Frederick J. Garbit

The Woman's Medical Companion and Guide to Health

Frederick J. Garbit

The Woman's Medical Companion and Guide to Health

ISBN/EAN: 9783337811624

Printed in Europe, USA, Canada, Australia, Japan

Cover: Foto ©berggeist007 / pixelio.de

More available books at **www.hansebooks.com**

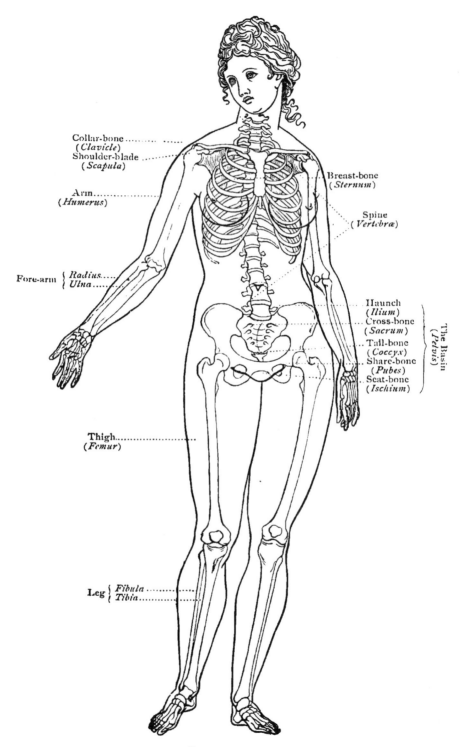

Collar-bone ....
(*Clavicle*)
Shoulder-blade ....
(*Scapula*)

Arm ....
(*Humerus*)

Fore-arm $\left\{\begin{array}{l}Radius .....\\ Ulna .....\end{array}\right.$

Breast-bone
(*Sternum*)

Spine
(*Vertebræ*)

Haunch
(*Ilium*)
Cross-bone
(*Sacrum*)
Tail-bone
(*Coccyx*)
Share-bone
(*Pubes*)
Seat-bone
(*Ischium*)

The Basin
(*Pelvis*)

Thigh ....
(*Femur*)

Leg $\left\{\begin{array}{l}Fibula ....\\ Tibia ....\end{array}\right.$

FRONTISPIECE.

THE

WOMAN'S MEDICAL COMPANION

AND

GUIDE TO HEALTH:

A PRACTICAL TREATISE ON THE

DISEASES OF WOMEN AND CHILDREN,

WITH

FULL AND DEFINITE DIRECTIONS FOR THEIR TREATMENT,

GIVING THE

CAUSES, SYMPTOMS, AND MEANS OF PREVENTION OR CURE,

WITH THE

LATEST AND MOST APPROVED METHODS OF TREATMENT ADOPTED
BY ALL SCHOOLS OF MEDICINE; THEIR DOSES AND MODES
OF ADMINISTRATION CAREFULLY PRESCRIBED.

Profusely Illustrated by the Profession.

BY

FREDK. J. GARBIT, M.D.,

Member of the Royal College of Surgeons, Fellow of the Royal College of Physicians, and for
many years Assistant at Bartholomew's and Guy's Hospitals, London, England.

BOSTON:

JOHN P. DALE & COMPANY.

1879.

INTRODUCTORY NOTICE.

THE author, having had a practical experience, extending over a period of nearly thirty years, in the study and treatment of the Diseases of Women and Children, in St. Bartholomew's and Guy's Hospitals, of London, and subsequently, both as Physician and Lecturer, in the Public and Municipal Institutions of New York City, has been brought into contact and practical acquaintance with every phase of disease to which women and children are especially liable, whether in the crowded cities or in the suburban districts. It is his purpose, therefore, in the "WOMAN'S MEDICAL COMPANION AND GUIDE TO HEALTH," to give young women, wives, and mothers the benefit of that varied experience; and, by the facts and principles brought to bear upon the daily life in the home, the nursery, the sick-room, and the factory or work-room, to afford them the means of *preventing*, as well as ameliorating and averting, many of the ills to which flesh is heir. The advice, observations, and directions as to symptomatology, treatment, hygiene, diet, regimen, etc., are deduced from a life-long experience, and a careful and critical analysis of the scientific researches of the most eminent authorities on matters relating to the public health. He therefore feels fully justified in his conviction that, in the work now submitted to the reader's notice, he is supplying a long-needed and universally-acknowledged want, — a reliable, comprehensive, plain-spoken Guide and Mentor, *specially* designed for, and devoted to, the physical and mental welfare of women. The author's chief aim is to maintain (or, rather, aid his readers in maintaining) the human organism in its pristine health and strength; and, where, from neglect or want of knowledge, it has been allowed to deteriorate and fall into a diseased condition, to point out the surest and the most effective means for its restoration, — thus securing to the women and rising generation that greatest of all human blessings, — a healthy mind in a healthy body.

PREFACE.

THIS work is specially designed for the use and guidance of Woman in the execution of her onerous and multifarious duties and responsibilities, as the most important and essential unit in God's grand scheme of creation. All will admit that "Knowledge is Power;"—and in the possession of that knowledge of the laws of her being, and their influence on herself and her offspring, alone lies safety for the human family,—for if she be weak and debilitated in mind or body,—if she be unacquainted with the anatomy and physiology of her sex, and of those principles which govern the existence of herself and offspring,—is not the entire human race affected *through* her?

From various causes a large majority of women have been debarred, hitherto, from obtaining an adequate knowledge of their physical life; and, in consequence, they have not been enabled to avail themselves of the vast influence and power they might wield, as the pivot on which the wealth and prosperity of the nation depend. In view of the fact that an accurate and comprehensive knowledge of their mental and physical functions, and their capacity for unlimited development, is all that is required to render Woman the true Archimedean lever of this nether world, we present this volume for her careful perusal, convinced that by following its counsels she would be enabled to fill her dignified and responsible position with gratification to herself and lasting benefit to the community.

(5)

It is our special purpose to faithfully portray the peculiarities of her structure and sexual functions; to demonstrate how those powers may be exercised, preserved, and maintained in their pristine vigor and normal health; and how they may be controlled and made a never-failing source of comfort, attraction, and usefulness to her fellow-creatures.

We have endeavored to point out the causes of those defects, deviations, and derangements from nature's standard which have been initiated and developed by ignorance, civilization, fashion's freaks, and the customs of " society ; " the diseases, suffering, and social disqualifications which, as a sequence, have fallen to her lot; and the method by which she may be effectually and permanently restored to that superior position to which she is entitled, giving her mental and physical strength to meet its requirements.

All technological or scientific phrases and terms will be avoided where practicable, or otherwise explained by the illustrations interspersed through the volume or in the glossary. *Fact*, and fact only, will be our standard ; theory and conjecture will be ignored. All the accidents, malformations, and special diseases to which the several sections of the female organism are subject will be truthfully delineated ; a *résumé* of the treatment by the various medical schools will accompany the diagnosis and symptomatology of each affection, to which we shall add our own experience and advice under the circumstances, defining what remedial treatment may be pursued by the nurse in attendance, and where the aid of a medical practitioner should be invoked.

In the Treatment of Diseases, special care has been taken to avoid all invidious preference of any particular method, to the disparagement of any other, for the reason that thousands of lives have been and are still sacrificed to the unreasoning bigotry and hostility existing between the various schools of medicine. We have, therefore, carefully selected the best and most efficacious remedies used by each of the schools, so that the patient can use the one most in accordance with her habits and ideas.

With the sincere and honest assurance that we are only actuated by a desire to benefit womankind permanently, by giving them a reliable guide, conscientiously abjuring every unworthy motive, and resolutely opposing every system of quackery or antiquated dogma, and endeavoring to keep pace with the onward stream of educational progress and reform in medical science, we therefore confidently commit ourselves to the good-will of our readers, trusting that our undertaking will merit the approbation and universal patronage of those on whose behalf and for whose service it is written, — the WOMEN OF AMERICA.

SYNOPSIS.

CHAPTER X.

CHAPTER XI.

CHAPTER XII.

WOMAN'S MEDICAL COMPANION.

CHAPTER I.

GENERAL DESCRIPTION OF THE HUMAN BODY.

The Body consists of a framework of bones, the *skeleton*, tied together at the joints by ligaments, and of the *soft parts*, the organs, vessels, nerves, muscles, and connective tissue, which collectively are called the *flesh*.

The parts of the skeleton are the skull and lower jaw, the spine, the ribs, the breast-bone, the collar-bones, the blade-bones, the basin, the arms and hands, the legs and feet. (See *frontispiece*.)

The Skull (*cranium*) consists of several bones all firmly locked together in the grown person, but some of which are separate in the infant to allow of the more easy passage of the head through the parts of the mother during birth.

The Spine, or backbone (vertebral column), extends from the back of the skull, which it supports, to the "basin," and is made up of short bones called "vertebræ," which are so shaped and placed upon one another as to form a continuous tube, in which lies the *spinal marrow* (spinal cord) ; the seven topmost bones are called *cervical*, or neck vertebræ ; the next twelve are called *dorsal*, or back vertebræ ; the next five are called *lumbar*, or loin vertebræ. The

(13)

column terminates with the *cross-bone* (*sacrum*), and *terminal-bone* (*coccyx*); these last form the back part of the basin.

To each dorsal vertebra, right and left, a RIB is attached, making in all twenty-four ribs. These ribs, with the exception of the two lowest, are united in front to the BREAST-BONE (*sternum*), so as to form a hollow box, the *chest*.

The Collar-Bones (*clavicle*), one on each side of the neck, are shaped like the italic letter *f*, and are attached by one end to the breast-bone and by the other to the blade-bone.

The Blade-Bones (*scapula*), whose triangular shape can be traced beneath the skin, are crossed obliquely by a ridge called the spine, which ends in a projection (*acromion*), forming with the end of the collar-bone the top of the shoulder.

The Basin (*pelvis*) consists of four sets of bones: in front the two *innominate bones*, which form the *share-bone* or the *pubes;* the *haunch-bones* at the sides, and the *seat-bones* below; the *cross-bone;* and the *terminal-bone.*

The Arm is attached by the shoulder-joint to the blade-bone or scapula. The upper arm consists of one bone, the *humerus*, jointed at the *elbow* to the forearm, which consists of two bones, the *ulna* on the inner side, the *radius* on the outer. The WRIST comprises eight bones. the HAND five, the FINGERS three each, the THUMB two.

The Leg is attached to the basin or pelvis by the hip-joint. The thigh has one bone, called the *femur*. The lower leg, connected by the knee-joint, and covered by the knee-pan (*patella*), has two bones, the *shin* or *tibia* on the inner side, the *small bone* or *fibula* on the outer. The ankle-joint connects the leg and foot. The under surface of the latter is called the *sole*, the upper is called the *back* or *dorsum;* behind is the *heel*, and in front are the *toes*.

The body may be divided into the Head, the Trunk, and the Limbs.

The Head consists of the face, the skull, and the lower jaw. Within the skull lies the *brain*, a mass of nervous matter continuous with the spinal cord. From the brain or spinal cord delicate threads called *nerves* pass to every part of the body. Motion and sensation are dependent on the healthy condition of these organs, and injury or disturbance of their action may be followed by pain, convulsions, or paralysis.

The Trunk consists of the neck, the chest, and the abdomen; an accurate idea of the respective organs contained therein and their location being most readily conceived by reference to the engravings on pp. 16 and 17.

The front or anterior part of the neck is called the *throat;* the back or posterior part of the neck is called the *nape.* Close up underneath the jaw the bone of the tongue (*os hyoides*) can be felt; a little lower is a projection, commonly called " Adam's apple," more prominent in men than women (the thyroid cartilage), the interior of which forms part of the *larynx,* or instrument by which the voice is produced; below this the *windpipe* (trachea) can be readily traced.

The Chest, or thorax, is formed by the twelve dorsal vertebræ, the collar-bones, blade-bones, ribs, and breast-bone with the flesh; all these bones are so jointed to each other as to allow of considerable alternate contraction and expansion of the chest during the act of breathing.

The. Breasts (*mammæ*). On either side of the breast-bone (*sternum*) are the Breasts.. These are made up of fat, connective tissue, vessels, nerves, and milk-glands. These glands have some resemblance to bunches of currants, and terminate by fine tubes or ducts in the Nipple or Teat (*mamilla*). The surface of the nipple is dark, and it is seated

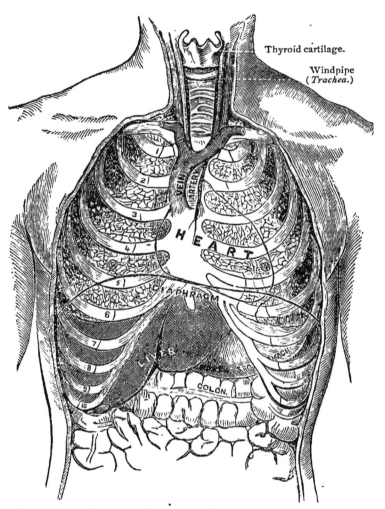

Thyroid cartilage.

Windpipe (*Trachea.*)

FIG. 1.—Front View of the Thorax. The Ribs and Sternum are represented in Relation to the Lungs, Heart, and other Internal Organs.

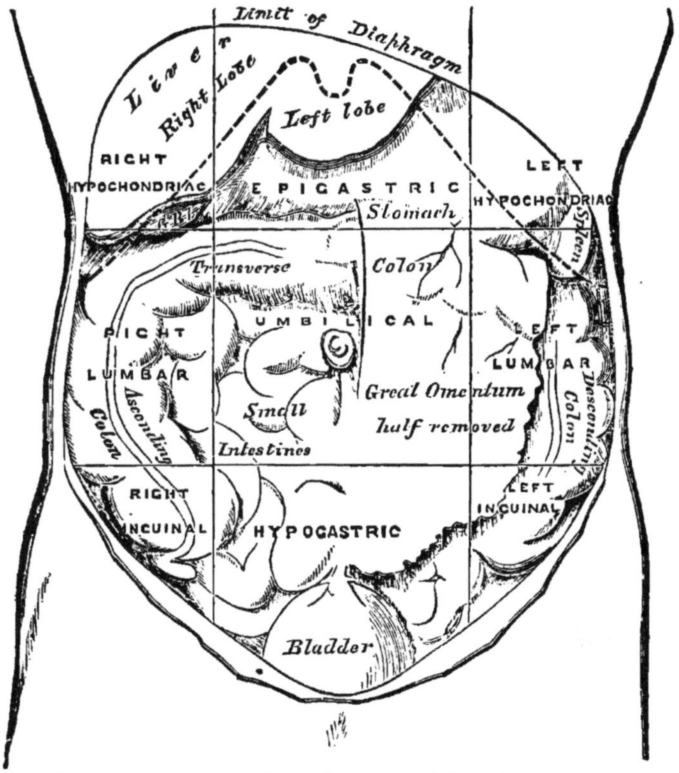

FIG. 2.—The Regions of the Abdomen and their Contents.
Edge of Costal Cartilages in dotted outline.

on a colored circle or *areola*, which in the virgin is usually of a rose-color, but becomes dark when pregnancy occurs, and never afterwards regains its former pink hue.

The hollow beneath the shoulder-joint is called the *arm-pit* (*axilla*). The space below the left breast where the heart is felt beating is called the *cardiac* region.

The interior of the chest is called the *thoracic cavity*. Its chief contents are the heart with its vessels, and the lungs with their bronchial tubes, in which terminates the wind-pipe. The office of the lungs is to expose the blood to the action upon it of fresh air admitted to them through the larynx, trachea, and bronchial tubes, during the act of *inspiration*, or drawing in the breath. If, then, the expansion of the chest is hindered by the clothing, as by tight-laced stays, or if the air is polluted, as by the breath of persons in close, unventilated rooms, or by the gases from drains, etc., the office of the lungs is interfered with, the blood becomes impure and poisoned, and the result is ill-health or disease. The blood flows to the heart by the veins from every part of the body ; is by it pumped into the lungs to be purified ; returns to the heart, and is thence transmitted by the arteries throughout the system. The *pulse* is commonly felt at the wrist, because at that part an artery lies near the surface of the body, and is therefore easily felt. The number of beats for an adult is ordinarily about seventy a minute, though it varies much, even in health, in different persons ; but it usually maintains its own rate in any individual.

The part of the chest containing the lungs and heart is separated internally from the belly by a fleshy partition called the *diaphragm*, or *mid-rib*, which thus forms the floor of the thoracic cavity.

THE BELLY (ABDOMEN)

Is divided into two parts, an upper and larger part, the *abdomen*, or belly, properly so-called, and a lower part, named the "basin," or the *pelvic cavity*. For convenience of description, the belly or abdomen is marked out into nine regions. A line is drawn across the body at the level of the pit of the stomach, and a second line at the level of the hips; these two horizontal lines are crossed by two vertical lines drawn each from the breast to the middle of the groin. The upper third is thus marked off into the epigastric region, or pit of the stomach, with on either side the hypochondriac regions (hypo, *under*, chondria, *the cartilages of the ribs*). The middle part of the belly, or abdomen, is divided into the umbilical or navel region, and right and left lumbar or loin regions, the flanks. The lower third is divided into the hypogastric (hypo, *under*, gaster, *the stomach*) or pubic region in the centre, and the right and left iliac or inguinal regions on either side. These three, the pubic and two iliac regions, are often together called the *lower belly*.

The chief contents of the belly, or abdomen, called the *viscera*, are as follows: in the upper third going from left to right are the liver, with its gall-bladder, portions of the intestines, the pancreas or sweet-bread, the stomach, and the spleen. In the middle and lower third are the small intestines or bowels, and the large intestine called the colon, which ascends from the right iliac region to a little above the level of the navel, then crosses and descends on the left side into the basin or pelvic cavity, where it is called the rectum or straight gut, and terminates at the fundament or *anus*. It is into this portion of the intestine that injections, clysters, or *enemata* are thrown.

KIDNEYS AND BLADDER.

In the lumbar or loin regions lie the *kidneys*, whose secretion, the urine, passes into the *bladder*. This vessel empties itself by a small canal called the *urethra*. The bladder, when distended with urine, can be felt like a ball rising out of the " basin" or pelvic cavity into the pubic region. The " basin" or pelvic cavity contains the bladder when empty, the womb with its appendages, and the rectum.

All these *viscera* — stomach, intestines, liver, pancreas, spleen, kidneys, bladder, womb, etc. — are covered, more or less, with a thin lining membrane, called the *peritoneum*. This membrane in certain parts is gathered into folds, which serve to tie or support certain of the viscera in their places, and these folds are called " ligaments."

THE NERVOUS SYSTEM.

Within the skull lies the brain, a mass of nervous matter similar to and continuous with the spinal cord. From this brain-matter or spinal cord, delicate threads called *nerves* pass to every part of the body. Some idea of the nature and beauty of their arrangement may be seen in Fig. 3, p. 21.

THE SPINAL NERVES,

Connecting with the cord, are in pairs, of which there are thirty-one. Each pair has two roots, — a *motor* root, arising from the anterior columns of the cord, and a *sensitive* root, springing from the posterior columns. A section of the cord is surrounded by its sheath. The spinal nerve is formed by the union of the motor and sensitive roots. After the union, the nerve, with its motor and its sensitive filaments, divides and subdivides as it passes on, and is distributed to the tissues of the several organs.

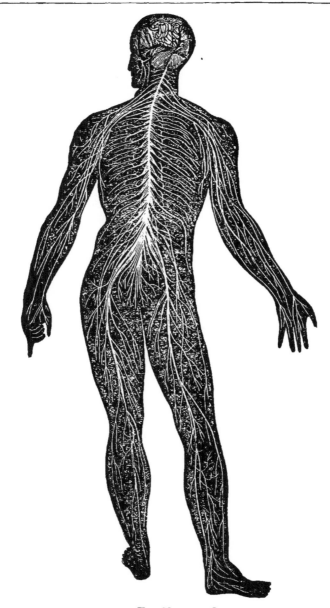

THE NERVOUS SYSTEM.

The thirty-one pairs of spinal nerves are divided into eight pairs of *cervical*, twelve pairs of *dorsal*, five pairs of *lumbar*, and six pairs of *sacral* nerves.

THE SYMPATHETIC NERVE.

Consists of a series of these ganglia, or knots, which extend down each side of the spinal column, forming a kind of chain throughout its whole length, communicating to both the cranial and spinal nerves, and distributing branches to all the internal organs.

These nerves, then, are undoubtedly the organs of feeling and sensation of every kind; through them the mind operates upon the body. The intelligent mind, whatever that may be, whose seat is in the brain, *wills* that a certain action shall be performed, and instantly through the main channel of communication, the spinal cord, the message flies, branching off here or there according to the direction in which the work is to be done, and setting in motion the muscles which form it.

CHAPTER II.

THE FEMALE ORGANS OF GENERATION.

THESE are the *ovaries,* or egg receptacles, in which the ovum or egg is secreted, whence it passes along the *Fallopian tubes* to the *uterus* or *womb,* whose office it is to contain the fecundated ovum during its growth, and then to expel it along the *front passage* or *vagina* into the world. These organs are all contained within a cavity, the walls of which are composed of bones and of soft parts, known as the *basin,* or *cavity of the pelvis.*

The *external* organs of generation are usually included under one name — THE PUDENDA, or the " Privates."

Passing backwards from the " privates " (vulva), we find the *perineum.* This bridge-like structure extends to the *anus.* It measures an inch and a quarter in its normal condition, but when, during birth, the child's head is pressing upon it, it is capable of extension to three or even five inches.

The *anus* is the circular opening into the bowel, and is a muscular structure capable of considable dilatation and contraction. Behind this, you can feel the terminal bone of the spinal column, or the coccyx.

THE INTERNAL ORGANS. (See cut 4, page 25.)

1st. **The Vagina,** or front passage. This is a canal or tube measuring four or five inches in its natural condition, and is extremely elastic. In its healthy state, the walls are close together, thus forming a substantial means of sup-

port to the womb.   In certain cases the walls become very
relaxed, when, as a natural consequence, the womb loses a
large proportion of its support, and it is apt to fall — an
affection known as *Prolapsus Uteri, Procidentia Uteri*, etc.
To the fore-part of its upper wall or roof the urethra and
bladder are attached ; farther back, about two-thirds of its
length, the neck of the womb projects into it.   The surface
of the vagina, when healthy, is only just moist, except during labor, when an abundant secretion or mucus is poured
forth, to aid the passage of the child into the world.   A
similar secretion is very frequently found in unimpregnated
females, and in very large quantities, but this does not
arise from the same cause or the same source.   This unnatural and unpleasant discharge comes from the womb, and
is known by the name of LEUCORRHŒA, or the *Whites.   In
these and all the other affections we shall have to mention,
medical advice should be at once obtained.*

Behind the vagina, in the hollow of the cross-bone, or
sacrum, lies the *rectum.*

THE WOMB (UTERUS)

In the virgin resembles a small flattened pear in size and
shape.   It is about two and a half inches in length, one
inch in thicknes, two inches in width, and weighs about
one ounce.   After child-bearing, these dimensions are permanently increased, so that the whole organ is larger and
heavier than in the adult virgin.   The bottom of the womb
is called the *fundus*, the middle third is called the *body*, and
the remainder is called the *neck* or *cervix.*   In the centre
of the body is a cavity that will contain an almond, lined by
a mucus membrane, which, during pregnancy, becomes
greatly thickened.   From this cavity a small canal leads
through the neck or cervix to the external mouth of the
womb.   There are also two minute openings near the fundus,
which are continued through the Fallopian tubes.

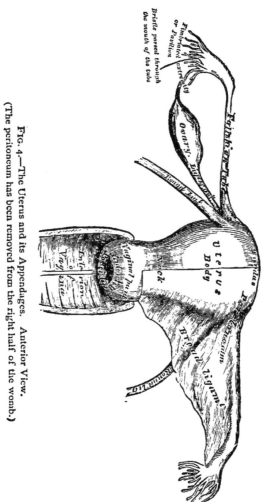

FIG. 4.—The Uterus and its Appendages. Anterior View.
(The peritoneum has been removed from the right half of the womb.)

The *neck* of the womb, or *cervix*, in a healthy woman, who has never been a mother, projects about three-quarters of an inch, or for two-thirds of its total length, into the " front passage," or vagina, presenting a smooth, conical surface, having a transverse or circular depression in its centre, the mouth of the womb (*os uteri* or *os cervicis*), with an anterior and a posterior lip. The opening will admit a large knitting-needle or a quill. In women who have borne children, the length and size of the neck or cervix vary greatly ; usually it becomes thickened, and the orifice (*os cervicis*) is often notched, and will sometimes admit the top of the forefinger.

THE FALLOPIAN TUBES

Are two pipe-like fleshy canals which pass off from opposite sides of the bottom of the womb (*fundus uteri*) ; they are about four inches in length, and as thick as a crow-quill, the passage through them hardly admitting a bristle. They end in a sort of trumpet-shaped mouth (the *pavilion* or *fimbriated*, that is, *fringed extremity*), which, at certain times, seizes the ovary in its grasp, and receives the ovum, or egg, which then passes along the Fallopian tube to the cavity of the womb.

THE OVARIES (EGG RECEPTACLES OR OVARIA)

Are two fleshy bodies, about the size and shape of a large almond, which lie half encircled by their respective Fallopian tubes, a little behind and about half an inch away from the bottom of the womb (*fundus uteri*), one on each side of the ovary to which they are connected by a ligament. Each womb, contains a number of vesicles, in which the " ova," or eggs, are formed, and which, as they become ripe, fall into the mouth or " pavilion " of the Fallopian tube to pass to

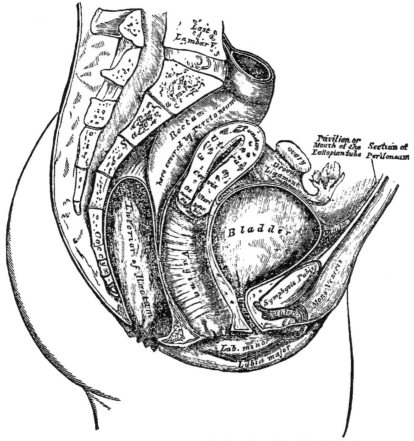

FIG. 1.—Pelvic Organs in position.   Bladder distended.   Womb virgin.

the womb.   The Fallopian tubes and ovaries are sometimes called *the appendages to the womb.*

The womb, Fallopian tubes, and ovaries are supported in their mid-centre position in the basin or pelvic cavity chiefly by a membrane, the *peritoneum,* in which they are enveloped, and which is attached by its outer border to the soft parts lining the side of the " *basin* " (*pelvis*), and to the other viscera, like a diaphragm.   This membrane forms a broad fold on the right and left sides of the womb, and these folds are called the *broad ligaments;* it forms two narrower and more cord-like folds behind the womb, which pass one on each side round the uterine (*rectum*) to the cross-bone, and these are called the *utero sacral-ligaments;* two other less distinct and slighter folds pass in front between the womb and the bladder, and are called *utero-vesical* (or womb-bladder) ligaments.   The bladder, womb, and bowel are thus all tied together, and consequently irritation or disease affecting one of these organs frequently involves one or both of the others.   The womb is further upheld by the vagina.   Two fleshy bands, called the *round ligaments,* arise from each side of the womb, a little below and in front of the Fallopian tubes, and pass downwards to the groins.

The entire aspect of the

UTERINE ORGANISM

(Exhibited on page 27, and again referred to in our chapter on Uterine Displacements, for the purpose of showing their *normal* position) will illustrate the intimate connection of the several functions with each other, and demonstrate the important fact that the healthy or diseased condition of one organ necessarily and inevitably involves the integrity of the whole.

CHAPTER III.

PUBERTY AND MENSTRUATION.

DURING infancy and childhood the breasts and organs of generation, both internal and external, remain undeveloped; but when the girl reaches the age of *puberty* — fourteen years or thereabouts — these organs take on growth, enlarge, and gradually become mature, and fit to perform the functions of reproduction; the girl, in the course of the next five years, becomes an *adult* woman.

MENSTRUATION.

There is no function of the female economy of which even females themselves are more ignorant than menstruation. That a process so vital to their general health, as well as to the fulfilment of their natural functions, as the source from which future generations are to derive their existence, should be so imperfectly understood by women generally, is a disgrace and a crime, for the existence of which there is not even the shadow of an excuse. Ignorance of this first and fundamental law of woman's nature is the cause of two-thirds of the demoralization, wickedness, and loss of virtue with which this world has been cursed since the creation. Fully three-fourths of those insidious and life-destroying maladies to which women are liable may be directly traced to their misapprehension on this matter. What wonder, then, that the medical fraternity should be so divided in opinion, and originate so many contradictory and speculative theories, within the last two centuries, on menstruation and the disorders arising out of the derangement of this function?

The fault, without doubt, lies primarily at the doors of the maternal parents themselves. From a *spurious* idea of modesty, from indifference, or pure carelessness, they have permitted their daughters to arrive at the age of puberty without the slightest intimation as to what they might expect to experience ; no word of counsel or caution has escaped their lips — the poor girls, affrighted at the appearance of the unexpected discharge, have endeavored to arrest it, and thus laid the foundation for an interminable train of painful and fatal diseases, making their life a prolonged and never-ending misery to themselves and those with whom they are connected.

The Ancients had many superstitious notions regarding menstruation. The wonderful periodicity and regularity of the flow once in every twenty-eight days led to the conviction that this flow was caused and governed by the moon, the same as the ocean tides. Were such the case, the whole of the race would be " unwell" at the same moment ; · but the contrary is the fact, for there is not an hour, or, indeed, a moment, in the whole year in which thousands are not undergoing that periodic visitation.

POPULAR ERRORS CONCERNING MENSTRUATION.

Many of our most eminent medical practitioners confidently assert that menstruation is inseparably connected with and dependent on the process of ovulation, or *conception.* But this also has been proved to be an error *in fact.*

The truth, as established and corroborated by the general experience of women is, that ovulation *can and does* exist *without menstruation,* and that menstruation frequently occurs without ovulation. Conception or ovulation doubtless has an exciting or stimulating influence on the menstrual function, but that the one is necessarily and inevitably the cause of the other is manifestly erroneous.

The physiological function and sole duty of the ovary is to mature and deposit its ova, or eggs, once every twenty-eight days, which it regularly does, in the majority of healthy females. The same principle regulates the occurrence of the menstrual function, but the simultaneous occurrence of the ovulating with the menstrual epoch is *a pure coincidence.* Many a female has become pregnant, not once only, but several times in succession, without even the slightest sign of menstruation.

In this country the sexual function is not assumed until the fourteenth year, as a rule ; in warm climates it appears somewhat earlier, and in colder regions, at a later period. Perhaps local causes and conditions have quite as much to do with the early or late appearance of the catamenia as the climate. It has been observed that those who are brought up luxuriously, and whose moral and physical training has been such as to exaggerate the susceptibilities of their nervous system, menstruate at an earlier period than those who are brought up roughly and are accustomed to coarse food and laborious employment. The appearance of the menses prior to the fourteenth year is much to be regretted, because it demonstrates a premature development of the generative organs ; and, on the other hand, a late or retarded first appearance is always to be regarded as an evidence of weakness or disorder. An undeveloped state of the uterine organs, indicated by a procrastination or non-eruption of the menstrua, always, in the mind of the skilled physician, excites apprehension for the welfare and security of the person in whom it is observed. In such cases we often find the body blighted, the mind dull and weak, with the chest and lungs insufficiently developed, all of which render the patient an easy prey to disease.

SYMPTOMS AND DURATION.

The first accession of the menses is usually preceded by headache, heaviness, languor, pains in the back, loins, and down the thighs, and an indisposition to exertion. There is a peculiar dark tint of the countenance, particularly under the eyes, and occasionally uneasiness or a sense of constriction in the throat. The perspiration from the skin has a faint or sickly odor, and the smell of the breath is peculiar. The breasts are enlarged and tender. The appetite is fastidious and capricious, and digestion impaired. These symptoms continue for one, two, or three, or more days, according to circumstances, and subside gradually as the menses appear. The period lasts for three, five, or seven days, according to the peculiarities of the constitution.

These monthly periods return with great regularity from the age of fourteen to about forty-five, when they usually cease. This period is ordinarily one of great anxiety to females, the symptoms which present themselves at the time, such as sickness at the stomach, capricious appetite, swelling and pain in the breast, etc., being frequently mistaken for pregnancy.

Menstruation very rarely ceases suddenly, but the intervals become irregular, eventuating in utter disappearance. During the menstrual period, especially in young persons, great care should be taken to ward off all influences, whether mental or physical, which may have the least possible tendency either to interrupt or increase the discharge; because upon the healthy and regular action of the discharge depends so much of the beauty, perfection, and security of the female. During this period there is an increased susceptibility and excitability of the system, and consequently a greater liability to derangements and to diseases of various kinds.

Serious and even dangerous results often follow a sudden suppression of the menses. Among the causes which produce trouble at this period, we may mention sudden frights, fits of anger, great anxiety, and all powerful mental emotions. Excessive exertions of every kind, long walks or long rides, especially over rough roads, dancing, frequent running up and down stairs, have a tendency not only to increase the discharge, but to produce falling of the womb. The discharge is not unfrequently morbidly increased, or entirely arrested, by taking purgatives, emetics, stimulants, and the various patent medicines recommended for female weaknesses. Cold and warm bathing, hip and foot baths, should be discontinued during the period. Care should also be taken not to expose the feet to cold or wet. Females subject to leucorrhœa, and who are taking vaginal injections, should discontinue them shortly before and during this period. During the menstrual period in a healthy person there is little required besides carefully avoiding the injurious mental and physical influences abovementioned. *If, however, the female be delicate and suffering from any of the numerous derangements of menstruation, or any other of the thousand and one sexual irregularities to which the sex are at all times subject, they should not attempt to medicate or prescribe for themselves,* BUT WITHOUT DELAY SEEK THE ADVICE AND AID OF A SKILFUL PHYSICIAN.

CHAPTER IV.

HYGIENE, OR THE LAWS OF LIFE AND HEALTH.

We have traced the development of the young female to the point where the first great constitutional change takes place, transforming her from a child to a *woman*—from a dependent and comparatively helpless being to an independent and responsible individual — responsible to herself and others for every act affecting her physical and mental condition, entailing either happiness or misery upon herself and those with whom she is connected socially or by family ties. She consequently needs the closest attention, and the most judicious and careful management, not only to counteract the tendencies of her semi-childish constitution, but also to control her in the enjoyment of her new privileges, and influence her in the selection of companions and pursuits. She has reached the spring-time of her life ; all her charms are budding forth like the opening rose ; she is now the delight and attraction, the life and soul, of the social circle, and henceforth is an indispensable integer of the species she is designed to perpetuate.

The first and most essential provision for the young aspirant to the privileges of society is, that her surroundings should be genial, health-inspiring, and calculated to expand her faculties, mental and physical, in order that she may *know*, and be thoroughly fitted to fulfil, her mission as a citizen of the world, and as the chief agent in moulding the character of the community in the immediate future.

The first law inculcated by nature and by our Divine Creator, is that of Cleanliness.

BATHING.

From the time of Adam until now, the importance of keeping the skin clean has been universally acknowledged and more or less acted upon by the members of every community, civilized, semi-civilized, or savage. Sanitary regulations formed a part of the civil and religious laws of the Jewish, Mohammedan, and Christian sects; and even amongst the so-called heathen, cleanliness was deemed a primary essential to admission even to the lowest grades of society. A *dirty* man, woman, or child, might be found, doubtless, in every community, but seldom in company with *intelligence* or *morality*.

It is not merely the hands and face which have to be kept clean, — every part of the body, seen or unseen, must be subjected to the daily operation of cleansing, for the reason that our skin is not merely a *covering* and *protection* from atmospheric and other influences; it is a huge collection of more than *seven million spiral canals* (or 2,800 to every square inch of the body), through which we receive a great portion of the supply of air by which we live, and from which the refuse of what we have partaken in our daily food, and the perspiration or vapor contained within our system, passes off. Any obstruction, therefore, to the free passage of the perspiration through these pores, by the accumulation of particles of dust, coagulated perspiration, etc., not only interferes very seriously with the health of the body at every point, but occasions the majority of those annoying and unsightly cutaneous eruptions which everybody looks upon with horror and apprehension. The pores of the skin are equal in importance to the *lungs* themselves. Now these pores cannot be kept open and perfectly free from impediment by the ordinary ablution from a small basin and a little soap. A sitz bath, or if that is not available, a large washing-tub, with a sponge, a good rough towel, and carbolized, Castile, or even ordinary toilet soap (so long as it is not *scented*), will answer the purpose. You will speedily see and experience the benefits arising from the use of the bath, if you only examine the condition of the person when it has been neglected for a few days, and the under-garments have not been changed sufficiently often. The insensible perspiration accumulates and dries upon the surface of the skin, mingling with the oily matter secreted by the oil-glands, and, with the shreds of the scarf-skin, forms a tenacious, gluey matter, which completely closes the pores. These pores being so closed, the perspiration, and other matter which is seeking an exit from the body, is retained to poison and embarrass the living current of the blood, or else to seek an outlet through the kidneys or lungs, which are already burdened with their own legitimate work. You will acknowledge then, dear reader, that a clear, purified, and healthy skin is one of the first and most imperative essentials to a healthy body.

COLD, WARM, AND SPONGE BATHS.

A great variety of opinions exists as to the advisability or propriety of *cold baths*. Any bath below the temperature of 75° is called a cold bath; and, if the body is in a condition to bear it, it acts as a decided and powerful tonic. Our own experience, as well as the experience of many thousands of our brother practitioners, has proved the efficacy and undoubted advantages of its use, promoting the compactness, solidity, and strength of the body, — a tangible proof of which is given in the almost instantaneous *reaction* which follows its application. The vessels immediately contract, and the blood retreats towards the internal organs, causing the bather to feel a genial glow all over the body, from the blood being forced back through the invigorated vessels from the crown of the head to the toes.

As we have said, some persons are so conditioned that the shock of *cold water* cannot be borne. In such cases, when a sensation of *chilliness* is felt, it is an evidence either that the bath has been too profuse, or that the system is too much enervated to produce the reaction we have spoken of; under these circumstances the body must be *gradually educated* to its use, by using *tepid water*, and reducing the temperature slowly until it becomes accustomed to the use of water in its natural state. A wet sponge, with or without soap, applied in turn to every part of the body, and immediately followed up by a brisk rubbing with a flesh-brush or rough towel, will soon accomplish the desired result.

With persons in a feeble condition of health it will be necessary to expose a part of the body at a time, quickly sponging and rubbing each part dry before proceeding with the other, — so subjecting the whole of the person, however feeble and delicate they may be, to the bracing influence of water and friction, without the slightest risk of shock, or cold arising from exposure. There is no form of bathing so universally applicable or so generally conducive to health as the sponge bath.

THE WARM BATH

Is usually of the same temperature with the body (blood heat), from 98° to 105°. It produces no shock, and is of special service to those who have passed life's meridian, or who are suffering from nervous and muscular debility, or whose systems have been prostrated by sickness or inactive life. In the last-named cases, great care should be exercised in regulating the temperature, so that sensations of heat or fulness, or increase of pulsation, should not be induced. As a rule, the warm, vapor, or shower bath should be used under the advice of the physician, as serious consequences might ensue from their injudicious or excessive use.

There has been, and is even now, a large amount of error and fanaticism about the *exclusive* use of water as a curative remedy. The " water-cure " as a remedy for all diseases, known and unknown, is simply an exaggeration and a ridiculous caricature, calculated to bring into derision one of the most valuable remedial agencies with which nature has provided us. But even this will effect it's own cure in time. People will learn that water, judiciously used, in the form of baths, is a potent moral and physical renovator of the race ; and that a community with clean hands, clean bodies, clean faces, and clean, healthy habits, will naturally appreciate and insist upon clean streets and clean cities — and eventually *clean consciences.* *Cleanliness* in physical matters naturally causes an irresistible affection for *purity* in every other form, until it pervades the moral as well as the physical nature.

CLEANLINESS OF PERSON,

Is not all that is required. The same principle must be rigidly carried out in all the domestic arrangements. No soiled clothing, no animal or vegetable refuse, no stagnant waters or decaying organic matter, no defective drainage or close rooms, no miasmatic or malarial poison, should be tolerated about the homestead or adjacent buildings under any pretext. With well-ventilated, well-aired, well-lighted, well-scoured apartments ; culinary utensils thoroughly cleansed ; free ingress for the balmy breezes of heaven in every room ; plain, wholesome food ; systematic and regular manual exercise, — walking, running, jumping, dancing, in *moderation,* — and an avoidance of all excess, dissipation, late hours, and bad habits generally, — acting out such rules, the person who has but just passed the threshold of adult life cannot fail to rejoice in the possession of a " sound mind in a sound body," and can safely look forward to a long, happy, healthful, and useful life, ending in a hale and honored old age.

A CAREFULLY REGULATED DIET

Is of all means the most appropriate for moderating the excitement and derangement resulting from the momentary plenitude of the circulatory system. The food of a young girl should consist mainly of vegetable substances, preparations of milk, of the tender, juicy meats, and of light and easily digestible substances, some few of the more succulent fruits, and puddings made of farina and other cereal products. Water, milk, broma, and cooling liquids should form the chief part of her drink, and on no account whatever should she indulge in candies, ice cream, and other confections, pickles, solid and highly seasoned meats, and made dishes, sour and unripe fruits, stimulating articles, alcoholic liquors, or the daily use of coffee and tea; they should all be studiously and resolutely avoided. Tepid, or nearly cold baths, sponge baths, as we have already described, taken occasionally, say twice or thrice a week, will contribute, together with regimen, to produce a general purifying, cleansing effect, and will have the advantage, moreover, of softening the skin, and dispersing the cutaneous eruptions to which girls are particularly subject at the period of puberty.

In order to maintain the generative organs in a normally healthy condition, and in a suitable state of preparation for the periodic exercise of the menstrual functions, moderate exercise in calisthenics or gymnastics, walking, riding, and running easily, the skipping-rope, jumping, horseback, etc.,

should be indulged in.   Special precaution should be taken
in regard to the underclothing ; judicious friction should be
kept up in the genito-urinary region by wearing Canton-
flannel drawers, etc. ; the wearing of corsets with busks or
whalebones, or anything which obstructs the motion and
free development of the pelvis, thorax, and neck, should be
absolutely and rigidly forbidden.   The subject of clothing
will, however, be detailed in another section.

We will now, however, pass from the consideration of
physical regulations, for a moment, to consider the mental
and moral influences which should be brought to bear on
the new candidate for the honors and privileges of womanhood.   The most important media for determining the
physical and social future of the young aspirant is

THE SENSATIONS.

A sensation is an effect produced on the mind through
a nerve.   Hunger is a sensation.   It is an effect produced
on the mind through the stomach.   Nausea is a sensation
produced by some injurious substance acting upon the coats
of the stomach.   In this way the various conditions of the
body, whatever their origin or exciting cause, have a pro-
portionate effect upon the mind ; and every mental emotion,
no matter how remote or apparently trifling, has a depressing or exhilarating effect upon the constitution.   It is a
demonstrated fact, that the mental faculties and the physical
functions are so intimately associated and absolutely identi-
cal, the one with the other, that the hygienic laws refer with
equal force to both ; and any neglect or infraction of those
laws would have an equally injurious influence on the mental
and physical development of the individual subjected thereto.

Sensations are either pleasurable or painful.   Pleasurable
sensations arise from the healthy and legitimate exercise of
some mental or physical function, or the coming in contact

with some genial and harmonious influence, and are a suitable and adequate reward for the control or self-denial exercised in keeping the desires and faculties within proper limits. For example, the sensations of freedom and satisfaction felt after a moderate amount of exercise, or the partaking of a wholesome, moderate meal, is a present and tangible reward which the gormandizer or the unmanageable romp never feels or knows. The muscles find a sort of enjoyment in action. Those who lead a sedentary life, either from choice or necessity, lose much mental and physical enjoyment. Hence there is pleasure in labor; and the working-women (the women who pass their time in household employment or in light and healthful labor of any kind), though frequently an object of pity with the wealthy and the lazy, are usually the happiest members of the human race. The eye and ear, when directed to agreeable sights and sounds, drink in their inherent beauty, and gradually become so thoroughly imbued with their spirit, that their possessor reproduces and transmits their beneficent influence on all with whom she may be associated. The mind is nourished and expanded, and by the irresistible tendency of sympathy, and the desire to communicate to others the advantages or blessings it has itself experienced, the blessings received by one intelligent and appreciative mind are distributed and dispensed throughout the community. The female organism is specially adapted for this mission. From her very birth, woman's nature is eminently susceptible of pleasant, joyous, agreeable impressions; her mind is, beyond all dispute, the most pure and truthful media for the transmission of Nature's eternal truths; her sympathies, her intelligence, and her imaginative powers, ever enlisted on the side of the true and beautiful; and it is only when that nature and those sympathies have been directly or indirectly warped, restrained, and arrested by impure or antagonistic associations, that we find her suffering from physical or mental disqualifications, which unfit her for the high and useful position in the social economy for which God and Nature destined her.

HOME INFLUENCES AND ASSOCIATIONS.

In view of the facts we have narrated, the influences and associations by which the young girl is surrounded should be of the purest and least exciting character. Kindness, sympathy, gentleness, cheerfulness, and broad benevolence and generosity of sentiment should be brought to bear upon her mind in every phase of her existence. Harshness, coercion, insincerity, unreasoning prejudice, sensational and inflated ideas of mankind and the world generally, are essentially repugnant to her nature, and should, therefore, be studiously and resolutely kept from her path. She is in the sunshine of her youth: let her see or hear nothing of the dark side of nature. To ensure this, *home influence* is undoubtedly the best. Many of our country boarding-schools and city work-rooms are nothing better than moral pest-houses. Another inexhaustible source of mischief is the cheap periodical literature of the day, sensational romances, highly-wrought novels and love tales, exciting a morbid taste for the marvellous, inspiring a desire to experience those imaginary scenes and sentiments which are never known or realized — *except on paper;* the already exalted and excited imagination having been raised to fever-heat by questionable books and unsuitable companions. Endowed with a beautiful, delicate, and impressionable nervous organism, she contracts baneful habits and thoughts, is tormented by an absorbing amorous melancholy, becomes sad, dreamy, sentimental, and languishing, and, like a delicate plant withered by the rays of a burning sun, she fades and dies under the influence of a poisoned breath.

As a matter of course, the remarks we have just made refer to those young persons whose buoyant, joyous nature, cheerful, open disposition, sanguine temperament, and genial, courteous, and agreeable manners, and attractive figure, render them the favorites of the circles in which they move. But there are many interesting and intelligent young women to whom these observations would be scarcely applicable. They are of a lymphatic, cold, retiring temperament, indifferent to the allurements of company, the fascinations of romantic scenery, the pleasures of perusing an interesting volume or witnessing a legitimate, well-wrought drama. Individuals evincing such a tendency should be encouraged to seek such sources of recreation, in order to dissipate their morbid conditions, infuse vigor into their mental and physical systems, and develop those latent powers which they possess in fully the same measure as their more vivacious sisters.

We have hitherto spoken mainly of the *outside* influences — those independent of the

DOMESTIC RELATIONS.

It is here, within the magic home-circle, that the *true*, the *life* character of the woman is formed. If the mother be morose, fretful, hasty, careless, slovenly, haughty, despotic, unreliable, loving, orderly, or thrifty, the daughters will, to a great extent, reflect, either exaggeratively or in miniature, the virtues or vices which they have more or less inherited from their parents, especially the mother. The parental peculiarities, mental and moral, as well as physical, are, as it were, *photographed* on the children, never to be entirely effaced. As with hereditary defects of the constitution, affections of the skin, and other congenital diseases, so it is with mental idiosyncracies, they are transmitted from generation to generation ; and, unless arrested and eradicated by some superior curative power, increase in intensity until the physical and mental faculties are utterly absorbed and transformed by the ruling power or agency.

Before we leave this department of our subject we would give a few general directions as to the hygienic regulations of a well-ordered household, in order to preserve the body and mind in a healthy and vigorous condition, sustain and develop their vitality, and enable them to avoid disease or derangement.

REST AND SLEEP.

The human frame resembles a clock ; it runs down and is wound up once in every twenty-four hours (or, rather, *it should be*). Were a female required to work on uninterruptedly, no matter what might be the nature of that employment, she would undoubtedly wear out in the course of a very few days ; her physical and mental powers would be prostrated beyond all possibility of recuperation.

It is a merciful interposition that periods of repose are allotted to us. Everything has its proper place. Rest is not less a luxury after exercise than exercise is after rest. They both confer happiness at the same time that they promote our well-being. But it must be remembered that, as nature has ordained *night* for rest, the turning of night into day either for pleasure or for labor must necessarily be attended with evil, physical as well as mental. The abridgment or the alteration of the hours of rest carries with it its own inevitable penalty. *Two hours' sleep before midnight* is more productive of benefit to both body and mind than six hours after that period. There is no more truthful axiom extant than —

> " Early to bed and early to rise
> Makes a man healthy, wealthy, and wise."

Of the health and wisdom derivable from the adoption of such a system every reasoning being may receive daily demonstration ; the *wealth* is of course a relative question, and may truly consist in the accession and maintenance of a sound mind in a sound body.

In regard to the period allotted to rest and sleep, nature is the most efficient arbiter. It depends solely on the physical condition of the body, the nature of the occupation, and the age and sex of the individual. Dr. Abernethy, an eccentric, shrewd, but cynical " Old School Physician," had some " hard and fast " rules, which, nevertheless, had a certain amount of common sense in them. He divided the day of twenty-four hours into three sections, — eight hours for recreation and rest, eight hours for sleep, and eight hours for work ; but he accompanied this suggestion by a rather invidious allusion in reference to *sleep,* saying that six hours' sleep were sufficient for a man, seven hours for a woman or child, and eight hours for a *fool.* You may rest assured that *none* of his patients voluntarily included themselves in the last-named category.

Practical experience demonstrates that more than eight hours' employment for a child or female is necessarily and absolutely injurious to their mental and physical organisms. Nor should this *eight hours* be continuous, — it should be relieved by intervals of rest and recreation — not *sleep.* Sleep should be reserved for night, when darkness covers the earth and all nature slumbers ; when the feathered songsters have ceased their song, and the flowers have closed their petals. *Ten o'clock* should, as a rule, find every one within the precincts of the bed-chamber, *not burning the midnight oil in reading sensational and prurient novels and tales,* but courting the embraces of the drowsy god, and gathering mental and physical strength by enjoying that sweet repose which a conscience devoid of reproach, a healthy, well-nourished body, and a well-balanced mind will always invoke and secure.

OUR SLEEPING-ROOMS.

Sleep and rest, however, will afford us but little benefit, if the room in which we sleep is small, unsuitably situated, or surrounded with conditions prejudicial to health, as the hours spent within the bed-chamber *permanently* affect our physical and mental well-being, for good or for evil. We necessarily breathe a large amount of air during the night, and our health becomes seriously injured if we breathe this air over and over again. The room should be tolerably large (not one of the smallest, as is usual in families), *dry, well-ventilated, well-aired, well-lighted,* and, when practicable, on the *top floor.* Just as much, and even more, attention should be paid to the size, situation, temperature, and cleanliness of the room as to the parlors or drawing-room. It should be especially provided that *two persons* (except in certain cases) should never occupy one bed in a *small* room, nor should there be more than two beds, each occupied by one person, in a large room; and again the young and the aged should *never* occupy the same bed. Any accumulation of refuse, decaying organic matter, or even the location of a *large number* of trees, near the sleeping-rooms is highly objectionable. The temperature of the bedroom should not be lower than 70 or 75 degrees; there should be ample means for the ingress of *light* as well as air, for in health, as well as in sickness, *light* and *air* are the chief and most effective restorative and preservative agencies.

Fires in sleeping-rooms depend simply on local considerations. If the occupant be an invalid, with feeble circulation, and ailing, a *small fire,* kept up for one or two hours prior to retiring to rest, may be necessary in cold weather; but a person in a normally healthy condition should never indulge in a luxury so perfectly unnecessary and prejudicial to robust health.

The Windows should, during the night, be let down about an inch from the top, but in such a manner as not to create a *cross-current* from the opposite sides of the room. Immediately on rising, every window (the lower sash) should be opened wide. The oxygen which the partially opened window admits is of the greatest benefit to the occupant of the chamber, and if the practice is prudently persevered in, will prove an invaluable preservative to health. The *open fire-place* is one of the best ventilators we can have for the passing off of the vitiated air, and no sensible person would, for a moment, think of hermetically closing this useful aperture.

BEDS AND BEDDING.

While we most emphatically disapprove the use of feather-beds and feather-pillows, we do not sanction the opposite extreme, the unyielding hardness of a closely-packed straw-mattress: there is in this, as in all other matters pertaining to our every-day life, a *happy medium* to be observed. The *hair mattress*, or what is termed the *Excelsior*, is the best, the most comfortable, and the most healthy that can be used. No one, after a fair trial of it, would ever return to the use of feathers.

In hot weather linen sheets are decidedly preferable to cotton, where persons are able to provide them, but in winter cotton is much more desirable to use, especially by persons peculiarly susceptible to rheumatic affections. *Bed-spreads,* or "comforters," as they are called, are objectionable, because they concentrate the insensible perspiration, and envelop the individual in a sort of vapor-bath. Blankets serve the purpose as an outer covering much better, as they are light, porous, and are excellent radiators of heat. There should be just as few clothes as possible on the bed, — only sufficient to prevent chill and to keep up the same temperature as that experienced in the daytime.

NIGHT-DRESS.

The *under* as well as the *outer* clothing worn during the day should be taken off on retiring at night, care being taken to have the night-dress of the same quality and thickness of material as the *day-clothes.* The underclothing should be subject to the same regulations.

THE AMOUNT OF AND PROPER POSITION FOR SLEEP

Can only be regulated by individual circumstances. The average time for sleep in the case of a healthy person (female) is from seven to eight hours, according to their occupation and constitutional peculiarities. There is no absolute standard for this, any more than for the amount and description of food. Nature is the best indicator in this matter. If the sleeper be in health, he should get up when he first wakes, whether it be five, six, or seven in the morning. The object of sleep is to restore the wasted energies ; the extent of that waste and the recuperative power possessed by the individual will measure the amount required. The temperament, constitution, amount of exercise, and the exhausting nature of the mental application, are the elements to be considered in the calculation.

The most natural position for rest is to recline on the *right* side, — as it gives perfect freedom to the internal organs, — of course, occasionally changing to the *left* side ; but by no means on the *back,* as in that position the stomach, bowels, etc., are pressed upon the larger blood-vessels in the neighborhood of the vertebral column, thus obstructing the circulation of the vital fluid. The hands should never be raised above the head during sleep, but should be placed in an easy and natural position on a level with the body.

INCENTIVES TO SLEEP.

The best incentives to sleep with a person in possession of a healthy mind and a healthy body are judicious exercise, a proper admixture of mental and physical labor, recreation and amusement, the partaking of food in moderate quantities and at regular hours, the avoidance of all excesses, and the possession of a good conscience. No person can expect to enjoy a good state of health who indulges in *late and heavy suppers*, or who spends the best part of the night in carousals, entertainments, and exciting associations. Fully *three hours* should be allowed to elapse between the evening meal and the hour of retirement. The best preparation for rest is a quiet hour spent within the happy and exhilarating influences of the home-circle. As we shall have to consider the important subjects of food and clothing in a separate chapter, we will hasten to the consideration of that essential item in our domestic life.

EXERCISE IN THE OPEN AIR

Is another invaluable adjunct in the preservation of health. Persons who take but little exercise leave the lower part of their lungs comparatively unemployed. As a consequence, the breathing is labored and unnatural. In the case of young persons, females especially, from fifteen to twenty-five minutes' exercise should be taken in the open air every morning before breakfast, when they should inflate the chest by long-drawn inspirations and respirations, and, after a few such exercises, in addition to the morning bath, they would eat their breakfast with a relish and satisfaction they never knew before.

But this is not all the exercise that is required. It is quite true that many persons who have dwelt in one spot all their lives, and never enjoyed a trip to other scenes or climes, have yet lived to a good old age. But one thing is certain, — that, as a rule, both mind and body tire of contemplating one set of objects for any length of time; the ideas get contracted, the routine of life becomes monotonous; and gradually, but surely, both body and mind sink into a lethargic, apathetic condition, incapable of enjoyment itself, or of imparting enjoyment to others. The physical frame droops, loses all energy, and becomes predisposed to sickness of various kinds; and, when the individual is actually suffering from illness, frequently sinks below all possibility of restoration, from want of change of scene and occupation. Travelling, if only for a day or two, tends to draw the thoughts of the sick and feeble from themselves, and awaken interest in surrounding objects. In the young and healthy it expands the ideas, enlivens the imagination, and furnishes them with sources of occupation which relieve the monotony of daily life and afford fresh incentives to exertion. It opens up new sources of gratification within them, and gives an abiding and constantly renewing interest in the world and its surroundings, which permanently disengages their thoughts from subjects of a personal or painful nature, until their very existence seems to bear a new aspect, and shadows forth a loftier and more expansive world in which they can exercise their faculties and progress toward the final fruition of their ambition. The nervous system has a miraculous power over the bodily health, and the pleasurable sensations evoked by mingling with new people, effecting new associations and visiting new scenes, often awaken in the constitution latent energies essential to recovery. The facilities which extended railway communication, and general reduction in cost of living, have given to every one, however humble in station, to make short or long trips to the coast or inland districts (according to circumstances), prompt us to recommend this means of health preservation and restoration, as one within the grasp of all, and as infinitely preferable to the most effective medicament the drug-store ever yet furnished.

It is a true, but common saying, that "All work and no play, makes Jack a dull boy." The great problem of the day is to make a judicious selection, age and sex considered, of—

LEGITIMATE AND HEALTHFUL AMUSEMENTS.

These must be adapted, not only to the sex and age of the individual, but to the time and season of the year, locality, etc. There is a time for everything. There are many amusements in which a young or elderly lady may participate with advantage and benefit, without derogation from her dignity or position. Of course the same amusements are not adapted to all persons.

THEATRICAL ENTERTAINMENTS

are specially gratifying and attractive to those who have a taste for art, and a love for works of genius and poetry; it appeals to a higher order of feelings, expands the sympathies, and gives a more accurate idea of our relations to the outer world, and in this respect is a prolific source of health and gratification. Many people, we know, think theatrical exhibitions are immoral and hurtful; but in this, as in other matters, the *evil* exists only in the person seeking it, not in the thing sought. Immoral persons will find evil in a church or in a grove, just as they would in a theatre, if their *motive* be evil. To the pure all things are pure ; they will find sermons in stones, sermons in running brooks, sermons in everything. Why, even the roaring farces and the laughable comedies have points of excellence ; the absurdity of the situations and the jokes produce laughter, and the laughter drives away the gloom and care which might otherwise fill our mind and make our lives a continuous misery. The only precaution to be taken in going to the theatre is to choose good companions, not to stay to a late hour, and *not to go too often* ; excess in *anything*, even in bathing or eating, is injurious.

DOMESTIC AMUSEMENTS

(such as Fox and Geese, Hunt the Slipper, Blind Man's Buff, etc.) have unhappily gone out of fashion, not only in town but in the country districts. These and an occasional dancing-party (*en famille*) are exceedingly healthful and enjoyable, in the fall and winter evenings, as a variation in the delightful atmosphere of the home-circle. These are recreations which would bear the morning's reflection ; and if these domestic games, interspersed with singing, dancing, and social intercourse, were generally revived in America (both in town and country), we honestly believe that one-half of the sickness, invalidism, nervous debility, and *positive insanity*, by which our worthy citizens (male and female) are afflicted, would disappear forever. Amusements are necessary to give a completeness to life. It is only when *all* the numerous mental and physical faculties with which we are endowed are exercised in due and proper proportion, that there is a harmonious beauty in them. The *customs* of society put us all out of shape, and rob us of every womanly and healthy quality. Adhere to Nature's simple dictates, and you will not only preserve your beauty of form and character, but develop it to an extent far beyond your present comprehension.

GYMNASTICS AND CALISTHENIC EXERCISES,

as a means of healthful exercise for young women, cannot be too highly extolled. We are glad to find that it is now becoming a part of the curriculum of education in the public and private educational institutions throughout this and other States, and that it is so admirably combined with intellectual instruction, that the mental and physical development of the scholars take place simultaneously, *the one helping the other*, and rendering them capable of meeting and surmounting any difficulty with which they may be brought into contact.

Our attention must now be directed to the

ARTICLES OF FOOD

specially suited to the requirements of the female organism. It would be absurd to suppose for a single moment that *all* articles were equally suited at *all* times for the *support* of *all*, and every individual, man, woman, and child. There are many points to be considered in dietary, — age, sex, condition of body, the occupation, time of year, the climate, etc. Food, however, may be first divided into two principal sections, — that which produces blood and flesh, and that which produces heat. Those producing blood and flesh comprise the vegetable and animal products which are ordinarily to be found in an American larder; the heat-producing articles include oils, sugar, starch, farina, arrow-root, tapioca, gums, etc.

The main feature to which we have to look is not so much the properties it contains, as its relative digestibility, and its suitability to the occupation and condition of health of the person concerned. The young woman just entering upon her life of usefulness and often thankless toil needs some counsel as to her diet at the most critical period of her life, and we shall commence by advising her *not to adopt any hard and fast rule* either as to the quality, quantity, or kind of food for her consumption. If she be of a lively, sanguine, nervous temperament, full-blooded, and excitable, no highly-seasoned or luscious dishes should form her diet. On rising in the morning a glass of milk with an egg beaten-up in it, prior to her morning's " constitutional," would impart additional vigor to her frame, and prepare her for the thorough enjoyment of her morning meal. This might consist of bread and milk, a little fish or oatmeal, a cup of broma, and home-made bread and butter. Her domestic or other duties might then be undertaken without any feeling of fatigue, tiredness, or *ennui*, and, the digestive powers not having been overtaxed, she would be ready to do justice to the mid-day refection, which should be confined to plain, juicy meats, well-cooked vegetables, and some light, farinaceous pudding. If liquid beverage of any kind were desired, it should *supplement* the meal, and not be taken *during* the time of eating, lest it should interfere with the gastric secretions, and should consist of pure *water*, with an occasional addition of lemon-juice. She would then resume her special occupation until the evening hour, when her physical wants might be supplied by a little wholesome fruit, bread and butter, light biscuits, and broma or milk. The habitual use of tea, coffee, pickles, condiments, candies, or any heavy, solid material, should be studiously avoided. Between that period and bedtime any recreation of mind and body suited to her taste and not involving any undue strain on her mental or physical faculties would close the day happily, and leave her with intellect unclouded, and her system in a condition to enjoy and derive positive benefit from the period of rest assigned her by Dame Nature.

For the purpose of general guidance, but by no means as an infallible or authoritative statement (for the digestive powers of each individual, even when normally healthy, exhibit considerable variation), we append a

TABLE OF THE COMPARATIVE DIGESTIBILITY OF STAPLE ARTICLES OF FOOD.

BOILED DISHES.	TIME. h. m.	BOILED DISHES.	TIME. h. m.
Rice	1 00	Mutton Suet	4 30
Pig's feet, soused	1 00	Fresh Beef Suet	5 30
Tripe, soused	1 00	Tendon	5 30
Salmon Trout, fresh	1 30	**FRIED.**	
Sago	1 45		
Codfish, cured dry	2 00	Salmon Trout, fresh	1 30
Milk	2 00	Fresh Eggs	3 30
Tapioca	2 00	Fresh Flounders	3 30
Wild Turkey	2 25	Fresh lean Beef	4 00
Parsnips	2 30	Animal Heart	4 00
Beans, pod	2 30	Pork, recently salted	4 15
Apple Dumplings	3 00	Fresh Veal	4 30
Fresh Eggs, boiled soft	3 00	**RAW ARTICLES.**	
Fresh Mutton	3 00	Sweet Mellow Apples	1 30
Chicken Soup	3 00	Sour Mellow Apples	2 00
Orange Carrot	3 15	Cabbage, with Vinegar	2 00
Fresh Eggs, boiled hard	3 30	Fresh Eggs	2 00
Irish Potatoes	3 30	Milk	2 15
Mutton Soup	3 30	Cabbage-head	2 30
Oyster Soup	3 30	Sour Hard Apples	2 50
Flat Turnip	3 30	Fresh Oysters	2 55
Beets	3 45	Pork, recently salted	3 00
Green Corn and Beans	3 45	Old, strong Cheese	3 00
Domestic Fowls	4 00	**BROILED.**	
Soup, Beef, Vegetables and Bread	4 00	Venison Steak	1 35
Salted Salmon	4 00	Beef Liver, fresh	2 00
Old Hard Beef, salted	4 15	Fresh Lamb	2 30
Cabbage, with Vinegar	4 30	Striped Bass, fresh	3 00
Pork, recently salted	4 30	Beef Steak	3 00

BROILED.	TIME. h. m.	WARMED.	TIME. h. m.
Fresh Mutton	3 00	Meat hashed with vegetables	2 30
Pork, recently salted	3 15		
Pork Steak	3 15	FRICASEED.	
Sausage, fresh	3 30	Chicken, full grown	2 45
Veal, fresh	4 00		
		BAKED.	
ROASTED.		Potatoes, Irish	2 30
Turkey, wild	2 18	Cake, sponge	2 30
Turkey, domesticated	2 30	Custard	2 45
Sucking Pig	2 30	Corn cake	3 00
Goose	2 30	Corn bread	3 15
Beef, fresh, lean, under-done	3 00	Bread, wheaten, fresh	3 30
Oysters, fresh	3 15		
Mutton, fresh	3 15	MELTED.	
Beef, fresh, lean, dry	2 30	Butter	3 30
Domestic Fowls	4 00		
Wild Ducks	4 30	STEWED.	
Pork, fat and lean	5 15	Fresh Oysters	3 30

The vitally-important and much-disputed question of

CLOTHING AND DRESS REFORM

now remains to be considered. It has been truly said, that " What is everybody's business is nobody's business ; " and, consequently, every one has a different idea on the subject, and it is *let alone*. *Fashions* have been set from time to time, but without any regard to suitability or comfort, or even the satisfaction of the wearer. The laws of physiology are entirely set aside and ignored ; and though every woman on the face of the earth readily acknowledges the evils, inconveniences, and injuries attendant upon the wearing of their garments as at present constructed, it would appear that they prefer to suffer the most indescribable torture, both now and in after-life, rather than disobey the slightest decree of their imperious mistress — Fashion.

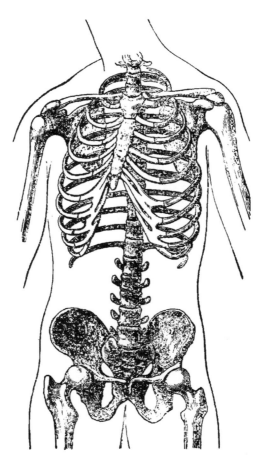

BONY FRAMEWORK OF THE BODY.

How strange it is that nature and the customs of society should be so constantly and persistently at variance! The bony framework, so symmetrical and perfect in its construction, so admirably adapted to every possible requirement of the human organism of which it forms a part, is infinitely more beautiful and serviceable in its normal condition, as detailed on page 57, than in the distorted and ungraceful mass of articulations presented in the waist of the fashionable " Miss " represented in the accompanying engraving. The inference is obvious.

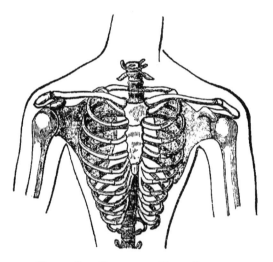

Fig. 7. The Results of Tight Lacing.

The freaks of Fashion are so numerous and erratic that they would scarce be worth noticing, were it not that the results were so dangerous and even fatal. From time immemorial, she seems to have run directly counter to Nature and Nature's laws. Dress should be made to suit the form, not the form distorted to suit the arbitrary rules of the costume. The Venus de Medici (the *beau ideal* of grace and symmetry in the female form) was never encased in a framework of whalebone and steel, with some twenty pounds' weight of dry-goods hanging upon the abdomen and hips. If you will just glance at the anatomical diagram on page 57, you will be able to judge of the normal shape of the ribs and "waist" of a symmetrical. well-formed woman. Now, compare it with the distorted "framework of a fashionable miss" of seventeen (Fig. 7), just emancipated from boarding-school, and you will be enabled to form some faint idea of the misery, suffering, and hourly martyrdom which the reckless votary of Fashion's follies has entailed upon herself.

DRESS-REFORM.

Prof. MARY J. SAFFORD BLAKE, in her admirable lecture on dress-reform, remarks, in reference to the absurdities of woman's dress, that, as a rule, from six to ten thicknesses of woollen or other material encase the region of the waist, while the lower extremities are covered with but *one* thickness, and that only of cotton. Under such circumstances, an effort to obtain proper warmth is usually made by adding an extra supply of skirts, although these garments contribute much more to pressure about the waist, weight upon the hips, and undue heat in the kidneys and abdominal organs, than to warmth in the lower extremities. Still it is in these lower parts of the body that heat is most needed, because there the circulation of the blood is less active, and an undercurrent of air around them is apt to produce chills.

"Let a woman step from a temperature of 70 degrees within doors to zero without, and stand on the street-corner five minutes for a car, while the breeze inflates her flowing skirts till they become converted into a balloon; the air whizzes through them and beneath them, and a wave of cold envelops the entire lower portions of her body. Then let her ride for an hour in the horse-car, with ankles wet from drabbled skirts, and exposed to a continual draught of air ; of course her whole system is chilled through, and it cannot be otherwise than that a severe cold will follow as the penalty for such exposure. A woman, accompanied by her husband, came to consult me, on one of the dreariest days of last winter. Her teeth chattered with the cold ; and you will not wonder at it any more than I did, when I tell you that she had on cloth gaiter-boots, thin stockings, loose, light cotton drawers, two short skirts of flannel, a long one of water-proof, another of white cotton, an alpaca dress-skirt, and an over-skirt. This made seven thicknesses, multiplied by plaits and folds innumerable about the abdomen. Each of these skirts was attached by a double band, and thus the torrid zone of the waist was encircled by fourteen layers. All this weight and pressure rested upon the hips and abdomen, and the result was — what it always will be if this pressure has been long continued — *a displacement of all the internal organs;* for you cannot displace one without in some way interfering with all the others. Here was this woman, with nerves as sensitive as an aspen-leaf to external influences, clad in such a manner that every breath of cold chilled her to the very marrow, the neck and shoulders protected by furs, the hands and arms pinioned in a muff, the head weighted down by layers of false hair, and the legs almost bare ; while her husband, the personification of all that was vigorous in health, was enveloped, as he told me, from head to foot in flannel. His every garment was so adjusted that it not only added to the heat generated by the body, but helped to retain it. I question whether that hale, hearty man would not have suffered twinges of neuralgia or of rheumatism, had he been exposed, as his wife was, to the severity of our atmospheric changes. Even in summer these changes are sudden and severe ; and then men are usually clothed in woollen garments only a trifle thinner and lighter than those worn in winter, while women are often decked in nothing but muslin, and are chilled by every sudden nor'-easter."

Through the courtesy of other physicians I have had the opportunity to be present at the autopsy of several unmarried women, of the class not compelled to labor unduly, so that most of the abnormal conditions of the generative organs could be rationally accounted for only by improper dress. In one girl aged twenty-two, whose waist after death was so slender that you might almost have spanned it with united fingers, there was an atrophied state of all the glandular organs. The death of women occurring under the influence of anæsthetics has in many instances been traced to impeded circulation, resulting from tight clothes. However loosely corsets are worn, the steels and bones must adjust themselves to the various curves and depressions of the body, and must be felt, or else the sure death that women so often declare would follow their abandonment would not be anticipated. As soon as the muscles give warning, by their weakness, that they are no longer adequate to the support of the body, it is high time they were given a chance to recuperate.

It does not require the foresight of a clairvoyant or fortune-teller to diagnose a chronic case of tight-lacing and heavy skirts. You know that when the abdominal muscular walls become inert, and almost wasted, one of the most important daily functions of the body is rarely, if ever, normally carried on. We might enumerate the ill-results that follow ; but these are only links in the long chain of disorders that have won the disgraceful distinguishing appellation of " Women's Diseases," when they should truthfully be termed " Women's Follies." There has been no blunder in the formation of women ; there would be harmony of action in each organ, and in the function assigned it, if Nature were not defrauded of her rights from the cradle to the grave.

FASHION'S PENALTIES.

"A few days ago I stepped into a large corset-factory carried on by a woman. I told her I was interested to know what women and children wear in this line, and asked to see her wares from the least unto the greatest. She began by showing me the tiniest article I ever saw in the shape of a corset, saying that was for babies. Then she brought forward another grade, and still another, and so on, till I think she must have shown me fifteen or twenty different sized corset-moulds, in which she runs the female forms that get into her hands. She informed me that all the *genteel* waists that I should meet on the streets in the fashionable part of the city she had made; and that the mothers brought their daughters in infancy to her, and that she passed them through the whole course of moulds till they were ready for the real French corset, when she considered them finished and perfect.

"Yesterday I visited the first class in one of our City Girls' Grammar Schools, consisting of forty-two pupils. I had five questions on a slip of paper, that I asked permission of the teacher to put to the girls.

"*First.* 'How many of you wear corsets?' *Answer.* Twenty-one.

"I asked them to stretch their arms as high as they could over their heads. In every instance it was hard work, and in most cases impossible to get them above a right angle at the shoulders.

"*Second.* 'How many of you wear your skirts resting entirely upon your hips, with no shoulder-straps or waist to support them?' *Answer.* Thirty.

" *Third.* ' How many wear false hair?' *Answer.* Four.

" *Fourth.* ' How many wear tight boots?' *Answer.* None (which I doubted).

" *Fifth.* ' How many do not wear flannels?' *Answer.* Eighteen.

" I went across the hall to a boys' class corresponding in grade, consisting of forty-four pupils. I asked for the number of boys without flannels, and found only six.

" Of course one hundred per cent. were without corsets, or weight upon hips, tight boots, or false hair. Every boy could raise his arms in a straight line with his body as far as he could reach, with perfect ease." — *From " Corsets vs. Brains," by Louise S. Hotchkiss.*

[Since writing the above, we have been informed by Professor Blake, that the young women of Boston have at last paid some little attention to the reform we have suggested, and that the *average* measurement round the waist now reaches *twenty-seven inches.* — ED.]

The Dress-Reform Committee, of 2½ Hamilton place, Boston, have done noble work in the breaking down of this absurd system of nature-distorting costume, to which the women of America have for so many years been voluntary martyrs. Desirous of inaugurating a *permanent* reform, they have not ventured upon revolutionizing the dress-making art suddenly, but have initiated and brought into practical use the suspenders and underclothing delineated in the following pages for the purpose of relieving the hips and abdominal region of the fearful burden they have hitherto had to bear. A glance at the simple but effective apparatus there portrayed and described will convey, in the most absolute and convincing manner, the strong common-sense, and practical utility which they evince, and the physiological advantages derivable from their use. They speak for themselves in language too eloquent to need any further comment from us. We cordially and sincerely commend those garments and apparatus to the attention and patronage of all our readers who desire to maintain their physical system in its pristine symmetry and health.

Fig. 1.   Fig. 2.

Fig. 1. UNION UNDER-FLANNEL. This is a garment worn next to the skin, and is made to cover the body, and impart uniform warmth, without ligature.

Fig. 2. CHEMELETTE. — A garment combining chemise and drawers in one, or it is made separate, with drawers to button on, called basque waist and drawers. These can be arranged to support skirts and stockings from the shoulder, leaving the lungs and other vital organs free and untrammelled in their action. This may be made of cotton, linen, flannel, or any fabric adapted to the habit of the wearer.

Fig. 3. EMANCIPATION WAIST. — This waist is made separate from the drawers, and may be made double, to take the place of the corset. It is made single for a corset-cover and skirt-supporter.

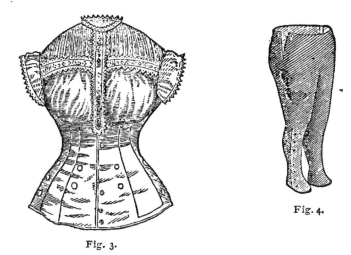

Fig. 3.

Fig. 4.

Fig. 4. DRESS DRAWERS. — The "Dress Drawers" may be worn in place of the underskirt, for extra warmth in riding or walking, and during extreme cold in and out of doors. This article is made of colored flannel, water-proof,

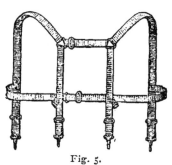

Fig. 5.

or of the dress material, as may be preferred, made to fit the ankle closely inside the boot, or with gaiters to go over the boot, and to fasten by buttons arranged for the purpose upon the "Emancipation Suit," or "Chemelette."

Fig. 5.   SKIRT SUPPORTERS AND SHOULDER BRACE.   Fig.
6.   SHOULDER BRACE, STOCKING AND SKIRT SUPPORTER.
Fig. 7.   STOCKING SUPPORTER AND SHOULDER BRACE.
Stocking-supporters separate are highly recommended.

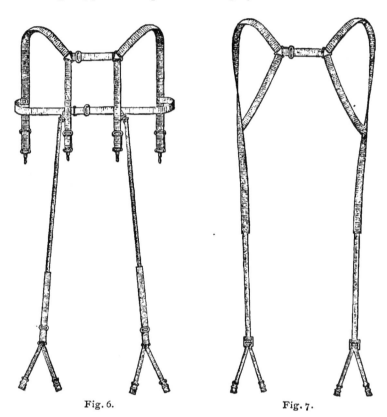

Fig. 6.   Fig. 7.

The principles on which these garments and suspenders
are made is —

1. That the vital organs in central regions of the body
should be allowed unimpeded action.

2. That a uniform temperature of the body should be
preserved.

3. That weight should be reduced to a minimum.

4. That the shoulders, and not the hips, should form the base of support.

We would only remark, in conclusion, that there is a considerable amount of error in the popular idea as to the analogy between the color of clothing and its suitability to the season or climate. White or light-colored fabrics make the coolest garments in summer and the warmest in winter, for the reason that in summer they prevent the sun's rays from passing inward, and in the winter they interrupt or arrest the heat of the body in its passage out. The contrary being the case with dark-colored fabrics, they are rendered less suitable for winter clothing than is generally supposed.

Another great evil prevalent among women is the fashion of wearing *thin shoes*, thus laying a sure foundation for consumption and a thousand other ills of similar character. If our ladies would in this matter only follow the example of English women, and wear the *hard* double sole (*half* the thickness of that of a gentleman's boot), they would live much longer, and leave a more hardy posterity behind them. We rejoice that the native common-sense of New England women has latterly much improved the *fashion* in this respect ; and there is every probability that, in the course of two or three years, our " ladies " will, as a rule, protect their feet from damp by the English " double-sole shoe."

CHAPTER V.

THE MARRIAGEABLE YOUNG WOMAN.

THE young ladies of America should be eternally grateful to that Divine Providence which has cast their lot in a land where, as a rule, woman is respected, independent, and the queen of the sphere in which she lives, moves, and has her being. The United States of America is, without dispute, the Utopia, the El Dorado, for womanhood. Morally, socially, and intellectually, she is the peer of, and in some respects superior to, masculinity generally, sharing their educational privileges and fully participating in all the amenities of "society." But here, alas, our eulogies and congratulations must, for a time, be suspended! Though they possess such manifest and solid advantages over their European sisters in educational advancement and social status, they are far behind them in the vitally important matter of health, *physical strength, and endurance.* The young graduate from the High School in Boston or New York may, *meteor-like*, far outshine her transatlantic sister in intellectual brilliancy, refinement of taste, social attraction, and mental development generally; but in symmetry of form, ruddy, robust health, freedom from constitutional ailment, and general power of endurance, she will not bear a moment's comparison. Like the comet in the solar system, she shines brilliantly, it is true, and becomes the cynosure of every eye; but, alas, for a period as brief as it is brilliant! The fragile, feeble, sparsely-nourished body is signally unequal to the strain brought to bear upon it by the prematurely-developed, delicately-organized mind which inhabits it. In place of the roseate, dimpled cheek and plump figure of the English girl beside her, you note the highly-wrought nervous organization, the pale countenance, fitfully illumined by the hectic flush, the anxious, enervated expression, the languid and exhausted air, the fretful, fitful restlessness of disposition so unmistakably indicative of physical inability to bear the mental pressure. The spirit is alive and equal to any and every emergency; but the flesh, the framework which contains that spirit, is unequal to its slightest effort.

Why should these things be? Nature is not to blame. God made of the same flesh and blood all nations of the earth. The same immortal, expansive soul inhabits European and American alike. Why, then, this great discrepancy, when the germ (material and spiritual) exists equally in every member of the human race? Where the *effect* is, there must the *cause* be found. Nature does not change or vary her laws, nor does the Almighty Creator permit any of his works to fall short of *absolute perfection.* The fault rests *entirely* with ourselves, a fact which we will now endeavor to demonstrate.

Five-sixths of the American women who arrive at the age of puberty are more or less afflicted with some congenital or hereditary defect or tendency of a mental or physical character, — an abnormality which, as a rule, will distinguish and materially influence the whole of their after-life.

PREMATURE DEVELOPMENT.

The influences which lead to this unhappy result are legion; but, fortunately, are all within our own control, if we will only exercise that discretion, care, and caution with which every human being is endowed. There is a time for everything. If a flower is forced into bloom two or three months before its time, it is proportionately deficient, either in fragrance, beauty, or length of life. The hardy perennial, which resists the snows and frosts of winter, knows not the atmosphere of the hot-house. So it is with the child brought up in accordance with Nature's laws. The smiling infant, nestling in its mother's loving arms, and fed from Nature's fountain *alone*, until dentition enables it to partake of stronger food; whose lithe and sinuous form is untrammelled by iron hoops and mummy-like bandages; whose rosy cheeks are fanned and braced by the pure breath of heaven, and illumined by the glorious sunlight; whose round and chubby limbs are daily bathed in waters from the limpid brook and crystal spring; whose beaming eyes are greeted with loving looks and cheering words from its tender nurses; whose little feet are cautiously trained to step from chair to chair until they gain sufficient strength to walk alone, — an infant the subject of such loving care will surely develop into a child able to combat successfully with all the ailments to which childhood is specially liable; its mental capacity will expand simultaneously with its physical growth, and, in all human probability, it will reach the close of the first epoch of its life in the full possession of its powers, and with the ability to press onward, happily and victoriously, to a youth of promise and practical usefulness.

CITY CHILDREN.

What a contrast, however, is presented in a walk down our crowded city thoroughfares! Fully fifty per cent. of the babies you meet you would not recognize as *babies*, — they seem more like old men and women in miniature; all the infantine loveliness has departed, if, indeed, it ever existed, and a really healthy, symetrically-formed, well-proportioned baby is the *exception;* and why? The reason is apparent: the surroundings of these hapless little beings are the exact reverse of the *pen-picture* we have just drawn; they are brought into the world, but only to drag out a miserable existence for a few years, and die an early death. Take the census of Massachusetts for 1875 as an example, and a comparison of the births and deaths in that State will reveal some facts worthy our most serious attention. The births to American mothers during the previous year were exactly *one-half* those to foreign mothers; that is, there was one birth to nine native-born mothers, while the foreign-born return one birth to four and three-fourths individuals. The percentage of births to American parents, furthermore, shows a constant diminution, while the foreign-born evince a corresponding increase. Nor is this the worst phase in the matter. Of these fully half die before they reach the age of thirty; three-fourths of the mortality occurring within the first five years (or the period of infancy).

DANGERS OF GIRLHOOD.

We will suppose, for argument's sake, that the young girl has surmounted all the dangers of infancy, and that the gayeties of girlhood life, the school-room, the play-ground, and young companions, loom out before her. Here are fresh dangers for *her*, fresh responsibilities for her guardians. Many an evil habit which has imprinted its sign-manual on the features, character, and future life of its victim, can be traced to her school-girl days, when the mind was fresh, fertile, and impressionable, and the physical functions in that peculiar condition of receptivity and susceptibility that made or marred their prospects and principles for the whole of her natural life.

Fresh from the nursery and the domestic hearth, the child's eyes open on new scenes and new surroundings;

its heart is opened to new impressions; its affections seek and welcome new companionships, — how vitally essential, then, that these scenes, impressions, and companionships should be pure, elevating, innocent, and congenial! The most prominent characteristic of child-nature is *imitation*. If vice, profanity, looseness, deceit, formality, fashionable folly, excess of any kind, characterize the surroundings of the young girl; if the senior members and visitors of the family display such tendencies, will not she follow but too readily the example set her? Will not she imagine that their example would fully justify her imitation thereof? But these *home influences* form but one moiety of her life. Her *books* and her *school associations* have quite as important an influence on her future. On the other hand, *undue restriction* — debarring the child from healthy and seasonable recreation; excessive taxation of the intellect by protracted study; exclusion from all associations of a genial, health-inspiring, vigorous nature, and the adoption of a rigorous, monotonous routine — will produce an effect quite as calamitous and unnatural as the course we have just depicted.

THE WOMEN OF THE FUTURE.

The only true and proper course of mental and physical culture is Nature's own way, — the *via media*, — in which the latent womanhood is *controlled*, not subdued; in which truthfulness, honesty, natural, unaffected demeanor are inculcated and encouraged; where the sanitary and hygienic laws of both mind and body are faithfully adhered to; where the hours of study are restricted to *six hours* as the outside limit, recreation another six hours, and the remaining twelve hours are devoted to food and rest. Combined with this, the moral and social elements must be of the purest, and the influence of sensational scenes and sentiments studiously avoided. The young girl who is just entering the confines of that mysterious physical change which is to transform her into an eligible candidate for the sacred duties of womanhood requires all the care and tenderness which love and sympathy can suggest.

THE AGE OF PUBERTY, OR WOMANHOOD IN ITS EARLY BLOOM,

has now fairly commenced. The chrysalitic period has passed. She appears for the first time on life's stage as " a thing of beauty ; " it rests with herself, her counsellors, and advisers, as to whether she is or is not to be " a *joy* forever." If the design of her Divine Creator be carried out to its full fruition, she will not only be a *joy*, but an unmitigated blessing to all future generations. How wonderful the change which has taken place, both outwardly and inwardly, physically and mentally, in the being now before us ; whether we take the sturdy, healthy, pleasant-looking dairywoman of the suburban farm, or the languid, pale, but graceful maiden of the city drawing-room. The awkward gestures, angularities, and innocent freedom of manner which characterized the school-girl have given place to a roundness and symmetry of form, a gliding gracefulness of gesture, a maidenly, gentle reserve, and a mellow, rich, melodiousness of voice, which, of itself, would indicate the fundamental revolution her constitution and her functions have undergone. Her childish playfulness and love of mischief is now superseded by a gentle dignity, and, in the full enjoyment of mental and bodily health, a kind of self-consciousness of power to please. A sense of the possession of new thoughts, new desires, and new relations to the world, vague though it be, pervades her whole being and imparts to her new fascinations and feelings which she scarcely understands. But if she has not been duly prepared and educated for this change, she seeks retirement, shuns society, has tearful paroxysms, is inattentive to her studies, her memory becomes treacherous, she exhibits a disinclination to mental exercises, loses all her playfulness of disposition, and listlessness and inertia pervade all her actions. This is the culminating point of her life, moral, physical, and mental. She is now destined to be the crowning glory of the human race, or a pitiful, miserable wreck of humanity, an aborted, distorted, and heterogeneous collection of noble elements run to waste, and rendered worse than useless. Every man possessing common sense and honesty holds in the highest and most sincere admiration the transcendental and glorious qualities of woman in her pristine condition of mental and bodily perfection — and, on the other hand, beholds, with most profound and unspeakable sympathy and sadness, the numberless departures from that normal state of beauty and completeness enjoyed by our first parents.

The only obstacles to this normal condition of health of body and mind are to be found in that one dread word — Disease — congenital or acquired.

THE DISEASES INCIDENTAL TO THE PERIOD OF PUBERTY.

At this critical period the seeds of hereditary and constitutional diseases manifest themselves. They draw fresh malignancy from the new activity of the system. The first symptoms of tubercular consumption, of scrofula, of obstinate and disfiguring skin diseases, of hereditary insanity, of congenital epilepsy, of a hundred terrible maladies, which from birth have lurked in the child, biding the opportunity of attack, suddenly spring from their lairs, and hurry her to the grave or the mad-house.

We propose, however, to take the several affections to

which young women are liable, in the order of their prevalence.

THE SYMPATHETIC ACTION OF THE UTERUS, OR WOMB,

with the other functions of the body is so marked, and is
so little understood by women generally, that a word or two
on that subject is necessary. In the earlier ages of medical science this sympathetic influence was much exaggerated ; but the recent researches of modern science have
demonstrated beyond all doubt that in many cases where
the breasts, stomach, bowels, brain, heart, vocal organs, or
facial nerves have been supposed to be the seat of the affection, ultimate experience has proved that the real location
of the disease was in the uterus — the action on the other
organs being only of a secondary or sympathetic character ;
as, for instance, in neuralgia, hysteria, etc.

The diseases of the uterine system may be divided into
three sections — FUNCTIONAL, ORGANIC, and MALIGNANT, —
the *functional* consisting of those deviations from the natural
menstrual secretion known as amenorrhœa, or suppression
of the menses ; dysmenorrhœa, or difficult and painful menstruation ; and menorrhagia, or excessive menstruation.
Nearly allied thereto is that annoying, depleting affection
known as leucorrhœa, the whites, or vicarious menstruation.
All these disorders, when uncomplicated with other affections, may continue for many years without evincing any
tendency to degenerate into organic disease of the womb.

ORGANIC DISEASES OF THE UTERUS.

The *organic* diseases of the uterus and its appendages are
frequently the result of inflammation, either of the mucous
membrane or of the muscular and vascular tissues, producing induration, softening, ulceration, and abscesses.

The secreting surface of that organ may also originate purulent discharges, or its cavity be distended with air, fluid, or masses of degenerated tissue called moles and hydatids. Certain organic changes not unfrequently produce *fibroid* tumors, of various degrees of consistence, which mainly arise from a change of nutrition.

The most insidious, dangerous, and fatal class of uterine troubles are those denominated *malignant diseases*, comprising fungous growths, malignant ulcerations, and morbid deposits. The fungous growths are called cauliflower excrescences, being nothing more than a mass of enlarged and diseased vessels, surrounded by cellular substances. *Malignant ulcerations* usually commence their attack at the neck of the uterus, rapidly spreading to the body of that organ, and even including the vagina in their operation. *Cancerous deposits*, on the other hand, may originate either in the neck or body of the uterus, or in the cellular tissue connecting the uterus with the adjacent organs, and may occur simultaneously in two or more places.

Besides these recognized diseases, the uterus is subject to various *accidents*, such as rupture, displacement, etc., of which we shall have to treat in their proper order.

The FALLOPIAN TUBES AND OVARIES are also liable to certain morbid changes of a similar character to those affecting the uterine system, the most common of which are obliteration of their canals, distention by serous effusion, tubercular or encephaloid matter, adhesion, and cancerous deposit.

But up to the period of commencement of the menstrual flow, diseases of the internal organs are of *very rare* occurrence. As we have already remarked, the assumption of the menstrual function opens the flood-gates of inherited, latent disease; it awakens the susceptibilities of the generative organs to the dominating influence, whether that influence be normal or abnormal in its tendency — whether the several functions be healthy and vigorous, or feeble and inert.

We will now consider the

AFFECTIONS OF THE EXTERNAL ORGANS OF GENERATION,

which may have had their origin prior to the establishment of the menstrual function. These are so constructed as to render them liable to a variety of complaints calculated to alarm the patient unnecessarily, from the prominence or urgency of the symptoms, though they are seldom attended with absolute danger.

The Labia Majora (or outer lips of the vagina), on account of their looseness of texture, sometimes become enlarged and inflamed from slight irritation, and necessarily a source of considerable annoyance. They are hard, red, and very sensitive to the touch; and more or less fever, accompanied by burning or shooting pains, is invariably present. It may arise from cold, excoriation or chafing, or from mechanical injuries, though in very many cases it is produced by neglecting to keep the parts properly cleansed *every day* during and after the periodic menstrual purgation. In the latter case, the accumulation of the acrid secretions causes an intolerable itching; friction or rubbing is consequently indulged in, and the result is excoriation. The same symptoms also occur in the Nymphæ, and from similar causes.

TREATMENT.

Frequent ablution of the parts with warm water, especially during the flow, is absolutely necessary in any case; which should be followed, if caused through cold, by administering belladonna, mercurius, or rhus; if by inflammation, rhus and belladonna in alternation; and if by mechanical injury, arnica, both internally and as a lotion, will prove effectual. (*See " Homœopathic Remedies."*)

ABSCESSES, TUMORS, ETC.

Abscesses not unfrequently form in the labia, and are extremely painful, and are usually caused by blows, falls, forcible intercourse, or other external injuries, though occasionally they may be traced to a general disposition to inflammatory action, without any other exciting cause. Where an abscess is present, a hard, throbbing pain is added to the previously described symptoms of heat, swelling, and redness, which involves the groin and a large portion of the thigh. The location of the abscess may be easily discovered by its hardness. Of course, rest and quiet are indispensable; a gentle purgative treatment must be adopted, and, if possible, suppuration be induced by means of warm poultices. If it will not then yield, incision will be necessary. (*See " Laxatives" in Remedies, and " Poultices" in Sick-Room Management.*)

Encysted Tumors, generally circumscribed and of varied circumference, are occasionally met with, the only remedy being incision or entire eradication.

Warty Tumors occur in this organ, both singly and in clusters, and are generally suspended by a pedicle from some part of the external surface. They vary in size from a pea to a turkey's egg, and have a tendency to spread to the internal surface, but are neither painful nor tender, and are only inconvenient on account of their bulk. In most cases these tumors are *venereal* in their origin; and if permitted to suppurate, form unhealthy sores. They must be taken away by knife, caustic, or ligature; the patient be kept quiet and have a moderately nutritious diet; and if syphilis be the origin, mercury, in one of its many forms, should be administered, with the occasional alternation of a gentle purgative to purify the system. (*See "Laxatives" and "Treatment of Leucorrhœa."*)

The Nymphæ, or Labia Minora, are subject, as we remarked, to the same contingencies as the labia majora. The same treatment is requisite, with perhaps additional attention to the local application of bread-and-milk poultices. It was the practice of the Arabs and Moors, in the earlier ages, to apply the rite of circumcision to their young women, by the excision of the part.

The Clitoris, adjacent to the labia, is subject to enlargement from inflammation and to cancerous growths. In the former case, cooling and astringent lotions outwardly, and belladonna, mercurius, or rhus inwardly, would give relief; but in the case of cancerous growths there is no alternative but extirpation, which should be performed at as early a period as possible. A curious popular error has arisen from the *malformation* and *enlargement* of the clitoris, which has deceived even the medical practitioner, and has been the cause of much scientific and legal controversy; viz., the possibility of the sexual organization of male and female being coexistent in one individual. We only mention the fact for the purpose of showing the source of error and dispelling an idea so absurd and erroneous from the minds of our readers.

Imperforation of the Hymen. — It is by no means infrequent that the hymen is found to be imperforate, or of such intense density that great difficulty is experienced in the performance of the menstrual function. In the case of imperforation, no inconvenience is felt until after the age of puberty, when the menstruous fluid is duly secreted, but, finding no outlet, is accumulated within the uterus, causing pains at those periods very nearly resembling those of labor, for which they have frequently been mistaken. The same symptoms, somewhat modified, attend cases of partial occlusion, though it does not entirely preclude conception, several instances having occurred in which safe delivery was ultimately effected. The remedy is necessarily the same in both cases; viz., the division of the confining membrane.

Inflammation of the Vulva. — The inflammation of the mucous membrane of the vulva is one of the most common, the most unpleasant, and the most obstinate of all the affections to which the female generative organs are subject. It occurs at all periods of life, though it differs in its outward manifestations. In children it is called *Infantile Leucorrhœa*, involving the whole of the mucous membrane, and extending to the vagina, and producing a profuse milky discharge. It is more generally found among the children of the poor, and is the immediate result of neglect, want of cleanliness, cold, mechanical injuries, etc. In the young woman or adult the discharge is thicker, more copious, and of a yellowish color, acrid, and excoriating the skin at the margin of the external organs. In the milder form, the treatment is simple and ordinarily successful.

TREATMENT.

The parts should be well fomented with camomile flowers or poppy-heads three or four times per day; and a weak dilution of merc. corr., hyoscyamus or hamamelis virginica applied. Internally, rhus, belladonna, or. aconite might be administered. If the disease has become obstinate and chronic, and assumes the form of Leucorrhœa, of which it is the type and origin, astringents and tonics of a more pronounced character should be given. (*See No.* 6 et seq.) If complicated by diarrhœa, pulsatilla or colocynth would remove the symptoms. In all cases extreme cleanliness should be observed, with a spare diet.

PRURITUS, OR ITCHING OF THE VULVA.

This distressing and troublesome complaint is the most dreaded and obstinate disease to which women are subject, and is generally symptomatic of some other disease in the uterus, bladder, or rectum. It takes away all rest or sleep during the paroxysm, and consequently causes extreme debility. It is by no means confined to the period of pregnancy, for young unmarried women are frequently troubled by the visitation. The irritating sensations are so overwhelming, and the desire to scratch so indomitable as sometimes to put decency to defiance; they make her melancholy, unfit for society, desirous of solitude, and subject to the most intense physical and mental excitement. It almost exclusively affects the abdomen, the private parts, and the fundament. The predisposing causes of pruritus are: uterine, vaginal, or urethral disease; pregnancy; depreciated general health; habits of indolence, luxury, or vice; uterine or abdominal tumors; want of cleanliness; constitutional syphilis, and severe exercise in one of sedentary habits. The immediate exciting causes are the contact of an irritating discharge and local inflammation or irritation, — leucorrhœa being by far the most prolific source. In every case the vagina should be carefully investigated for evidence of leucorrhœal discharge, unless some other sufficient cause is apparent.

TREATMENT.

The treatment should be mainly directed to the disease of which this is a leading symptom, — the internal remedies being accordingly chosen from conium, kreosote, bryonia, arsenicum, rhus, pulsatilla, silica, sulphur, lycopodium, and graphites, in accordance with the originating affection.

A very efficacious and simple external application is made by dissolving one ounce of borax (biborate of soda) in a pint of rose-water or soft rain-water, and washing the affected part therewith several times a day. We have found a solution of mercurius corrosivus extremely effective as an outward application, in combination with myro-petroleum soap. (*See No.* 252.)

We are now brought, in the logical sequence of events, to consider those diseases to which the young woman is especially liable prior to or independent of the menstrual function, the most prevalent being

CHLOROSIS, OR THE "GREEN SICKNESS,"

which derives its designation from the peculiar pale green-ish hue it gives to the complexion. As a rule it manifests itself about the age of puberty, and is characterized by the hue of the countenance just described, deficient warmth, perverted, depraved appetite, with occasional nausea or sickness, great physical and mental weakness, impaired digestion, palpitation of the heart, and general derangement of the sexual function.

Among the middle and higher classes a predisposing condition of the system (the very opposite of the robust, full-blooded condition which should exist) to chlorosis or general debility is unhappily far too general, so much so that more than three-fourths of the young women in the large cities of Europe and the Western Continent are the subjects of uterine trouble from the very commencement of their womanly career. Chlorosis usually accompanies re-tarded or suppressed menstruation; but though this is, and rightfully so, a subject of great anxiety to parents, yet there is no justification, under any circumstances, of adopting means for the purpose of hastening menstruation, nor for attempting to remove local sexual ailments by the adminis-tration of astringent, detergent, and other remedies to produce the menstrual flux. It is only when all other indications demonstrate beyond doubt that some abnormal obstruction alone retards this natural process, that it becomes absolutely necessary to invoke artificial auxiliaries; and even then the most skilful *medical advice* should be sought, and not the numberless patent and quack nostrums which beset the innocent and ignorant on every side, and only serve to still further derange and frequently destroy the healthy action of the sexual functions altogether.

SYMPTOMS.

Chlorosis is *always* a chronic disorder, and commences slowly. The patient is at first languid and listless, disinclined to amuse herself, and is easily fatigued by ordinary mental or bodily occupation; her face gradually becomes pale, and the skin assumes a sallow appearance; the bowels are constipated; she loses her appetite, and has an unnatural craving for certain articles of food; the tongue is white, the breath fetid; and if menstruation has already been established, the discharge loses its red color, and gradually diminishes in quantity until it is entirely suppressed.

In the confirmed or chronic condition there is often considerable emaciation; the flesh loses its firmness; the lips, tongue, gums, and inside of the mouth are unnaturally pale or whitish; slight swelling in the eyelids and face is observed in the morning, which wears off during the day, and at night the feet and ankles are swollen; the urine is pale and limpid, the abdomen is frequently enlarged from flatulency, particularly after eating; there is sometimes nausea or vomiting in the morning, heart-burn, and other symptoms of indigestion. The appetite is in many cases morbidly capricious; sometimes there is an insatiable desire to eat pickles, chalk, lime, pipe-clay, cinders, etc. The shortness of breathing, which, in the first stage, was only slight, is now exceedingly oppressive, and accompanied with palpitation of the heart on ascending the stairs, attempting to walk quickly, etc. The pulse is feeble and small; there is great difficulty in keeping the feet warm; sometimes there is cough, periodical headache, and a variety of nervous or hysterical symptoms.

Females of the lymphatic temperament and of weak constitution are most frequently attacked with chlorosis. It is developed under various debilitating causes, as frequent exposure to a cold, moist atmosphere, watery or poor diet, more especially when conjoined with fatigue and long watching, the various depressing passions, as grief, unrequited love, etc.

This disease seldom proves fatal; but, when left to itself, or badly treated, it may be prolonged during many months or even years, and leave traces of its injurious effects on the constitution in after-life.

TREATMENT.

There is no disease in which the administration of iron, in some one of its many forms, is attended with such uniformly favorable results. The tincture or chloride of iron should be taken in doses of fifteen drops three times a day, about half an hour before each meal; the diet must be nourishing, as fresh meat and bread, and easy of digestion. The patient should walk out in the air and sunlight, take plenty of sleep, and frequently sponge the body and rub thoroughly dry with a coarse towel. Care is necessary, however, to avoid attempting too much at first. The bowels have to be regulated before taking a tonic course, and the stomach thus become accustomed to the medicine. The use of tight corsets and the prevalent method of dress among women, viz., overloading the abdominal region and hips with a superfluity of clothing, while the lower extremities are exposed to the inclemency of the weather, should be entirely abjured; the bed should neither be too warm nor too soft; sleep should not be protracted beyond eight or nine hours, — and all food, drinks, or amusements calculated to cause undue mental and physical excitement, such as wine, highly-spiced condiments, and made dishes, vivid emotions, frequenting balls and other entertainments of an exhausting character, the reading of highly-wrought romances, etc., should be absolutely forbidden. (*See Nos. 7, 156, 157, 158, and 159.*)

HYSTERIA, OR HYSTERICS.

Hysteria has in many respects a close resemblance to epilepsy, and is supposed by many physicians to be a species of that disease. Several well-marked symptoms, however, distinguish these disorders from each other. In hysteria the face is not nearly so much distorted, nor does it ever acquire a livid color, as in epilepsy; and in the former affection the patient generally hears what is said to her and seldom becomes entirely insensible; froth does not appear at the mouth, there is no grinding of the teeth, nor is the tongue ever injured; the breathing is not stertorous or snoring; and the hands remain open.

SYMPTOMS.

A paroxysm or fit of hysteria is generally announced by headache, restlessness, cramps, coldness of the feet, yawning. and sometimes by immoderate fits of laughter, or crying and laughing alternately; the patient experiences a peculiar sensation, as if a ball were moving about with a rumbling noise in the abdomen. This, after some time, rises to the stomach, and thence to the throat, where it fixes itself, causing a most intolerable feeling of choking or strangulation. The breathing now becomes hurried, the heart palpitates; giddiness, sickness at stomach, and dimness of sight follow. The patient then falls down, seized with convulsions; she screams, perhaps tears her hair and beats her breast; her body is writhed to and fro, and her limbs assume a variety of postures. The convulsive movements are not constant; a succession of fits take place, with longer or shorter intervals between them. Sometimes the urine is discharged involuntarily; and during the absence of the convulsions the patient laughs wildly, cries, or screams, and sometimes a distressing hiccup comes on. The abdominal muscles may be irregularly contracted, the abdomen drawn inwards towards the spine, or tense and distended with air; the veins of the neck are greatly distended, and the carotid arteries beat with unusual violence. In delicate females the face is pale and flushed alternately; in the more robust, it is flushed, and appears fuller than usual. The patient having remained in this state during a longer or shorter period, often for twenty-four hours, and sometimes considerably longer, at length begins to recover gradually. The spasms abate; wind is freely discharged from the stomach; there is frequent sighing or sobbing; she complains of severe headache, with a feeling of soreness over the whole body and limbs, and lies in a languid and listless state for some time before she is able to rise. The recovery in some cases is sudden, and accompanied with a loud fit of laughter or immoderate crying; and there is often a copious discharge of pale urine.

SIMILARITY TO OTHER DISEASES.

This disease simulates so many others, and assumes such a variety of symptoms, that a concise description fails in conveying an adequate idea of it; but we do not see any necessity for giving a minute account of all its various forms and relations, because, however formidable in appearance, it is never attended with positive danger.

A point, however, of considerable importance with regard to hysteria is the difficulty of distinguishing it from other diseases; indeed, it has such a close resemblance to hypochondria in males that medical men are often embarrassed by the variety of symptoms which occur in hysterical females; and in many cases considerable experience and judgment are required in order to be able to discriminate between functional or even organic disorders and the endless variety of forms which this affection presents. An hysterical female sometimes complains of great pain and tenderness in the abdomen, and even screams if it be touched; she may have headache at the same time, and remain in bed during several days; but the pulse continues tranquil, and the skin is not hotter than natural. Many girls, however, have been bled repeatedly while in this state, under the idea that some inflammatory action was going on.

Pains about the region of the heart, accompanied with palpitations and occasional fainting fits, constitute another form assumed by hysteria, and may at first lead any one ignorant of the use of the stethoscope to suppose that organic disease of the heart exists.

Females from fifteen to thirty years of age are most liable to hysteria, and it is generally observed in those of a highly nervous temperament, with spare habit of body; or in plethoric and fat persons with soft and relaxed muscles, who are subject to irregularities of the menstrual discharge.

EXCITING CAUSES.

The most common exciting causes are disappointed love, jealousy, undue excitement, ungratified desires, and all powerful mental emotions which act strongly on the nervous system and tend to induce derangements of menstruation. Hysteria, in fact, depends almost entirely on the education, social position in life, mode of living, and moral training of females; many, from having been over-indulged when children, become irritable, wayward, capricious; and, in a word, are so self-willed that the slightest disappointment or opposition brings on a paroxysm. The eminent physician Sydenham remarked, some years since, that "upon the least occasion they indulge terror, anger, jealousy, distrust, and other hateful passions; and abhor joy, hope, and cheerfulness, which, if they accidentally arise, as they seldom do, quickly fly away, and yet disturb the mind as much as the depressing passions do; so that they observe no mean in anything, and are constant only to inconstancy. They love the same persons extravagantly at one time, and soon after hate them without a cause; this instant they propose doing one thing, and the next change their mind and enter upon something contrary to it, but without finishing it. So unsettled are their minds that they are never at rest." People in general are not much inclined to sympathize with hysterical females, however formidable or alarming the fits may appear, because it is well known that this affection is, in a great measure, under their own control; and, in fact, in nine cases out of ten the paroxysm is the result of a fit of bad temper, or of some excitement which could not have arisen in a well-regulated mind.

TREATMENT.

Two indications are to be attended to in the treatment of hysteria: the first is to shorten or moderate the violence of the paroxysm; the other, to prevent the return of the fit. The application of cold water to the head and neck, and of pungent salts to the nostrils, is sometimes practised with advantage, though in mild cases the fit may be allowed to run its course with safety. When the paroxysm is severe, the first thing to be done is to prevent the patient from receiving injury by the violence of her struggles. She should be placed in bed in a well-aired apartment, her shoulders raised and her dress loosened. If she be capable of swallowing, a teacupful of cold water or the following draught may be given : Camphor mixture, two ounces ; sal volatile (aromatic spirit of ammonia), a teaspoonful ; mix. Or a teaspoonful of ether may be given in a little cold water. Should the face be flushed and the head hot, cloths moistened with ether are to be placed on the forehead, or wet towels or pieces of linen may be applied to the same part.

In order to effect a radical cure of this affection, attention must be paid to the general health of the patient, and to the state of the digestive organs and womb. If the habit of body be full and plethoric, low diet and exercise are proper ; but if the patient be delicate, and her stomach debilitated, tonic remedies, such as small and repeated doses of quinine and preparations of iron, are the most suitable remedies. Medicine, in order to have any decided effect in this disorder, must be directed towards improving the condition of the digestive and uterine functions. (*See also Nos.* 13, 102, 231, 232, 233, 234, 235.)

Hysteria may attack any part or organ of the body, and resemble organic disease. Like hypochondria, it is really a disease, and is to be treated accordingly. Moral influence can do much, but it cannot do everything. It is a nervous disease, and is to be managed on the same general principles as other nervous diseases. It should always be remembered that hysteria is to woman what hypochondria is to man. Both are really diseases ; are probably symptoms of some disturbance of the central nervous system ; gradually increase in frequency ; are often premonitions of actual insanity; may occur at any time of life after puberty; and both diseases may often be relieved and cured by appropriate treatment.

ILIAC PASSION, OR ILEUS.

This disease consists of excessive vomiting, with obstinate constipation of the bowels. It is so dangerous and erratic that it may commence suddenly and terminate fatally within four or five days, though cases of this description are fortunately very rare. It usually commences with acute griping pain, obstinate constipation of the bowels, retraction of the navel, and the usual symptoms of severe colic, which, not being relieved by any mode of treatment, a still more distressing state supervenes. The patient is racked with severe pain; the abdomen becomes swollen and tender to the touch; the pulse is weak, small, and quick; thirst is urgent; face anxious and shrunk; fecal matter is vomited; cold sweats, hiccup, and frequent fainting-fits follow, and death generally puts an end to the patient's misery. In some cases acute pain is felt at a particular part of the abdomen, accompanied with heat of skin, quick pulse, thirst, and the ordinary symptoms of inflammation; in others there are no symptoms of fever; in the latter case, life may be prolonged a considerable length of time.

CAUSES.

Ileus may arise from various causes, the principal of which are: ruptures, one portion of the bowels passing within another and becoming entangled; contraction or stricture of the bowels; obstruction from cancerous or other morbid growths; bands, formed by false membranes, strangulating or compressing a portion of intestine; paralysis or torpor of the bowels, arising from hardened fæces, impacted in some part of the intestinal canal ; or it may be a symptom of inflammation of the bowels.

TREATMENT.

In every case, the first thing to be done is to ascertain whether or not the disease is the result of hernia or rupture. A hernial tumor is sometimes so small that the patient is ignorant of its existence, or may not consider it worthy of notice ; and women are often ashamed or unwilling to admit that they have any complaint of this nature. The bare statement of the patient is not by any means sufficient in these cases : the most skilful and careful medical examination for rupture should be made. The necessity for procuring the best professional assistance at an early stage of the disease, in order to avoid intense suffering and death, cannot be too absolutely insisted on.

Another essential point to be attended to before having recourse to any remedial means, is to ascertain whether or not the disease is accompanied by inflammation, the signs of which are a constant, acute, and burning pain in the abdomen, which is distended, tense, hot, and acutely sensitive to the slightest pressure, urgent thirst, and high-colored urine. In this case, instead of administering opiates and strong purgatives, which would soon destroy the patient, recourse must be had to judicious blood-letting, and the means usually adopted to subdue inflammation of the bowels, of which the ileus may be only a symptom, which will, of course, be removed along with the inflammation. If the disease does not depend on hernia, and if no inflammatory symptoms be present, the careful administration of purgatives and opiates will be necessary.

NYMPHOMANIA, OR FUROR UTERINUS,

is a disease frequently confounded with pruritus vulvæ,

from the similarity of some of the symptoms, and also from the fact that they are sometimes found to coexist in the same patient. The diseases are, however, widely different in character, and require special treatment. The source or seat of this affection, nymphomania, has been a much-disputed point for many years, some placing it in the brain and others in the genital organs. We believe, however, with M. Columbat and other modern authors, that it is a simultaneous irritation of the brain and sexual organs. As its name implies, it is an exaggerated voluptuous sensation, accompanied by irresistible and insatiable venereal desire. It is less a disease than a symptom, for the concurrence of the brain and the sexual organs is essential to the erratic manifestations. In the former case, the disease is developed under the influence of mental causes, which secondarily irritate the genital organs; and in the latter, it is a primary irritation of the organs of generation, reacting sympathetically upon the brain, and especially on the cerebellum.

SYMPTOMS.

This affection, which may appear at all ages, generally attacks women of an original *uterine* temperament (that is, of sanguine temperament, and suffering extreme irritability of the abdominal viscera). Women of this constitution have certain characteristics: their stature is small, their skin dark, and their complexion highly colored; and the breasts, and all the other attributes of womanhood, are fully developed and extraordinarily sensitive. Young widows, women of ardent nature, fond of pleasure and excitement, and, finally, those attacked with some chronic cerebral or uterine affection, especially those inhabiting warm climates, are especially liable to this form of disease, from the vivid character of their passions and the super-exaltation of their imagination.

We will not pursue the frightful train of symptoms attached to this disease, but only remark in general that the patient gradually loses all self-control, completely forgetting her native modesty, at last falling into a furious delirium. The generative organs are red, excoriated, much swollen, and the seat of a purulent, fetid discharge ; there is ardent thirst, grinding of the teeth, spasm of the œsophagus, and, in some cases, what are called hydrophobic symptoms, or rabies. The attacks are frequent, periodic, and often fatal, especially where organic disease of the uterus or its appendages is present.

TREATMENT.

In general, much greater benefit is derived from *hygiene* than from drugs, which are not so successful in opposing a disease the origin of which lies more in the mental and moral than in the physical region. Some means of diversion should therefore be devised to turn her imagination from seductive or improper thoughts, such as constant mental and bodily occupation of some kind, travelling, new scenes and companions, etc. Everything capable of producing erratic excitement should be studiously avoided. Frequent and prolonged warm baths, with cold affusions to the head, cold and sedative drinks, and light diet of cereals, poultry, milk, vegetables, and total abstinence from all stimulating drinks and food, from soft featherbeds, and all the enervating luxuries of modern civilization, must be rigidly enjoined.

HYSTERALGIA, OR IRRITABLE UTERUS.

This disease is also known as *neuralgia* of the uterus, from the fact that those afflicted with irritability of the uterine region are more or less affected with neuralgia of that organ.　Dr. Gooch defines it as " a tender and painful state of the uterus, neither attended by nor tending to produce change in its structure."　This disease occurs in females of every temperament, and may appear at any time within the menstrual age; the young and middle-aged, however, are most subject to it, the aged being rarely attacked. This affection, at times, is one of extreme suffering.

SYMPTOMS.

There is pain in the lower part of the abdomen and loins, which usually comes on a few days before or after menstruation, and from which the patient is never quite free.　It is subject to aggravations from mental excitement and bodily exertion; hence the patient is induced to give way to the relief afforded by repose.　The result is, that, in consequence of the pain, want of exercise, and fresh air, the general health is broken, and a languid circulation, constipation, and dyspepsia are superinduced.

Upon examination, the uterus is found tender from slight pressure.　Sometimes the neck and body are slightly swollen, but not hard; the mouth of the uterus and vagina are usually healthy.　The disease may continue for months or years; it may be subdued by medical treatment, or it may subside spontaneously.　It is a sure cause of barrenness while it exists, but it does not terminate in organic disease of the uterus or endanger life.

Leucorrhœa sometimes, though not always, accompanies this disease.　It may be distinguished from acute and chronic inflammation of the uterus by the absence of heat and throbbing, and by the long continuance and non-progressive character of the symptoms, without apparent invasion or degeneration of the organ itself.　It is distinguishable from " painful and difficult menstruation" by the non-cessation of the pain, in greater or less degree, throughout the intervening periods. It differs from displacements of the uterus and vagina by the maintenance of those organs in their natural position during this affection. From a comparison of this malady with long-continued and painful affections of other parts of the frame, it must be inferred to be nervous. Thus, the breasts, the spine, and the various joints may be the seat of acute pain, which may endure for many years without being accompanied by organic change.

The CAUSES of this disease are often obscure. The most prominent are : bodily exertion during menstruation or at too early a period after delivery or abortion ; excessive coition, and an improper use of astringent injections. It may also supervene on extreme fatigue, such as long journeys, dancing, dissipation, late hours, etc.

TREATMENT.

There are few diseases so tedious of cure and liable to relapse as this. The indications are : first, to relieve the pain ; second, to restore the constitution to its normal condition. The violence of the pain must be allayed by absolute rest during the paroxysm, and narcotics, such as opium, hyoscyamus, etc. (*Nos.* 231, 256, 257), either alone or in combination with assafœtida. If the stomach be too irritable, they will be found equally as efficacious by injection into the vagina or rectum. Opium or belladonna plasters, or anodyne fomentations to the sacrum and abdomen are also useful. Counter-irritation should be practised by means of small blisters to the loins. Much relief will be afforded by vaginal injections of warm water alone, or aconite and belladonna if the pain be severe. The warm hip-bath will often give relief. The bowels should be kept open by the mildest possible means, as all active purgation in this disease is invariably injurious.

Although, during a severe paroxysm of pain, close confinement is indispensable in the horizontal posture, this should not prevent the patient from often being carried into the open air, and taking moderate carriage exercise, particularly as soon as there is an abatement of the most violent symptoms. A generous diet, — but not so as to burthen the stomach, — fresh air, and a gradual course of tonic medicines (preparations of iron are preferable) are the best means of treating the constitutional condition accompanying this obstinate disorder. (*See Nos.* 317, 318, 319.)

There is another affection of the uterus, much resembling this, which might be, perhaps, more properly named *rheumatism of the womb.* The plan of treatment is the same as above recommended.

There is a painful condition of the vagina frequently met with by the physician, analogous to the affection of the uterus we have just described, and which may be termed *Irritable Vagina.* It is characterized by extreme tenderness when the lining membrane is touched by the finger, and a hysteric attack not unfrequently follows coitus. In some marriage develops this weakness; in others, child-bearing, or both, become agents of development. The treatment is mainly the same as directed for hysteralgia.

LEUCORRHŒA.

Perhaps there is no single term in the whole catalogue of " woman's diseases " more undefined, more extensively misunderstood, or about which a greater amount of practical ignorance exists than this. It is derived from two Greek words, signifying " white discharge ; " and *fully five-sixths of the female population of all our large cities,* youth and adult, are afflicted with it. By the people generally the term is vaguely and indifferently applied to *all non-sanguineous vaginal discharges,* no matter what their character or the diseased conditions of the organism from which they spring ; the natural consequence being much confusion and loose treatment on the part of the physician, and an untold amount of suffering, annoyance, and anxiety on the part of the patient. It will, therefore, be apparent to our readers that, in the treatment of this insidious and mysterious affection, it is primarily essential that all conclusions should be based on an accurate knowledge, not only of its symptoms, but of its seat.

In popular phraseology, " leucorrhœa," " fluor albus," and " female weakness" mean the same disease, which is described as a light, colorless discharge from the genital organs, varying in hue (according to its virulence or origin) from the *white* or colorless, mucilaginous, mucous discharge to the yellowish, light green, or slightly red and brownish exudation, differing in consistency from a thin and watery fluid to a thick, tenacious, ropy kind of substance, and fluctuating in quantity from an almost imperceptible increase of the healthy vaginal secretion to from one to six ounces during the twenty-four hours.

Leucorrhœa may occur at any time of life, from early infancy to old age, but it is most frequently prevalent between the ages of fifteen and forty-five, or the *change of life*, seldom continuing after that period, except when it has its origin in some organic disease of the womb. In children afflicted with hereditary scrofula, it is often present prior to puberty, and even in infancy, materially interfering with, and in all probability causing retardation or suppression of, the menstrual function. As a rule, the leuchorrhœal discharge is much more profuse at the menstrual period than at other times.

Whether taken as a disease or as a symptom, it may, like all other diseases connected with the reproductive organs, be divided into two principal forms, — the acute and the chronic ; the acute being nothing more nor less than an attack of inflammation of the mucous membrane lining the parts, whatever may have been its cause. In a large majority of cases it will be found to be nothing more than a catarrhal inflammation, occasioned by taking cold, which, if promptly and resolutely treated, will speedily disappear. If neglected, unnoticed, or improperly treated, it necessarily passes into the chronic form, which is simply a continuation and aggravation of the acute form, and is frequently complicated with severe and obstinate inflammation of the adjacent tissues and ulceration of the neck of the womb.

CAUSES.

Many individuals erroneously look upon leucorrhœa as the direct and immediate result of " general debility," and most persons suffering from the disease vainly imagine that if they could only get *something to strengthen them* they would be cured. Under this delusion many a woman has dallied away valuable time and money in taking successively, but by no means successfully, the many strengthening bitters, universal panaceas, and nice-tasted, pretty-named nostrums, advertised to regenerate and rejuvenate suffering humanity generally. But this is a great, a sad, and a fearful mistake, by which human life has been risked, and the period of restoration to health indefinitely postponed. The debility is not the *cause* of the disease ; neither is the discharge. Both the discharge and debility are simply the outward manifestation or result of some morbid action going on in some portion of the uterine organism. What that disease is, or has been, which has given rise to these symptoms, can generally be ascertained by the nature of the discharge and the peculiarities which each particular case presents. For instance, there are three distinct varieties of discharge from the vagina alone : *mucous, purulent,* or *mattery,* and *watery* ;

and there are morbid conditions capable of producing each of these evacuations. Inflammatory action is not absolutely necessary for the secretion of mucus, though it must exist where pus or matter is present. As both pus and mucous secretion is frequently found in the leucorrhœal deposit, it is therefore essentially necessary that the true source of the various forms of the malady should be accurately ascertained, in order that it may be treated intelligently, and not blindly.

In far too many instances leucorrhœa has been treated and looked upon as a vaginal disease, pure and simple. With the very limited knowledge of its pathology which many physicians possess, it is no wonder that they have fallen into the habit of treating it upon routine principles, and that the success of that treatment has been anything but flattering. *Nine out of every ten cases of leucorrhœa have their origin in congestion, inflammation, or ulceration of the neck of the uterus;* and when the cause is removed — when this condition of the organ is terminated — the discharge entirely ceases, and the nervous and muscular derangements of the uterine apparatus soon give way to generous diet and tonic treatment, the organs being restored to their normal health and vigor.

That leucorrhœa is *hereditary* is a fact too well substantiated to need comment; many women, more especially those possessing a lymphatic, nervous constitution, with soft flesh and pale in color, being *hereditarily predisposed to uterine affections generally.* With such persons, a cold, unsuitable diet, nightly dissipation *of any kind,* tight lacing, and the various other indiscretions and reckless indulgences prevalent among women of all classes and stations in life, would inevitably result in a more or less aggravated attack of leucorrhœa, as a symptom and sequela of some other affection; while those differently constituted might commit the same imprudences, but would be affected in a totally different manner.

The immediate *exciting causes* of the disease include severe colds, sudden changes of atmosphere, sitting upon cold, damp seats, such as stone, grass, earth, etc., exposure of the upper and lower extremities to cold air, violence, excessive indulgence, tight lacing, encumbrance of the abdominal region and hips with weighty and tightened clothing, irritation from stimulating injections, inflammation of the rectum, hemorrhoids, miscarriages, abortions, uterine displacements, ulceration of the womb, tumors of various kinds, purgatives and emmenagogue medicines given for the purpose of hastening the menstrual discharge, warm injections, the abuse of warm baths, late hours, excessive indulgence in fish, flesh, or acid and watery fruits, the intemperate use of tea, coffee, and other so-called harmless, non-intoxicating beverages, and the uncontrolled indulgence in the depressing passions, — fear, grief, bad temper, etc. It is quite common in cold, damp climates, and especially in the two extremes of society, the *high* and *low*, — those who are compelled to live in narrow lanes and alleys and in basements, where the atmosphere is damp and fetid and loaded with noxious gases exhaled from decomposing masses of refuse, — and those who, living in the highest strata, revel in every luxurious indulgence : beds of down, soft cushions, highly-spiced food, deprived of all its nutrition, and social habits and amusements calculated to exhaust all their vitality and energy, have ruined their constitutions and secured to themselves quite as luxuriant a crop of uterine diseases as their less fortunate sisters who drag out a miserable existence at the opposite extreme of the social circle.

A leucorrhœal discharge is not unfrequently produced in young female children by the presence of pin-worms, which find their way from the rectum to the vagina. In such cases, a removal of the worms is speedily followed by an abatement or disappearance of the discharge. It is always advisable, when little girls are troubled with a discharge from or itching of the vagina and neighboring parts, to make a vigilant examination. I have frequently known children to be kept awake night after night from the irritation caused by the presence of two or three little worms just within the lips of the vagina. They can be easily removed with a little piece of cloth.

SYMPTOMS.

In very rare instances the vaginal discharge is the only symptom noticed. As a rule, however, there is a miserable catalogue of aches and pains extending over the entire system. In fact, the constitutional symptoms, in many cases, are so distinctive that they are mistaken for the *cause* of the disease. The color of the skin, the bloated appearance of the face, the shortness of breath, constipation, moral and intellectual obtuseness, general debility, etc., will all be mentioned by the patients in the enumeration of their ailments, as separate *ailments*, but not one word as to their origin — the *leucorrhœal discharges* — will escape their lips. The duty of the physician as to vaginal examination under such circumstances is self-evident.

As we have before said the color, quantity, and consistency of the discharge varies proportionately with the other constitutional indications, and also with the intensity of the affection which originates it. It is sometimes so copious as to require the same provisional protection as during the menstrual flow, and in some cases amounts to an actual *flooding* of the peculiar secretion, and is accompanied by the uneasiness, pain, and prostration felt at severe menstrual periods.

In slight cases of catarrhal leucorrhœa, there is no irritation or excoriation of the parts ; the discharge is extremely mild, and but little inconvenience is felt. In acute cases, those arising from colds, the usual indications of catarrhal inflammation — a sense of heat and soreness in the parts, a feeling of weight or heaviness, or a bearing-down pain, with languor and general feeling of weariness — are present. These symptoms are sometimes accompanied with slight chills, pain in the back, quick pulse, thirst, high-colored urine, and other febrile indications. Under these circumstances, if proper treatment is at once instituted, there will be but little difficulty in subduing the symptoms, and the organ speedily assumes its normal health and strength.

If improperly treated, and especially if astringent injections and cathartic medicines have been used, the disease will most certainly become obstinate and chronic, the languor and debility increase, the discharge become still more profuse, pain and a general sense of heaviness in the abdomen be continuously felt, the digestion become impaired, and nausea, loss of appetite, headache, vertigo, palpitation of the heart, weariness upon the slightest exertion, and a host of dyspeptic symptoms, will soon manifest themselves. Ultimately, the disease will extend to and penetrate the womb ; congestion, inflammation, or ulceration take place ; the tissues become relaxed ; prolapsus uteri, or some similar affection of the uterus, follow, and a general increase of constitutional derangement supervene, which will declare itself by a loss of flesh and color, a quick, small pulse, the dryness or partial coating of the tongue, a constant aching pain in the small of the back and about the hips, great exhaustion and general debility, eruptions of small blackheaded pimples on the face and forehead, and the sinking of the eyes with a dark circle around them. The intellectual and moral faculties are always more or less weakened.

When the discharge is purulent, fetid, and stained with streaks of blood, the affection causing it is evidently of a serious character. As we have previously observed, simple leucorrhœa is, in itself, seldom, if ever, serious ; but there are so many uterine affections producing a discharge similar to that of leucorrhœa, that the person suffering therefrom should, for her own safety, comfort, and ease of mind, without delay place herself under the care of some intelligent, experienced medical practitioner. It is a disease, which, at the best, is difficult of cure, and the longer it is permitted to continue, the more obstinate it becomes, though a well-directed and persistent course of treatment seldom fails to afford permanent relief.

TREATMENT.

We have endeavored to show that this disease has its origin in deep-seated causes, — nothing less than a positive abuse of the fundamental laws of nature. The intelligent physician well understands the absurdity of attempting by the administration of medicine alone to change the long-continued and unhealthy action of organs that have for years failed to perform their natural functions, while the original cause of their derangement remains still in action. It is evident, therefore, that there is no alternative but that the woman must first perseveringly retrace her steps and reform her errors of habit, etc., before there can be a gleam of hope of her ultimate restoration to health. She must first place herself, physically and mentally, in a receptive condition for the remedial influences which are to be brought to bear upon her debilitated and disorganized system. All the surrounding circumstances which may in any way tend to excite or aggravate the disease must be promptly removed. All *indulgences* and *luxuries*, or *excessive use* of necessaries, must be absolutely forbidden. Bodily and mental excitement of all kinds must be avoided; the diet must be regulated and strictly adhered to. The *laws of nature* as to hygienic and sanitary matters must be strictly observed and implicitly obeyed; the food must be simple, nourishing, and as little stimulating as possible, and be taken at regular intervals; and tea or coffee, acid and watery fruits, absolutely abandoned.

Moderate exercise in the open air will be most effectual in promoting the cure, though great care must be exercised in the avoidance of fatigue; the clothing must be adjusted so as to admit of perfect freedom of motion, — the waist, especially, being free from all ligature. All exciting and depressing emotions should be avoided as far as possible. If the patient resides in a damp, low, or unhealthy district, she should be removed, at any rate for a time, to a dry, open, and healthy region, when such an arrangement is possible.

These requisitions being complied with, the medical remedies take their proper places, as alteratives, in arresting the disorder — being, of course, chosen in accordance with the indications given of the origin of the discharge. The most effective at present known are pulsatilla, sepia, alumina, calcarea carb., kreosotum, nitric acid, mercurius, cocculus indicus, conium, sulphur, and silicea. (*See Nos.* 6, 256, 257, 258.)

Most physicians speak highly of *water* as a remedial agent in this disease, and experience has taught me that too high encomiums can hardly be awarded to it when *judiciously* employed. Under the head of "*Causes* of Leucorrhœa," it will be remembered, I asserted that the disease was not unfrequently occasioned by the use of water. This is true Some over-fastidious females are not content with cleansing their persons in the ordinary way, but think it necessary (*while in the most perfect health*) to syringe themselves out

once or twice per day with tepid or cold water. Vaginal
injections during health are not only uncalled for, but, in
some instances, positively injurious; for, in its natural,
healthy condition, the lining membrane of the vagina is
kept constantly moistened by a mucous secretion. Now, an
injection, even of simple water, washes away this secretion,
and leaves the surface dry, in a condition easily irritated
and prone to disease. Water is, unfortunately, not the
only injection made use of, and cleanliness not the only
pretext for its use, prevention of conception and abortion
being frequently attempted by this means. With this matter, however, we do not propose to deal at present, but
simply to protest against the use of any injections *in the
healthy uterus.*

In *any* form of leucorrhœa, vaginal injections are of the
greatest importance; and they are especially beneficial in
cases where the discharge is acrid and causes excoriation
of the parts. The water dilutes the secretion, and thus
renders it less irritating, besides having a decided tendency
to reëstablish a healthy functional action, as has been
abundantly proved by prompt recovery following its use in
many cases. Cold hip-baths are also beneficial. Some
extremely severe and obstinate cases have been entirely
cured by the injection of a decoction of hamamelis virginica into the vagina, the cavity having previously been thoroughly cleansed with injections of warm water and castile
soap, repeating the operation some three or four times
per day.

The imperative necessity of *immediate attention to leu-
corhœal* or any vaginal discharge, no matter how slight, is
self-evident. Nine-tenths of our bed-ridden and chronic
invalids of the female sex may *directly trace their ailments
to neglected leucorrhœa.* It saps their very life-blood, destroys their ambition, prostrates them *mentally* as well as

physically, and renders their existence a positive calamity to themselves and to those by whom they are surrounded.

The Effects of Leucorrhœa,

if neglected or permitted to become chronic, are so disastrous and fatal to the comfort, health, and happiness of the sufferer, that we take this opportunity to again impress upon the minds of our readers the imperative necessity for immediate attention to its first symptoms. As we have remarked, it saps the very life-blood of the system, impoverishes the tissues, reduces the once hale, healthy, lively girl to a walking shadow, pale, emaciated, without vitality, weak, fragile, hysterical, and peculiarly susceptible to consumption, fever, contagious diseases, etc. It deprives her of every particle of physical or intellectual enjoyment in this life ; it extinguishes all hope or ambition, and makes her wish for death as the only avenue of escape from her pain and trouble.

ANÆMIA.

Anæmia means *poverty of the blood*. There are a variety of causes which produce this condition, such as hæmorrhage, exhausting diseases, blood-poisoning, and confinement within doors, from sedentary employment, etc. The symptoms are paleness of countenance, debility, nervousness, and nervous palpitation of the heart. This disease is very apt to be confounded with *purely* nervous diseases, such as *chlorosis* or *neurasthenia*. Anæmia is essentially a disease of *the blood;* chlorosis is an affection of the nervous system. One may cause the other, and they are frequently associated. Both are especially liable to occur to young girls and boys at the age of puberty.

TREATMENT.

The treatment of anæmia is important, for the reasons we

have given. Electrization, a course of tonics, and a generous dietary, calculated to build up the constitution, are all
needed, and produce the most satisfactory results. But
above and beside all these restorative and recuperative remedies, *air*, *sunlight*, and *exercise* are absolutely essential to
the restoration of the patient to a vigorous and healthy condition. Abundant and nourishing food, such as meat, beef-
tea, fresh eggs, bread, etc., and *plenty of sleep*, are the
most effective auxiliaries to nature's efforts. (*See Nos.*
156, 157, 158, 159.)

ANGINA PECTORIS, OR NEURALGIA OF THE HEART,

is an affection to which young women (especially those
brought up in our large cities) are peculiarly liable. It is a
strictly nervous disease, and begins with a sense of pain
and constriction in the region of the heart. This is accompanied with more or less pain and numbness in the left
arm. In women it is not uncommon for it to be attended
with great sensitiveness and pain of the breasts. When
the attack is violent, the pain in the heart is excruciating
and even terrific. There is attending this a feeling of great
oppression in the chest, amounting, in the severest cases, to
a sense of suffocation. The heart palpitates violently, the
brain is oppressed, and fainting sometimes occurs.

The disease is brought on, in nervous persons, by over-
excitement. Walking up-hill against a strong wind, may
bring it on. If walking at the time of the attack, the patient is compelled to stop and stand still till the pain subsides. The disease is often connected with organic changes
in the heart's structure, such as ossification and other alterations.

TREATMENT.

When the affection is connected with organic disease of
the heart, remedial measures must, of course, be directed

to the cure of that disease. To relieve a severe attack, the patient must be placed in a recumbent, quiet position ; wind in the stomach, if present, must be expelled by ginger, peppermint, ether, or some aromatic. If there is acidity or sourness of the stomach, it must be corrected by a tea-spoonful of soda in a half-tumbler of water ; and if the stomach is full of undigested food, let the patient take a tablespoonful of ground mustard, stirred up with a teacup-ful of warm water. This will cause almost instant vomit-ing. These things having been done, give some quieting or antispasmodic medicines. Great relief is sometimes afforded by sending a magnetic current through the region of the heart, by applying one pole of the machine in front and the other on the vertebral column. During the inter-vals the general health is to be improved by a wholesome, nourishing diet, gentle out-door exercise, and a careful con-trol of the passions. (*See Nos.* 302, 303, 304, 305.)

CHAPTER VI.

DERANGEMENTS OF THE MENSTRUAL FUNCTION.

AMENORRHŒA, OR SUPPRESSION OF THE MENSES.

THERE is naturally a considerable amount of confusion and error engendered in the minds of our women by the commingling of the two affections or conditions; viz., the retardation and the suppression of the menstrual flow. The menstrual function may be retarded in its operation by imperfect development, by local obstructions of a temporary character, by a chronic affection of some adjacent organism, or by congenital malformation, in either of which cases the exertions of the patient herself, her guardians, and her medical adviser must be directed to the invigoration of the system in general, and to the development of the uterine system in particular. Where the obstruction arises from malformation, surgical interference will obviously be necessary; but on no account should the artificial hastening of the flow be attempted by the administration of emmenagogues, such as ergot, savin, etc. But we will now proceed to the consideration of amenorrhœa.

Amenorrhœa is that condition wherein the menstrual function, after having being established for a longer or shorter period, has been arrested, without being interrupted by pregnancy, nursing, or old age. This may happen at any period of menstrual life, and it may take place suddenly or very gradually; in other words, it may be acute or chronic.

CAUSES.

Acute Amenorrhœa is caused by sudden exposure to the cold and damp air, immersion of the feet or hands in cold water; cold ablutions of the genital organs; sitting upon the ground, etc., during the menstrual term; the eating and drinking of ices and very cold drinks, especially while the body is in a state of violent perspiration; violent bleeding; the application of a large blister; the employment of purgatives or emetics, and strong perfumes or odors; the act of coition during the flow, or the setting-up of any fever or severe disease at that period. Also, a severe bodily or mental shock received just previous to or during the period of the discharge, or the exercise of vivid emotions, such as anger, love, jealousy, immoderate joy, sudden fright, disappointment or grief; all circumstances tending to sudden determination of the blood in a contrary direction to the uterus, may give rise to this condition. The most common cause, however, is the application, in some way or other, of cold to the feet. The majority of women pay so little regard to this period that they are continually, even hourly, exposed to derangements from this cause alone by standing in thin shoes, sitting upon damp, cold ground, going too lightly clothed, and dabbling in water, etc. Some indeed are so utterly reckless as to deliberately bathe their feet in cold water during their menstrual flow, in order that, by so arresting it, they may be enabled to keep a prior engagement to a ball or party. Such criminal acts as these, however, bring speedy and severe punishment with them.

SYMPTOMS.

These are extremely variable, but most commonly there is fever, with a sensation of heat, weight, and pain in the pelvis, uterine colic, disagreeable tension of the loins and upper part of the thighs, enlargement of the breasts and abdomen, lassitude, nausea, vomiting, dizziness, headache, ringing in the ears, frequent palpitations, scalding urine, etc. Or there may be symptoms of local inflammations either of the lungs, brain, intestinal canal, or of the uterus itself. Occasionally, instead of inflammation there are severe neuralgic pains of the womb, or a species of hysteria simulating inflammation, and changing from one organ to another as soon as remedies are brought to bear upon it. Attacks of apoplexy and paralysis have also been known to result from sudden suppression, as also aphonia, or loss of voice, curious derangements of vision, and cutaneous disorders. These secondary attacks may be mitigated in severity by the occurrence of *vicarious menstruation*, or discharge of blood from some other part of the uterine region, by which the temporary plethora is removed without any return of the uterus to a normally healthy condition; or by *uterine leucorrhœa*, which appears to afford relief; and more naturally, since there is a sort of action of the uterus which, though faulty, seems preferable to a condition of perfect indifference and lethargy. On account of the secondary attacks to which it gives rise, sudden suppression is a much more serious disorder than any other form of menstrual derangement.

TREATMENT.

There can generally be no difficulty in ascertaining the fact of suppression : if in any cases there should be any doubt as to their nature, they should at first be treated as simply inflammatory. Attempts are sometimes made by unmarried females, with the view of avoiding exposure of character, to deceive physicians in reference to one of the causes of menstrual suppression. The experienced physician needs no caution on this point, but we will, for the information of our readers, give a more detailed account of its dangerous and malefic influence, in our chapter on Abortion and Miscarriage.

The *acute* form is more easily cured than the *chronic*. Our first endeavor should be to effect a resumption of the menstrual discharge; and for this purpose the patient should take a warm bath, or put the feet into warm water, and take a bowl of warm gruel. Rest in bed is absolutely necessary, and warmth should be applied to the legs and thighs by means of hot flannels or bottles filled with warm water, or cloths saturated with turpentine and warm water. *Gentle* purging will also be found useful; but if it be induced to any extent, it will necessarily defeat its own object, as copious discharges of any kind, by relieving the constitution, to a certain extent supersede menstruation, and prevent effort on the part of the uterus. Should this course fail after reasonable perseverance, and a state of congestion prevail, relief will be obtained by loss of blood. If adopted in the earlier stages of the malady, it will, in all probability, prevent the local disorders to which we have referred; but when they do arise, they must, of course, receive the treatment usual for the specific diseases. The hysterical affection of the different organs should be combated with what are called antispasmodic medicines, such as assafœtida, musk, camphor, etc. (*See Nos.* 213 or 214.) If colic be present, the most certain relief will be found in aloes and myrrh, in their several combinations, till the bowels are opened. (*See Nos.* 118, 119, 120, 121.)

Upon the approach of the next period, great attention should be given to the patient, and every means used to facilitate the natural secretion. The bowels should be kept open, the surface comfortably warm, and the hip or foot bath used alternate nights. The strength, if necessary, must be supported by a generous, but not stimulating diet.

If, at the proper time, the menses should be reëstablished,

our object will be accomplished; but if merely a white discharge appear in its stead, we must again, during the interval, put into action all those means mentioned in the treatment of the tardy appearance of the menses. If the white discharge persist during the interval, it must be treated simply as leucorrhœa. But if no discharge appear, neither red nor white, recourse must be had to those means and remedies which we shall prescribe in our next section for chronic suppression of menses, according to the condition of the patient.

CHRONIC SUPPRESSION OF THE MENSES

may be the consequence of an acute attack, arising from neglect or improper treatment, or it may be caused by gradual depression of the bodily powers, giving rise to delicate health. It may also arise from diseases of the ovaria, uterus, and other parts of the body, as inflammation, induration, ulceration, hydatids, etc., of the uterus and its appendages; anteversion, retroversion, and complete prolapsus of the womb. It may also be the sequela of pulmonary consumption, disease of the heart, scrofula, particularly of the bones, inflammation of the brain or spinal marrow, the stomach, spleen, liver, lungs, pleura or peritoneum, or any irritation in the system which retains the blood and prevents it from being directed upon the uterus.

The quantity of the secretion may gradually diminish, and the time become irregular and uncertain, till at length the uterus altogether ceases to act; or, which is perhaps more frequent, the menses diminish in quantity and become of a paler color, with shorter intervals, until leucorrhœa becomes permanently established in its stead.

SYMPTOMS.

The SYMPTOMS which arise from *chronic suppression* are

various, being often the same (though less marked) as those mentioned under *acute suppression*. The features of the young woman, heretofore brilliant with freshness and health, are observed to assume the impress of feebleness, depression, and languor; the roses upon the countenance fade, the fire of her eyes is extinguished, and a dark circle surrounds them; finally, the most frequent symptoms are habitual headache, dyspnœa, dizziness, pains in the back, sides, limbs, and joints; deficient appetite, and a general failure of the vital powers, ending in a confirmed state of ill-health. The moral alteration is equally great; sometimes there is an excessive sensibility, which renders the sufferer impatient and irascible; at others, her ideas are sad, her imagination sombre; sometimes the patient seeks for solitude and sheds causeless tears; again, on the contrary, she becomes passionately fond of music and amusements of all kinds. The general health rarely suffers before three or four successive periods have passed, unless it be accompanied by considerable leucorrhœa.

If the menses do not occur after suckling, and the health of the individual appears to suffer, an examination into the parts should be instituted to ascertain their condition. There may be an obstruction or obliteration of some portion of the canal in the neck or mouth of the uterus or of the vagina, in consequence of inflammation following delivery. The introduction of the finger will satisfy as regards the vagina, but the permeability through the neck of the uterus can only be determined by passing up a moderate-sized bougie (a slender gum elastic instrument).

Some care is necessary to distinguish between *chronic suppression* and *pregnancy*, as the patient, if she be in a position to have children *creditably*, may mistake the suppression for the first symptoms of pregnancy. The arrest of the menses, when occasioned by conception, is shortly followed by the morning sickness, and an alteration in the breasts, etc.

The remedies for this form of suppression will vary according to the cause which has given rise to it, and the state of the system generally. When it is consequent upon disease of the genital organs, or some other part of the body, we shall find that, upon the patient's recovery from such disease, the menstrual flow will generally return. When the menses have been superseded by leucorrhœa the proper treatment of it will generally restore the uterine functions.

The reader will readily perceive, from our previous remarks, that every deviation from menstrual regularity does not necessarily demand medical interference, for in many cases of early menstruation, and indeed with young girls generally, there will, at first, be considerable irregularity, which must not be mistaken for disease. This may also be the case with hearty, robust young women, in whom a temporary suspension may be caused by cold, excessive passion, or mental emotion, — the flow voluntarily resuming its proper course, after a time, without artificial aid. The rule with the medical practitioner on this point should be, *never to interfere, unless there be some tangible evidence that the health is suffering from the absence of this discharge.*

TREATMENT.

In *chronic suppression* the general health is rarely invaded for the first three or four months of the affection; up to that period the treatment should be purely *hygienic;* that is, strict obedience to the physiological laws, in the proper regulation of diet, exercise, proper clothing, well-ventilated apartments, avoidance of all excess in passions, amusements, and food and beverages; strict cleanliness, the avoidance of all those modern luxuries and indulgences which predispose the system to disease, and, more than all, a careful selection of associates and pursuits, and a strict adherence to the higher law of morality, without which the most careful physical and intellectual development is worse than useless. If we find that the pulse manifests a tendency to excess of action, the treatment should be commenced with such remedies and regimen as will reduce it to a proper standard, before we proceed to the use of medicines which have a direct tendency to produce the menstruous discharge, such as slight bleeding, the application of leeches to the vulva, etc., purging, and a strict vegetable diet. Under such circumstances, colocynth will answer admirably in doses of two teaspoonfuls twice per day. After these remedies have been persevered in for a time, tincture of cantharides or guaiacum, with a small quantity of opium, may be administered, not more than ten drops, in a tablespoonful of water, at a time, and applications of the electro-magnetic battery may prove serviceable in cases where all other means have failed. This last is, however, a powerful remedy, which should be employed very cautiously, lest it may depress the nervous system, and thus protract the disease ; when moderately applied, it often rouses into activity the energy of torpid organs and functions, but when used to excess, it may altogether destroy their excitability. It should not be employed in cases of local congestion or general plethora, nor during pregnancy, and should seldom, if ever, be used alone.

In many cases of suppression, the flow may be reëstablished by the application of simple counter-irritants, stimulating poultices, liniments, etc., to the groins, over the ovaries ; with scrofulous constitutions the sirups and tinctures of iodine and the various preparations of iron have proved effective, and aloes, myrrh, turpentine, savin, and the sulphates, carbonates, and citrates of iron (as given in Prescriptions 6, 8, 118, 119, 120, 306, 307, 308) are extremely useful when judiciously applied.

Very few of the remedies we have just enumerated, except the iron and iodine, are applicable to patients of feeble, nervous, or lymphatic temperaments, or who are constitutionally scrofulous, until their general health has become reëstablished. Change of scene and occupation, the enlivening influences of the social circle, music, and a judicious variation of the scenes and surroundings, so as not to create abnormal excitement, will prove the most efficient auxiliaries in the restoration of the patient to health.

As a last resource for the girl or young woman laboring under obstinate suppression of the menstrual flow, *marriage* has been recommended by many eminent physicians from the time of Hippocrates to the present day ; but we should have considerable hesitation in indorsing the recommendation or advising such a course ; as, apart from the moral aspect of the question, the *physical* uncertainty of the result would not compensate for the risk. We should advise that, whenever practicable, the sufferer from chronic suppression should remove to a fresh and dry atmosphere (preferring mountainous localities) ; we should also recommend the use of nourishing food, such as rich soups, light meats, etc., bitter infusions and gently excitant beverages, mineral waters, dry friction, flannel underclothing, cold bathing and swimming, pedestrian and equestrian exercise, healthful games, calisthenics, gymnastics, etc. In conclusion, we would remark, that the moral condition of the patient merits as close an attention as the medical treatment. If we neglect to attend to the state of the mind, menstrual disorders, depending upon certain mental conditions, such as profound dejection, resist all the resources of the healing art ; they are generally powerless in remedying the malady in a young woman who is tormented by a disappointed or unfortunate love. It is to the persuasive eloquence and counsels of friendship, the consolations of a prudent, enlightened mother, and especially the gratification of the affections by marriage, *when there is not extreme prostration*, that we are to look, in these cases, for the arrest of the disease at its source.

VICARIOUS MENSTRUATION.

In cases of suppressed menstruation, where the monthly menstrual effort occurs without secretion on the part of the uterus, and where the system generally is suffering from the consequent plethora or irregular distribution of blood, an attempt is made by the natural powers to afford relief by a discharge of blood from some other part, generally one which is already diseased or enfeebled. This is called *vicarious menstruation*, and has been known to take place from the nostrils, eyes, ears, gums, lungs, stomach, armpits, bladder, nipples, toes, fingers, from the stumps of amputated limbs, from ulcers, and from the surface of the skin generally. The mucous membranes, however, of the the lungs, stomach, and intestines, are the most common seats of the discharge. This discharge generally comes on suddenly, and continues at intervals for several days. In most cases, it seems to relieve the constitutional distress consequent upon suppression, but does not promote the natural establishment of the function during the interval. In general, it is not followed by more serious consequences than those resulting from the loss of blood. The most dangerous form is when it proceeds from the lungs.

TREATMENT.

The same may be said of vicarious menstruation as of suppression, — that medication for the *symptom* is often worse than useless ; the real object being to rid the system of the *cause,* and not merely the *effect.* Our desire is not so much to compel the return of the menstrual flow, as to cure the patient of that disordered condition of her system which led up to the suppression. In the words of the eminent Dr. Wm. Hunter: " With regard to the management of the menses, my opinion is that you should pay no regard to them, but endeavor to put her to rights in other respects. If you cure the other disorders, you can cure the irregularity of the menses, which is the consequence and not the cause of her complaints." Taken as a whole, the remedies prescribed, and the hygienic and other regulations given for amenorrhœa, are equally applicable and efficacious in this form of menstrual derangement. ◄(*Nos.* 118, 119, 120).

MENORRHAGIA, OR EXCESSIVE MENSTRUATION.

Menstruation may continue too long, or occur too often, or be too profuse while it lasts ; or all these irregularities may be experienced by the same person. Any one of them will prove a serious irritation and a drain upon the constitution : the whole together, if not arrested, will undermine and destroy it.

The cause of this, as of all other female diseases, is in a great majority of cases overlooked. It is not to be attributed, as so many suppose, to a congested state of the womb, but is usually the result of the inflammatory or ulcerated condition of the uterine neck. In a still larger number of cases it arises from a succession of *ovarian abortions.* When the blood has run low, and nutrition is defective, as in persons of a consumptive habit, the ovarian vesicles fail to reach maturity. Like other products of the uterine economy, they become blighted and abort. And, as these blights occur often, nature is busy every two or three weeks in casting them off. Hence, the excessive and frequently repeated flow. They come and go irregularly, and without order, because they spring from processes directly in contravention of nature's laws.

It is not easy to explain how inflammation and ulceration of. the uterine neck should in one case produce suppression and in another profuse menstruation. Yet it is a self-evident truth that such opposite results do come from one and the same apparent cause. Probably the explanation is to be found in the different degrees of inflammatory action, in the varieties of constitution, and in the various degrees of tenacity with which the vessels hold the blood. Bleeding from the genital organs may be produced by a variety of causes which have nothing to do with menstruation. Such hemorrhages are properly *uterine* or *vaginal hemorrhages*, and *not* profuse menstruation. They are the result of inflammations or tumors within the uterine neck, or of constitutional weakness. The womb may bleed for days, or even for months, from pure debility.

As the normal quantity of fluid discharged at the monthly evacuation varies in different women, *menorrhagia* only exists where there is a disproportion between the loss and the power of replacing it. It is the *relative*, and not the *absolute* quantity lost which constitutes the disease, so that seeming derangement should only be considered *excessive menstruation* when it has an injurious effect upon the general health.

We shall include every variety of menorrhagia under two great divisions, principally in reference to their severity. In the *first division* are those characterized by a sudden gush from the vagina, suddenly ceasing for a few hours, and then again recurring, pursuing this intermittent course throughout the entire period of menstruation. On the other hand, sometimes the discharge goes on regularly, but, instead of being over in three or four days, lasts for ten days, a fortnight, or even three weeks ; or it may return, in its usual quantity, every two or three weeks ; this last variety, more frequently than the other, being connected with that condition of the lining membrane of the uterus which gives rise to leucorrhœa. In this condition, there may be more or less discharge of *clots* of blood along with the proper secretion, though it rarely occurs in young or unmarried women. The subjects of it are generally women of a torpid or sluggish temperament, whose constitutions have been impaired by disease or frequent childbearing. One or two small clots appear at first, at short intervals, returning each time in increased quantity. It is not known, in these cases, whether the discharge is altered in quantity or quality.

SYMPTOMS.

These are : languor, exhaustion, weakness across the loins and hips, paleness of the countenance, headache, ringing in the ears, and giddiness ; these occur, to a greater or less extent, in the slighter cases. If the disease continues, and especially if leucorrhœa be present, all these symptoms become very much aggravated — the languor increases, the face becomes sallow, there is an aching pain across the loins, extending around the abdomen, repeated and severe headaches, and derangement of the stomach and bowels ; and, finally, there is extreme exhaustion, with a feeble pulse, melancholy, nervous symptoms, ending in dropsy and even epilepsy. A prominent cause, in this physic-taking community, of this as well as many other ills to which woman is heir, is the enormous amount of quack pills and patent medicines of every variety of form and name. These *so-called medicines* contain, for the most part, ingredients that operate violently upon the lower part of the intestines adjacent to the uterus, thus determining an excessive quantity of blood to that organ, besides debilitating the bowels and frequently producing piles.

At a more advanced period of life, the most frequent exciting causes of menorrhagia are childbearing and protracted nursing of children, a practice which is extensively carried on among the poorer classes for the avowed purpose of preventing a too rapid increase of the family. This result it achieves very effectually, when it gives rise to this disorder, but at the untold expense of great loss of health and extreme suffering on the part of the unhappy woman. *Excessive coition* is another very prolific source of menorrhagia, and always aggravates the disease. In the severer cases, conception does not take place, though it not unfrequently does in the milder forms. Menorrhagia may or may not return after parturition, but it necessarily predisposes the woman to abortion and miscarriage in subsequent pregnancies, and, from the excessive relaxation of the parts, is largely productive of falling of the womb and vagina.

TREATMENT.

The primary object is, of course, to remove the cause of the affection if possible. If it arise from excessive nursing, the child should at once be weaned, and marital intercourse should, for a time, be suspended. In any case, she should be kept perfectly quiet upon her back ; cloths, wrung out in cold water, should be laid over the uterus, vulva, and thighs ; cold, acidulated drinks, such as iced lemonade, solution of elixir of vitriol in ice-water, etc., should be given freely, and the introduction of all warm fluids into the system strictly forbidden. The bed-chamber should be kept cool, and the foot of the bed raised about ten inches or one foot above the level; opium, or some similar soporific administered, and all conversation prohibited ; and gallic acid, ergot, or (especially) cannabis indica (in 5 to 10 drop doses), be given internally. If local pains of a continuous or sharp, energetic character exist, belladonna or opium plasters may be placed over the seat of distress with great advantage. Salt-water sponge-baths, daily, specially applied to the lower extremities, will do much towards relieving the weakness in the loins and spinal column. Tonics, especially preparations of iron, should be freely given. If the patient be of sanguine temperament, the diet should be almost exclusively vegetable. Tea and coffee, feather-beds, foot-stoves, and stimulating food or drinks must be entirely avoided. If of a *lymphatic* temperament, the diet should be generous; but everything tending to excitement of the uterine system should be carefully prohibited. The invalid should live in a dry atmosphere, and very moderate exercise should be taken throughout the interval. No melancholy companions should be allowed to visit her, and all mental emotion should be strictly avoided. (*See also No.* 272.)

METRORRHAGIA.

In the second division of menorrhagia, or *Metrorrhagia*, as it is usually called, the discharge is more profuse, and its effects more severe than in the first; and it is accompanied by alterations in the condition and size of the mouth of the uterus, occurs at a later period of life, and is much more difficult of cure.

SYMPTOMS.

The attack commences much in the same way as in the milder form, but not so suddenly, and is not confined to any particular temperament, though it is more frequently found in the sanguine than in the debilitated or melancholic. It rarely appears before the fortieth year, or after the cessation of the menstrual flow or change of life. There is, for some time previous to the attack, irregularity in the time, quantity, and duration of the menstrual periods, with occasional attacks of leucorrhœa during the intervals. When the menses have flowed naturally for about twenty-four hours, the bloody discharge appears; large clots are expelled, and there is a considerable increase in the fluid discharged. At the inception of the affection, the duration of the attack is not more than eight or ten days; but in long-standing cases, it not unfrequently extends throughout the entire intra-menstrual period. The quantity lost is sometimes very large, producing excessive exhaustion and weakness of the loins, a sense of weight or pain in the pelvis, difficulty in discharging urine, great depreciation of general health, excessive constipation, blanching of the countenance, and nervous and muscular prostration or anæmia.

The exact change in the uterus and its appendages resulting from this condition is not known in detail, but examination reveals the mouth low down in the pelvis, the neck more or less swollen, and tilted forward so as to press upon the bladder, thus producing the affection we have described. There is sometimes tenderness upon pressure in the neck and body of the uterus, the vessels of which are evidently very much congested or engorged with blood, so that the discharge is not the result of secretion, but of the rupture of some of the vascular twigs which ramify the lining membrane of the uterus. The disease, *unless the organic changes are very considerable*, is not usually obstinate; it may subside by skilful treatment, or possibly spontaneously in three or four months, or may continue as many years.

CAUSES.

These are much the same as in the milder affection already noticed, those most subject to the disease being women who live indolent lives, yield too readily to the passions and emotions of the mind, dance inordinately, and keep late hours, who are in the habit of tight lacing, and adopting other fashionable absurdities of costume, who take little or no exercise, are intemperate in the use of stimulants and hot drinks, as tea, coffee, etc., and who are too prodigal of marital privileges.

This description of excessive menstruation is easily distinguished from the flow of blood arising from organic diseases of the uterus, as corroding ulcers, cancers, polypi, etc., by the irregularity of its occurrence in these diseases, and its persistence after the usual period of this excretion has expired.

TREATMENT.

At first, the remedial measures we have already mentioned should be perseveringly employed ; and if they are found ineffective, as is frequently the case, the use of ergot and other similar remedies must be resorted to ; it may be given in doses of five to ten grains, three times per day. During an attack the patient should be kept in a state of perfect rest, on a hard mattress, covered rather lightly with bed-clothes. All her drinks should be cool and unstimulating, unless she become faint, when a small quantity of wine or brandy may be permitted. At the same time that the ergot is given, cold must be persistently applied to the lower extremities by means of the douche or wet cloths. The precaution of keeping the feet warm should always be observed. If the discharge is not arrested, it is a matter of serious consideration how far the use of injections of cold water and astringent solutions may be permitted prior to its entire cessation, considerable danger attending their use during the period of attack. As soon as the discharge has ceased, curative measures should at once be taken to eradicate the disease of which this exudation is so formidable a symptom. We have found the repeated application of blisters to be very efficient in diminishing or suppressing the discharge ; and, in our later experience, belladonna plasters have proved specially effective in the removal of pain from the affected region, the swelling of the uterus rapidly diminishing, the leucorrhœa disappearing, and the parties being able to walk about with comparative ease. (*See No. 272.*)

The patient is always liable to a relapse, consequently she should permit two or three menstrual periods to elapse, taking the same precautions and adopting the same medicinal and hygienic regulations, so that her health shall be thoroughly reëstablished, before she resumes her ordinary routine of duties.   Should a relapse unfortunately occur, the symptoms must be met again by the same treatment and regimen as before.   During the intervals, the treatment directed under the primary or milder form of the affection must be strictly observed.   The bowels must be kept free, *without purging*, and tonics, mineral waters, liberal, nutritious diet, moderate exercise in the open air, and freedom from undue mental and bodily excitement must be rigidly adhered to as the rule of life.

DYSMENORRHŒA, OR PAINFUL MENSTRUATION.

This is one of the most distressing and agonizing forms of menstrual derangement to which women are subject, its most prominent feature being *the pain*, which begins three or four days in advance of the evacuation, and continues uninterruptedly throughout the term.   These pains extend through the loins and over the entire surface of the abdomen, and run down to the thighs.   They vary considerably in character and intensity, according to the constitution of the individual ; they may be moderate and lasting only a few hours at a time, or they may be so protracted, intense, and agonizing as to cause fainting, spasms, or convulsions, and even catalepsy. They are occasionally so violent as to resemble the pains of labor. The stomach and bowels become irritable, producing vomiting or diarrhœa, with scalding of urine, and, in the severer forms, utter prostration of the nervous system supervenes. This disorder, therefore, may be divided into two varieties, according to the constitutional peculiarities of the patients : —

1st. The *inflammatory*, when it occurs in women of full habit and sanguine temperament, and,

2d. The *neuralgic*, when it is confined to those of a nervous temperament and of a thin, delicate habit of body.

A *third* kind might be added, viz., where there is a *mechanical* difficulty arising from some impediment existing in the passage leading to the uterus ; but these distinctions are only important as regards the treatment, the causes are generally the same.

CAUSES.

The causes of this complaint are very numerous, many of which we have enumerated under the preceding forms of menstrual irregularity. There is, beyond all doubt, such an affection as rheumatism or neuralgia of the womb, though this is much more rare than usually supposed. Pains at the monthly period are frequently caused by displacements of the uterus, occasioning an unnatural pressure upon the nerves, which, when so greatly distended with blood as they must be. at those periods, cause exceedingly painful sensations during the passage of the menstrual flood. In these cases the neck of the uterus is bent at right angles, and the canal which passes through it is of course *strictured*, so that the evacuation is made with great difficulty and pain, the neck being, from inflammation and other causes, almost closed. An increased flow of blood to an inflamed part always causes pain, from the fact that the inflammation ren-

ders it so much more sensitive by the undue compression
of the sensory nerves. Congestion of the lining membrane
of the womb itself is a frequent cause of painful menstruation. The condition of the membrane of the womb is
similar to that of the larynx in membranous croup. There
is the same pouring forth of what we call *coagulable lymph,*
which forms itself into a membrane. This membrane the
womb strives, by strenuous contractions, to throw off, and
finally succeeds in expelling it, not whole and entire, but
in *shreds and particles.* These shreds, which women sometimes call *skinny substances,* are characteristic of the disease, and the efforts to expel them cause pains very much
like those of natural labor, and sometimes almost as severe.

SYMPTOMS.

This disease may attack women at any age, more especially unmarried women, and women who have never borne
children. For a few days previous to the monthly paroxysm
there is a general sense of uneasiness, feelings of cold and
headache, alternating with pains in the lower part of the
back, often of a severe and bearing-down character. In
plethoric constitutions, there is frequently a flushed face,
hot skin, and full pulse. When menstruation is established,
the eruption is slow and scanty, or in slight, intermittent
gushes. The discharge is frequently paler than usual, or
mixed with small clots of blood. In many cases there is
the peculiar membranous matter discharged of which we
have spoken, and which has frequently given rise to unjust
suspicions of conception when occurring in the unmarried.
Dr. Denman regarded the existence of this membrane as a
sign of sterility, asserting that he never knew a female in
whom this secretion existed, to be the subject of conception. Our experience and observation has taught us, however, that though conception seldom takes place under such

circumstances, it is not by any means impossible. Dysmenorrhœa does not appear to us as necessarily connected with
any derangement of uterine structure. The most common
cause is, undoubtedly, cold taken during menstruation, miscarriage, or delivery, which, giving rise to sudden contractions, in some cases, especially when unrelieved, produce
hardening and alterations in the neck and mouth of the
uterus, thus giving rise to barrenness and a disposition to
cancer. The majority of cases are curable, although a few
resist all the known means of alleviating the malady, and
are only cured when the function of menstruation ceases.

TREATMENT.

If the dysmenorrhœa be caused by displacement of the
uterus, the trouble will, of course, be entirely overcome
when the organ is restored to its proper position ; if by
stricture of the canal through the uterine neck, it can only
be rectified by artificial enlargement of the passage, an
operation which requires the manipulation of a careful and
skilful physician. As in the other forms of menstrual irregularity, the first effort must be to assuage the pain and
lessen the annoyance to the patient, the chief reliance being
placed on sedatives. Afterward, the entire energies of the
physician must be directed to the prevention of its return
by the administration of appropriate remedies during the
interval. The same remedies as previously directed, with
the addition of electro-magnetism to the inert organs ;
belladonna ointment to the neck of the womb, and
belladonna plaster to the sacral region (or base of
the spinal column) will prove eminently beneficial.
(*See No.* 271.) During the interval, the means we have
previously recommended to strengthen the patient and
lessen the local and general irritability, should be adopted.

The diet should be generous, accompanied by daily exercise in the open air.

The treatment of painful menstruation of the *mechanical* species, as we have previously remarked, is necessarily confined to *mechanical* remedies, — the use of the catheter. The instrument should be allowed to remain only a few minutes at a time, the frequency of application (not more than twice or thrice a week) depending entirely upon the nervous irritability of the patient. It will be advisable in such cases to use daily vaginal injections of warm water, the patient being kept under the influence of sedative medicines, with an occasional mild aperient.

CHAPTER VII.

WOMEN AS WIVES AND MOTHERS.

HAVING considered the ordinary diseases to which the young women of our large cities are especially liable prior to marriage, we will leave the consideration of hereditary, congenital, and general diseases for a time, and proceed to the contemplation of woman in her relations to the community as a wife and mother, and the duties, responsibilities, and contingencies attaching to that condition.

We will assume, for argument's sake, that her choice has not been trammelled by the dictation of parents or guardians; that she has not been influenced in that choice by mere outward appearance, or the glare and sheen of the ball-room, but *pure affection*, based on a practical and personal knowledge of the husband's habits and tendencies, his capacity and willingness to contribute to and secure her permanent happiness and comfort, and his absolute freedom from any malady, hereditary or acquired, which might inflict suffering or injury on herself or her offspring. She has now assumed new honors and new responsibilities. She is now a monarch in her own right, — the queen of a household which shall prove to her an earthly paradise or a lifelong dungeon, the grave of all her hopes and aspirations, the scene of untold woes and miseries, to the description of which language is utterly inadequate.

PROSPECTIVE MOTHERHOOD.

Every young married woman in the prospect of becoming a mother is, naturally, deeply interested in her novel and bewildering position. It is new to her, and she feels its importance in her inmost soul. She is all anxiety to know whether it will affect her own health, if there is any plan she ought to pursue to preserve it, and whether a strict adherence to such a plan will tend to secure a vigorous constitution to her expected offspring. Unfortunately, however, in too many instances, these reflections lead to no useful result; sometimes from ignorance of the importance of the subject, but with the majority of newly-married women, from an unwillingness *to ask* for the necessary information. From whatever cause it may proceed, there can be no doubt of the fact that, hitherto, there has been a lamentable want of self-management during the period of pregnancy. I say lamentable, because of the importance of the interests involved, and the melancholy consequences often resulting from such neglect.

It is undoubtedly true that when pregnancy occurs in a woman of sound constitution, who habitually observes the ordinary laws of health, and regards also those which are demanded by the new condition in which she is placed, in general her health will not be much affected by it. But then this presumes a previous careful and constant observance of those hygienic and sanitary regulations by which alone the health is maintained; for no woman can be indifferent to those laws, or violate them during this period, without paying the cost in suffering, and occasionally in positive danger, and *inevitably* in endowing her innocent unborn offspring with the same maladies and agonies she herself endures. And here we feel it to be our painful but inexorable duty to specially call the attention of our fair readers to the vital importance and life-long results, for good or for evil, of

PRE-NATAL INFLUENCES

on the immortal being she is so soon to introduce into this

nether world. It must be remembered that the individual characteristics of each parent (mental and physical) are strikingly reproduced in the children, and are necessarily transmitted by the same law from generation to generation, in proportion to the intensity of the peculiarity in the original stock. The structural formation, the mental and moral proclivities, and even the psychological features, are reproduced and perpetuated from one generation to another, until the tastes, the pursuits, and the features of a particular line or family become as familiar as a household word. In America, as well as in England, the social records show that the members of certain families have been noted for many generations as athletes, gymnasts, dancers, scientists, mechanicians, inventors, financiers, statesmen, etc.; for we are daily met with the remark, in our social intercourse, when the conversation turns upon the excellencies or peculiarities of certain individuals, that such a thing "runs in the family." This observation refers with equal truth to disease, anatomical and mental peculiarities, and habits of thought and conduct. These family features may be transmitted by either or both parents, and they may be shared equally by the children of both sexes. Instances of the absolute truth of this dogma multiply so thickly, day after day, that we need only call attention to a few illustrations, to enforce the fact upon our readers' minds.

HEREDITARY PECULIARITIES.

Any one who has read the history of Frederick the Great and his gigantic body-guard, so famous for their immense physique, will recollect that, in order that this commanding stature should be perpetuated through succeeding generations, he decreed that they should not be allowed to marry any women but those of corresponding size. While this law was rigidly adhered to, the physical status of this body-guard remained unaltered: it was only when obedience to it ceased to be enforced, that the stature of this famous regiment became reduced. The practice of the ancient Greeks of putting away their dwarfed, diseased, or maimed children, that their physical defects might not be perpetu-ated, shows how early this law of inheritance was recognized and acted upon. The natural consequence was that for many ages the Greek nation was universally known and acknowledged as the type of physical perfection and sym-metry; and it was only when, enervated by luxury and indolence, they abandoned the simplicity, frugality, and temperance of their Spartan forefathers, that they declined in physical and intellectual development, until they became the miserable and degraded beings of the present century. In our own quarter of the globe, we have only to visit the Mexican territory, and to compare the native inhabitants of that region with the men and women of Vermont and New Hampshire (the beau-ideals of the typical "image of the Creator") to form an idea of the universal operation of hereditary characteristics.

TRANSMISSION OF DEFORMITIES.

The law of heredity receives still more forcible illustra-tion in the transmission of malformations and deformities (or arrest of development). For instance, harelip, horny excrescences, and other peculiarities of formation in toes, fingers, etc., have been reproduced for four or five succes-sive generations. Nor has this transmission been confined to the anatomical structure, for even the color of the hair has been faithfully transferred from father to son and grand-son, thus proving that, though the coloring pigment is specially influenced by the nervous organism, every department of nature is equally susceptible to and dependent upon the law of heredity. As with the body, so with the mind. The mind has a constant and abiding influence on the body, and *vice versâ*. Now, from the very moment of conception until the instant that the child opens its eyes on the scenes and surroundings in which it is to perform its part, it is virtually part and parcel of the maternal system, — it participates in every physical and mental change experienced by the woman ; every indiscretion of diet, of undue exposure to the inclemency of the weather, every error in dress, every physical ailment, every fit of despondency, ill-temper, or excess of joy, grief, or hilarity, has a corresponding influence on the living occupant of the womb, and leaves its lasting impress on the little being whose *life* commences with the period of impregnation. The eminent authoress George Eliot remarks, in reference to mental hereditary influence : —

> What! shall the trick of nostrils and of lips
> Descend through generations, and the soul
> That moves within our frame like God in worlds,
> Imprint no record, leave no documents
> Of her great history?

To this interrogatory let answer be in the lives of Vernet, Bonheur, Teniers, Caracci, Titian, Beethoven, Bach, Goethe, Schiller, the Herschels, father and son. Thousands of bright and shining lights in the constellation of science and art, in all ages and countries, owe their eminence and intellectual superiority to the beautiful, elevated, and heaven-inspired nature of their mothers. It was their maternal inheritance, the development of the germ implanted by their maternal progenitor. In our own beloved country, where woman has enjoyed an intellectual equality with man, these exquisite mental endowments have been transmitted with marvellous impartiality to the female branches of our leading families, as witness the Brownings, Beechers, Careys, Sigourneys, and others, whose works will live in the hearts of posterity till time shall be no more. Had women in past ages been permitted the same opportunities for mental development as their more favored colleagues of the sterner sex, their roll of honor and fame would have been as great and glorious as that of their masculine co-workers. The very desires and yearnings women of earlier ages have had for increased knowledge, and for the privilege of imparting and bringing into practical use that knowledge, have mirrored themselves upon the minds of their offspring, and *created*, by reflex action, the heroes and patriots of the present and past century.

HEREDITY OF HABIT, DISEASE, AND CRIME.

But there is a sad, a heart-rending reverse to this picture. The principle of heredity of taste, of habit, and of intellect, necessarily involves an heredity of vice, of disease, of crime, of sensuality, of dissipation, — of all those abnormalities of moral and mental force which have, in the present as well as the past, transformed this world from a Paradise to a Pandemonium. The *passions*, in all the fearful intensity of their original paroxysm, are permanently and indelibly photographed on the fœtus. The poor mother, whose period of pregnancy has been marked by a succession of domestic vicissitudes and griefs, cannot wonder that her dear new-born infant is ushered into existence with a prostrated nervous organism, and bears the impress of premature old age on its infantile countenance. What wonder that the offspring of a mother who, during the nine months she has borne her child, has indulged persistently in the use of the dram-bottle, should, before it is able to walk, exhibit an unnatural craving for intoxicating beverages, and positively refuse to take its natural nutriment! Two-thirds of the drunkards of this present generation have been *made* so, and *born* with that hereditary tendency to dissipation; they are only following out the behests of their *nature*, by the development of the germ implanted in them by their progenitors. In their case, it is a *disease*, and not a *crime*. Their parents are the only criminals, — and the parents' crime has culminated in an hereditary *curse* on the children and children's children. If we desire a proof of the law of heredity in reference to *vice*, a visit to our State prisons will show that fully 75 per cent. of our youthful criminals owe their irresistible passion for the commission of crime to the moral depravity and physical and mental degradation of their parents. In fact, so firmly established is this truth, that most of our juvenile offenders feel a sort of *family pride* in tracing their pedigree to the most notorious law-breakers our criminal records can furnish. Suicidal tendencies are frequently inherited, and sudden deaths at a certain age have been known to run through successive generations of families.

Education may do and doubtless does much to transform and modify, but it never *creates*. It may make a good musician, but it cannot make a musical genius. It may and does *modify* hereditary mental and physical defects, but it cannot eradicate them.

TRANSMISSION OF TRAITS OF CHARACTER.

When considered in all their varied relations, the scales of influence are evenly balanced between the parents and their children; neither can evade their responsibilities and be found guiltless. A large share of the influence coming from the father is communicated through the mother to the child by the impressions she receives from him during the period of gestation. If there is unison of spirit, a harmonious blending of their natures, there is more likely to be an equal mingling of the traits of both parents; while, on the other hand, if the husband is brutal, if she loathes his presence, then will the child, in all probability, be stamped for life with his most undesirable characteristics. Likes and dislikes, of persons, places, and even articles of diet (frequently of the most erratic nature), possess the mother's mind during this period; she tries to overcome the feeling, but in vain, and the consequence is that these unpleasant peculiarities are reproduced in the child. These indisputable facts show us the imperative necessity of making the surroundings of the mother as genial and pleasant as possible, if we would have the child all it might be, — bright, happy, and beautiful.

HOW PARENTS SHOULD LIVE AFTER CONCEPTION.

The life and surroundings of both parents, before, at the time of, and after, conception, are conditions that directly affect the child. The importance of physical perfection, both in the individual and in the race, cannot be overstated, for upon a sound physical basis rests strength of mind and soul, and all its multiple mental and moral outgrowths. The sickly in body, the depressed in mind, by becoming fathers and mothers, run the fearful risk of multiplying and intensifying the maladies and defects from which they suffer, and which they have, most probably, received from the neglect or ignorance of their parents. It is incumbent upon the parents, therefore, to make constant effort after purity of body and culture and strength of mind, that these desirable qualities may become the birthright of their offspring. This can only be attained by a simple but nutritious dietary, partaken at regular hours, and selected with due regard to its digestibility and suitability to constitutional peculiarities; healthful and congenial employment, both mental and physical; judicious and frequent exercise in the open air; the cultivation of music as a recreation; and, above and before all, the inculcation and development of a spirit of purity and cleanliness in thought, act, and deed. Human beings must learn, and adopt as an inviolable principle, the fact that sexual appetite was *only* given as a means of perpetuation of the race, not as a vehicle for lust, reckless passion, and indiscriminate indulgence in the lowest and most degrading species of sensual gratification. Until both man and woman have been enabled to bring their passions thoroughly under the control of the moral and intellectual faculties, they will never secure a progeny worthy of their high and holy mission, as the propagators of the highest race of beings known in God's creation.

DUTIES OF THE HUSBAND DURING THE PERIOD OF GESTATION.

During the last twenty-five years the attention and interest of the public has been constantly drawn to the important subject of improvement in stock; and farmers, stock-breeders, and those who devote their time and attention to the raising of cattle, take especial care in the selection of the finest and most healthy animals, and those possessing the highest qualifications for the purposes required. This having been done, the efforts of the owners are concentrated on the preservation of the health of the animals so selected; their food, their general surroundings, their exercise, and their periods of rest, so that their constitutions shall not be deteriorated or undermined by overwork, neglect, irregularity, etc.; for the experienced stock-breeder is fully cognizant of the fact that, just in proportion as the animal is cared for, and its welfare vigilantly watched, so will he be rewarded by the superior physical properties, health, and consequent market-value of the stock so raised. Now, if this be true in regard to the lower order of animals, how much more emphatically so in relation to the human race, in which the maintenance and improvement of the intellectual as well as the physical faculties are of the most vital importance in the exercise of the reproductive function! The instances we have cited in the first part of this chapter demonstrate the fact most indisputably, that these physical and mental qualifications or peculiarities which distinguish either or both of the parents are reproduced and often intensified in the children.

It must be presumed, as a matter of course, that the husband and wife have been attracted to each other by an affinity of taste, intellect, and mental and physical culture, far above the mere physical perfection sought for by the cattle-breeder or stock-raiser; but yet, the same principles must be acted upon, and the same care observed in the one case as in the other. Some of our first physiologists have remarked, and every-day experience has tested its truth, that if you want to make your child great and gifted in an intellectual point of view, you must first make it a " powerful animal;" that is, it must be amply endowed with physical vigor and health. This desirable object can only be effected by the adoption of a strict and inflexible course of conduct.

YOUR CHILD BEGINS ITS LIFE AT THE MOMENT OF CONCEPTION.

The instant that you have reason to suspect that a new organism has been brought into existence, the united efforts of yourself and your husband should be directed to the maintenance of your health, comfort, and happiness. Your husband should bear in mind that *the future of your unborn infant*, morally, intellectually, and physically, entirely depends upon the course of life pursued by himself and you *during the nine months of gestation.* You are to live a natural, healthful life ; not one of seclusion, melancholy, or abstemiousness, or, on the other hand, of wild excitement, indulgence, and neglect of those wholesome, health-producing home-duties which ordinarily devolve upon the wife in the daily routine of the homestead. But, as the condition of gestation, though perfectly natural, renders the woman susceptible of a constant succession of influences, alike unknown and incomprehensible to her or those with whom she is surrounded, it is the husband's paramount duty to evince towards his wife that self-denial, affectionate solicitude, and vigilant watchfulness and anxiety to anticipate her every wish, which an earnest, ardent love would prompt. He must treat her as an intelligent, reasoning being, not as a mere vehicle for the gratification of his passions. He must keep her mind free from all trouble, anxiety, or disturbing influences. Her diet and exercise must be regulated with the utmost care and discretion ; all indigestible or highly-seasoned dishes, stimulating beverages, late hours, exciting amusements, or arduous labors being especially avoided ; in a word, he must, as a rule, allow her to have *her own way* in these matters ; for her own common-sense and inclination will usually prompt her in the right direction. And, lastly, though by no means least, he should seize every opportunity of evincing his affectionate solicitude for her welfare, by those little unobtrusive kindnesses and attentions which are so grateful to, and heartily appreciated by, every true woman. A strict observance of this course of conduct cannot fail to produce the result so ardently desired by both parents, — an offspring in the full possession of all the mental and physical faculties which go to make up the perfect man or woman.

SIGNS OF PREGNANCY.

The diagnosis of early pregnancy is no easy task : it frequently baffles the most experienced physicians ; therefore great care and discrimination should be exercised before venturing upon a positive assertion. The general condition of a pregnant woman is a plethoric habit of the body, a quick, full pulse, and an apparent increase in the circulation of the blood. The sympathetic action of the several organisms (the brain, stomach, etc.) with the uterus is markedly evinced ; variations in temper and disposition are of frequent occurrence ; the appetite is curiously capricious ; and the skin occasionally becomes sallow, or shows discolored patches in various parts of the body. The special deviations from the normal physical condition, or unvarying signs by which pregnancy may be determined, are : —

1. Cessation of Menstruation. — The non-appearance of the catamenia at the proper time is one of the first circumstances which leads a woman to suspect her pregnancy ; and if a second term passes by without their appearance, it is usually looked upon as conclusive. But, strictly speaking, it is not so, for menstruation may be arrested by various diseases which we have already described ; or, on the contrary, it is by no means infrequent that menstruation will continue its course uninterruptedly for several months after conception or during the whole period of gestation. Nevertheless, although exceptions of this kind do occur, when menstruation ceases without any perceptible cause, the woman otherwise remaining perfectly healthy, we take it as pretty good evidence that conception has taken place.

2. Morning-Sickness, combined with other symptoms, is of considerable value, though, of itself, it is extremely unreliable, because pregnancy frequently occurs without the slightest indication of sickness in the morning; while, in other cases, morning-sickness may present itself, from various causes, and yet the patient not be pregnant. This irritability of the stomach, arising from sympathy with the uterus, commences soon after conception, and ceases shortly after the third month.

3. Salivation. — This is sometimes, though far from invariably, present. When it exists, however, it differs materially from mercurial salivation, inasmuch as there is a total absence of the peculiar odor of mercurialization, and the sponginess and soreness of the gums produced by that metal.

4. Enlargement of the Breasts. — About two months after conception, the woman's attention is called to the state of the breasts. She feels an uneasy sensation of fulness, with a throbbing and tingling pain in their sub-stance, and at the nipples. They increase in size and firmness, and have a peculiar, knotty, glandular feel; the areola (the colored circle about the nipple) darkens, and after some time milk is secreted. But it must be recollected that the breasts may enlarge from other causes; this happens with some women at each menstrual period when the cata-menia are suspended, or after they cease; and at such times a milky fluid may be secreted.

5. Enlargement of the Abdomen. — The gradual enlargement of the abdomen, taken in connection with the symptoms already mentioned, enables us to esti-mate with considerable certainty the period of pregnancy at the time the examination is made. Distention of the ab-domen, however, sometimes takes place from other causes than pregnancy; therefore, this sign alone is not sufficient to warrant us in pronouncing upon a case.

6. Quickening. — This term is applied to the first movement of the child within the womb, or rather to the first perception of such movement on the part of the mother. Some women labor under the erroneous idea that the child does not commence its life until the fourth month, — the time about which the quickening is usually felt. The fact is, however, we have just as much reason to believe that the child is quite as much alive at the *fourth week*, and, indeed, *from the very moment of conception.* Quickening sometimes takes place at an earlier period than the fourth month, while occasionally it is delayed until the sixth or seventh month. The sensation is, at first, like a feeble pulsation, and, though so slight, is often accompanied by sickness at the stomach, a feeling of faintness, and, sometimes, complete syncope or swooning. By degrees it becomes stronger and more frequent, until the movements of the extremities are plainly distinguishable.

PRESERVATION OF HEALTH DURING PREGNANCY.

To the vitally important question, " How shall perfect health be secured?" we reply, " By vigilant attention to *dress, diet,* and *exercise.*" These items are of the highest importance under ordinary circumstances, but in pregnancy this importance is increased a hundred-fold.

1. Dress. — Lycurgus, the great Spartan lawgiver, decreed that all pregnant women should wear wide, loose clothing. A similar law also prevailed among the Romans. The dress should be warm, loose, and light during the whole period; and, at this day, were there such a law, and the proper power to enforce it, you would hear fewer complaints of " bad gettings up," " fallings," " pro-lapsuses," " broken breasts," " weaknesses," and other complaints which do so much to undermine the constitutions of married women. · Let out your dresses early ; no part of the dress should be tight; even garters should be abandoned ; everything should be loose, so as to allow a free circulation of the blood. Tight lacing is highly injurious: how can it be otherwise? While nature is gradually increasing the capacity of the abdomen to accommodate the steady development of the child, the absurdity of compressing the chest with stays, or girding the abdomen with skirts, would seem patent to any one possessed of common intelligence, for they must know that it cannot fail to have an extremely injurious effect on both mother and child.

CARE OF THE BREASTS.

Special care must be taken that the dress is loose about the breasts. This is highly important, for not unfrequently the breasts and nipples are so flattened out by direct pressure that after confinement there is nothing that can be properly called a nipple left. Sometimes they are almost entirely obliterated, from compression during girlhood, and a continuance of that pressure during married life. But this pressure does not affect the nipple only : the secretory structure of the breast itself is permanently injured, and the important function of lactation never attains that state of perfection which it otherwise would. The suffering resulting from this state of things to both mother and child is by no means trifling. The breasts should be effectually protected from compression of any kind, and should be subjected to careful but gradual development, especially the nipple, their *most sensitive part*, by the medium of which alone the infant can obtain its natural nourishment. Dr. Tracy, one of the most eminent obstetricians in the United States, suggests the following method for keeping the nipples permanently prominent after they have been once drawn out : "Wind a bit of woollen thread or yarn two or three times around the base of the nipple, tying it moderately tight, but not so tight as to interfere with the free circulation of the blood."

Retraction, or *deficient development*, is not the only difficulty to which the nipples are subject. The most common affections to which they are liable are excoriations, cracks, inflammation, scaly eruptions, and small abscesses. These usually arise from the extreme sensitiveness of the skin, occasioned by the nipple being kept folded down upon the breast by the clothing; in this way, the skin around the base of the nipple, being folded upon itself, becomes very delicate and thin, and unfitted for the purpose for which it was designed. The natural result is, as soon as the child begins to nurse, the skin becomes irritated and inflamed; cracks, fissures, or abscesses form, and the mother is subjected to untold misery every time the child is put to the breast.

Now, the main object to be attained in preparing the breasts during early pregnancy for their future important function, is to thicken and toughen the skin upon and at the base of the nipple. For several weeks prior to delivery, the entire breast and chest should be bathed in cold water daily, and afterwards well dried and rubbed with coarse towels. Some recommend bathing the breast and nipple with brandy, or various decoctions of herbs; but I should infinitely prefer the cold-water treatment, or simply rubbing the parts upon all sides, and in every direction, with the palm of the dry hand. This rubbing should be commenced soon after the establishment of pregnancy, and continued until confinement. Should there be tenderness, excoriation, or soreness, the parts may be bathed in a weak solution of arnica.

DERANGEMENTS DURING PREGNANCY.

The undermentioned *derangements during the period of pregnancy*, are not considered in detail, for the reason that they cannot fairly be classed under the head of *diseases;* but remedies for each of these ailments will be found at the following numbers in the Appendix : —

Continued menstruation, 8, 272.
Headache and vertigo, etc., 9, 216, 217, 218.
Morning-sickness, 10, 214, 331, 332.
Constipation, 11, 179, 180, 181, 182, 183.
Diarrhœa, 12, 190, 191, 192.
Hysteria, or fainting-fits, 13, 231, 232, 233, 234, 235.
Palpitation of the heart, 14, 275, 276.
Toothache, 15, 321, 322, 323, 324.
Neuralgia, 16, 281, 282, 283.
Pains in back and side, 17.
Cramps in limbs, back, or abdomen, 18.
Varicose or swollen veins, 19.
Hemorrhoids, or piles, 20, 225, 226, 227, 228.
Jaundice, or icterus, 21, 236, 237, 238, 253, 254.
Incontinence of urine, 22, 280.
Difficult or scanty urination, 23, 280.
Flooding, 24, 219, 220, 221,
Miscarriage, or abortion, 25, 259.
False pains, 26.

FALSE PAINS.

Some time previous to delivery (varying from two weeks to a few days) women are frequently much annoyed with what are termed *spurious* or *false pains*. These pains sometimes so closely resemble true labor-pains, that it is exceedingly difficult to discriminate the one from the other. From this close resemblance arise what are called "false alarms."

Now, in view of all this, it becomes quite essential that both patient and nurse should fully understand the difference between true and false pains. False pains usually differ from labor-pains in the irregularity of their occurrence; in being entirely unconnected with uterine contraction, and being chiefly confined to the abdomen, which is peculiarly sensitive to touch and movement; and in their not increasing in intensity as they return. True labor-pains commence low down, and are first felt in the *back*, extending gradually to the front, recurring with regularity and increasing in intensity with each return.

Spurious pains arise from various causes, such as over-fatigue, indigestion, cold, mental emotions, constipation, errors in diet, and frequently by the active motions of the child.

TREATMENT.

As these spurious pains, when they come on early in pregnancy, are liable to bring on premature labor, or, when at full term, occasion great distress and loss of rest, it is always desirable to relieve them as speedily as possible. This may generally be effected by one of the following remedies, each of which is appropriate to the before-mentioned causes, in the order named: Bryonia, pulsatilla, nux vomica, dulcamara, and aconitum. Twelve globules of either of these remedies, dissolved in twelve teaspoonfuls of water, taking one teaspoonful of the solution every half-hour until the symptoms are relieved, will be found effectual. (*See No.* 26.)

METHOD OF CALCULATING THE TIME OF CONFINEMENT.

The *time* when confinement may be expected, particularly if it be a *first* pregnancy, is naturally a matter of considerable importance and interest to the young married woman; and it is certainly very desirable on all accounts that it should be as *accurately* determined as possible. It is impossible, however, by what is called *reckoning*, or by any other means, to ascertain the *exact* day upon which labor will take place. There are many circumstances which prevent this; among others, the uncertainty connected with the duration of pregnancy itself. The nearest approach to the actual time, and the most effective way of meeting the difficulty is to allow 280 days as the full period of gestation — a fact which is proved by the great weight of experience all the world over.

For the purpose of facilitating reckoning, the following tables have been prepared. The mode of using them needs but little explanation : Suppose the lady to be taken unwell on the 28th of December, and continue so until the 31st, the reckoning must then commence on the day following, — the 1st of January. Look for this date on the first column of the January table, and the corresponding dates of quickening and labor will be found in the same line ; that is to say, she will quicken about the 20th of May, and be confined about the 8th of October.

JANUARY.

Date of becoming Pregnant.		Date of Quickening.		Date of expected Confinement.	
JANUARY	1	MAY	20	OCTOBER	8
....	2	....	21	....	9
....	3	....	22	....	10
....	4	....	23	....	11
....	5	....	24	....	12
....	6	....	25	....	13
....	7	....	26	....	14
....	8	....	27	....	15
....	9	....	28	....	16
....	10	....	29	....	17
....	11	....	30	....	18
....	12	....	31	....	19
....	13	JUNE	1	....	20
....	14	....	2	....	21
....	15	....	3	....	22
....	16	....	4	....	23
....	17	....	5	....	24
....	18	....	6	....	25
....	19	....	7	....	26
....	20	....	8	....	27
....	21	....	9	....	28
....	22	....	10	....	29
....	23	....	11	....	30
....	24	....	12	....	31
....	25	....	13	NOVEMBER	1
....	26	....	14	....	2
....	27	....	15	....	3
....	28	....	16	....	4
....	29	....	17	....	5
....	30	....	18	....	6
....	31	....	19	....	7

FEBRUARY.

Date of becoming Pregnant.		Date of Quickening.		Date of expected Confinement.	
FEBRUARY	1	JUNE	20	NOVEMBER	8
....	2	....	21	....	9
....	3	....	22	....	10
....	4	....	23	....	11
....	5	....	24	....	12
....	6	....	25	....	13
....	7	....	26	....	14
....	8	....	27	....	15
....	9	....	28	....	16
....	10	....	29	....	17
....	11	....	30	....	18
....	12	JULY	1	....	19
....	13	....	2	....	20
....	14	....	3	....	21
....	15	....	4	....	22
....	16	....	5	....	23
....	17	....	6	....	24
....	18	....	7	....	25
....	19	....	8	....	26
....	20	....	9	....	27
....	21	....	10	....	28
....	22	....	11	....	29
....	23	....	12	....	30
....	24	....	13	DECEMBER	1
....	25	....	14	....	2
....	26	....	15	....	3
....	27	....	16	....	4
....	28	....	17	....	5

MARCH.

Date of becoming Pregnant.		Date of Quickening.		Date of expected Confinement.	
MARCH	1	JULY	18	DECEMBER	6
....	2	....	19	....	7
....	3	....	20	....	8
....	4	....	21	....	9
....	5	....	22	....	10
....	6	....	23	....	11
....	7	....	24	....	12
....	8	....	25	....	13
....	9	....	26	....	14
....	10	....	27	....	15
....	11	....	28	....	16
....	12	....	29	....	17
....	13	....	30	....	18
....	14	....	31	....	19
....	15	AUGUST	1	....	20
....	16	....	2	....	21
....	17	....	3	....	22
....	18	....	4	....	23
....	19	....	5	....	24
....	20	....	6	....	25
....	21	....	7	....	26
....	22	....	8	....	27
....	23	....	9	....	28
....	24	....	10	....	29
....	25	....	11	....	30
....	26	....	12	....	31
....	27	....	13	JANUARY	1
....	28	....	14	....	2
....	29	....	15	....	3
....	30	....	16	....	4
....	31	....	17	....	5

APRIL.

Date of becoming Pregnant.		Date of Quickening.		Date of expected Confinement.	
APRIL	1	AUGUST	18	JANUARY	6
....	2	....	19	....	7
....	3	....	20	....	8
....	4	....	21	....	9
....	5	....	22	....	10
....	6	....	23	....	11
....	7	....	24	....	12
....	8	....	25	....	13
....	9	....	26	....	14
....	10	....	27	....	15
....	11	....	28	....	16
....	12	....	29	....	17
....	13	....	30	....	18
....	14	....	31	....	19
....	15	SEPTEMBER	1	....	20
....	16	....	2	....	21
....	17	....	3	....	22
....	18	....	4	....	23
....	19	....	5	....	24
....	20	....	6	....	25
....	21	....	7	....	26
....	22	....	8	....	27
....	23	....	9	....	28
....	24	....	10	....	29
....	25	....	11	....	30
....	26	....	12	....	31
....	27	....	13	FEBRUARY	1
....	28	....	14	....	2
....	29	....	15	....	3
....	30	....	16	....	4

MAY.

Date of becoming Pregnant.		Date of Quickening.		Date of expected Confinement.	
MAY	1	SEPTEMBER	17	FEBRUARY	5
....	2	....	18	....	6
....	3	....	19	....	7
....	4	....	20	....	8
....	5	....	21	....	9
....	6	....	22	....	10
....	7	....	23	....	11
....	8	....	24	....	12
....	9	....	25	....	13
....	10	....	26	....	14
....	11	....	27	....	15
....	12	....	28	....	16
....	13	....	29	....	17
....	14	....	30	....	18
....	15	OCTOBER	1	....	19
....	16	....	2	....	20
....	17	....	3	....	21
....	18	....	4	....	22
....	19	....	5	....	23
....	20	....	6	....	24
....	21	....	7	....	25
....	22	....	8	....	26
....	23	....	9	....	27
....	24	....	10	....	28
....	25	....	11	MARCH	1
....	26	....	12	....	2
....	27	....	13	....	3
....	28	....	14	....	4
....	29	....	15	....	5
....	30	....	16	....	6
....	31	....	17	....	7

JUNE.

Date of becoming Pregnant.	Date of Quickening.	Date of expected Confinement.
JUNE 1	OCTOBER 18	MARCH 8
.... 2	 19	 9
.... 3	 20	 10
.... 4	 21	 11
.... 5	 22	 12
.... 6	 23	 13
.... 7	 24	 14
.... 8	 25	 15
.... 9	 26	 16
.... 10	 27	 17
.... 11	 28	 18
.... 12	 29	 19
.... 13	 30	 20
.... 14	 31	 21
.... 15	NOVEMBER 1	 22
.... 16	 2	 23
.... 17	 3	 24
.... 18	 4	 25
.... 19	 5	 26
.... 20	 6	 27
.... 21	 7	 28
.... 22	 8	 29
.... 23	 9	 30
.... 24	 10	 31
.... 25	 11	APRIL 1
.... 26	 12	 2
.... 27	 13	 3
.... 28	 14	 4
.... 29	 15	 5
.... 31	 16	 6

JULY.

Date of becoming Pregnant.		Date of Quickening.		Date of expected Confinement.	
JULY	1	NOVEMBER	17	APRIL	7
....	2	....	18	....	8
....	3	....	19	....	9
....	4	....	20	....	10
....	5	....	21	....	11
....	6	....	22	....	12
....	7	....	23	....	13
....	8	....	24	....	14
....	9	....	25	....	15
....	10	....	26	....	16
....	11	....	27	....	17
....	12	....	28	....	18
....	13	....	29	....	19
....	14	....	30	....	20
....	15	DECEMBER	1	....	21
....	16	....	2	....	22
....	17	....	3	....	23
....	18	....	4	....	24
....	19	....	5	....	25
....	20	....	6	....	26
....	21	....	7	....	27
....	22	....	8	....	28
....	23	....	9	....	29
....	24	....	10	....	30
....	25	....	11	MAY	1
....	26	....	12	....	2
....	27	....	13	....	3
....	28	....	14	....	4
....	29	....	15	....	5
....	30	....	16	....	6
....	31	....	17	....	7

AUGUST.

Date of becoming Pregnant.		Date of Quickening.		Date of expected Confinement.	
AUGUST	1	DECEMBER	18	MAY	8
....	2	....	19	....	9
....	3	....	20	....	10
....	4	....	21	....	11
....	5	....	22	....	12
....	6	....	23	....	13
....	7	....	24	....	14
....	8	....	25	....	15
....	9	....	26	....	16
....	10	....	27	....	17
....	11	....	28	....	18
....	12	....	29	....	19
....	13	....	30	....	20
....	14	....	31	....	21
....	15	JANUARY	1	....	22
....	16	....	2	....	23
....	17	....	3	....	24
....	18	....	4	....	25
....	19	....	5	....	26
....	20	....	6	....	27
....	21	....	7	....	28
....	22	....	8	....	29
....	23	....	9	....	30
....	24	....	10	....	31
....	25	....	11	JUNE	1
....	26	....	12	....	2
....	27	....	13	....	3
....	28	....	14	....	4
....	29	....	15	....	5
....	30	....	16	....	6
....	31	....	17	....	7

SEPTEMBER.

Date of becoming Pregnant.	Date of Quickening.	Date of expected Confinement.
SEPTEMBER 1	JANUARY 18	JUNE 8
.... 2	 19	 9
.... 3	 20	 10
.... 4	 21	 11
.... 5	 22	 12
.... 6	 23	 13
.... 7	 24	 14
.... 8	 25	 15
.... 9	 26	 16
.... 10	 27	 17
.... 11	 28	 18
.... 12	 29	 19
.... 13	 30	 20
.... 14	 31	 21
.... 15	FEBRUARY 1	 22
.... 16	 2	 23
.... 17	 3	 24
.... 18	 4	 25
.... 19	 5	 26
.... 20	 6	 27
.... 21	 7	 28
.... 22	 8	 29
.... 23	 9	 30
.... 24	 10	JULY 1
.... 25	 11	 2
.... 26	 12	 3
.... 27	 13	 4
.... 28	 14	 5
.... 29	 15	 6
.... 30	 16	 7

OCTOBER.

Date of becoming Pregnant.		Date of Quickening.		Date of expected Confinement.	
OCTOBER	1	FEBRUARY	17	JULY	8
....	2	....	18	....	9
....	3	....	19	....	10
....	4	....	20	....	11
....	5	....	21	....	12
....	6	....	22	....	13
....	7	....	23	....	14
....	8	....	24	....	15
....	'9	....	25	....	16
....	10	....	26	....	17
....	11	....	27	....	18
....	12	....	28	....	19
....	13	MARCH	1	....	20
....	14	....	2	....	21
....	15	....	3	....	22
....	16	....	4	....	23
....	17	....	5	....	24
....	18	....	6	....	25
....	19	....	7	....	26
....	20	....	8	....	27
....	21	....	9	....	28
....	22	....	10	....	29
....	23	....	11	....	30
....	24	....	12	....	31
....	25	....	13 ·	AUGUST	1
....	26	....	14	....	2
....	27	....	15	....	3
....	28	....	16	....	4
....	29	....	17	....	5
....	30	....	18	....	6
....	31	....	19	....	7

NOVEMBER.

Date of becoming Pregnant.	Date of Quickening.	Date of expected Confinement.
NOVEMBER 1	MARCH 20	AUGUST 8
.... 2	 21	 9
.... 3	 22	 10
.... 4	 23	 11
.... 5	 24	 12
.... 6	.... · 25	 13
.... 7	 26	 14
.... 8	 27	 15
.... 9	 28	 16
.... 10	 29	 17
.... 11	 30	 18
.... 12	 31	 19
.... 13	APRIL 1	 20
.... 14	 2	 21
.... 15	 3	 22
.... 16	 4	 23
.... 17	 5	 24
.... 18	 6	 25
.... 19	 7	 26
.... 20	 8	 27
.... 21	 9	 28
.... 22	 10	 29
.... 23	 11	 30
.... 24	 12	 31
.... 25	 13	SEPTEMBER 1
.... 26	 14	 2
.... 27	 15	 3
.... 28	 16	 4
.... 29	 17	 5
.... 30	 18	 6

DECEMBER.

Date of becoming Pregnant.		Date of Quickening.		Date of expected Confinement.	
DECEMBER	1	APRIL	19	SEPTEMBER	7
....	2	....	20	....	8
....	3	....	21	....	9
....	4	....	22	....	10
....	5	....	23	....	11
....	6	....	24	....	12
....	7	....	25	....	13
....	8	....	26	....	14
....	9	....	27	....	15
....	10	....	28	....	16
....	11	....	29	....	17
....	12	....	30	....	18
....	13	MAY	1	....	19
....	14	....	2	....	20
....	15	....	3	....	21
....	16	....	4	....	22
....	17	....	5	....	23
....	18	....	6	....	24
....	19	....	7	....	25
....	20	....	8	....	26
....	21	....	9	....	27
....	22	....	10	....	28
....	23	....	11	....	29
....	24	....	12	....	30
....	25	....	13	OCTOBER	1
....	26	....	14	....	2
....	27	....	15	....	3
....	28	....	16	....	4
....	29	....	17	....	5
....	30	....	18	....	6
....	31	....	19	....	7

PARTURITION, OR CONFINEMENT.

You have now arrived at the third great epoch of your life, on the happy consummation of which all your hopes of future happiness, comfort, and usefulness depend. In a few short hours an entirely new world will have opened to your vision, a new and inexhaustible source of joy, affection, responsibility, and fond anticipation will spring up in your heart, and the little being whom you are now about to introduce into the world will awaken within you feelings to which you have hitherto been a stranger, and which language would utterly fail to describe. We will, of course, suppose that your physician is in attendance, and that the nurse, — whom you have selected on account of her experience, genial temperament, and skill in the execution of the critical and onerous duties of her position, — has made all the necessary preparations for the comfort and safety of her patient and the " little stranger" now about to make its advent. Under these circumstances, it is but natural and proper that you should look forward to a happy and successful termination of your present trouble, for past experience demonstrates that, where both parents are in a normally healthy condition, and ordinary care has been taken, not more than one case in two hundred has a disastrous or unfavorable termination, either for the mother or child.

CHILDBIRTH.

Labor, as we have said, generally takes place at the end of two hundred and eighty days from conception; but it is not absolutely certain, for it sometimes occurs prematurely, or may be prolonged to the two hundred and ninetieth day, especially in first cases. The commencement of actual labor is usually preceded by some of the following premonitory symptoms: agitation, nervous trembling, lowness of spirits, irritability of the bladder, with frequent desire to urinate, nausea and vomiting, flying pains through the abdomen, followed by an increased mucous discharge.

The occurrence of true labor-pains may soon be looked for after the premonitory symptoms we have described. The pains usually commence in the back, but sometimes they are first felt at the lower and front part of the abdomen, and extend to the loins and lower part of the back. They are not constant, but periodical or intermittent, coming on at regular intervals of longer or shorter duration. At the commencement they are not actual pains, but rather a feeling of uneasiness. When active pains first begin, they are slight and of short duration, lasting but a few moments, and with intervals of rest lasting from half an hour to an hour or more. By degrees they become more and more frequent, gradually increasing in intensity until labor is completed, which usually takes from four to six hours.

Anxiety on account of the length of labor should never be indulged in. If the position of the child is right, protracted labors are no more dangerous than short ones. First labors are generally longer than subsequent ones. Your medical attendant and nurse having attended to the requirements of yourself and your new-born babe, your own course of action is self-evident; viz., to render implicit obedience to their instructions, banish all anxiety or thought from your mind, and court that rest and sleep your exhausted frame so urgently requires.

INSTRUCTION TO NURSES IN PARTURITION OR CONFINEMENT.

It is of primary importance that every woman should be more or less conversant with the necessary cares and duties of a lying-in chamber, for the purpose, not of making them poor physicians, but of fitting them to become competent, efficient, and trustworthy nurses, so qualified that in cases of emergency they may render intelligent assistance.

It not unfrequently happens, especially in quick cases, that the medical attendant may be detained, and delivery take place before he arrives. In such a case how important that the nurse should be capable of meeting the emergency, and securing the safety of both mother and child !

Though, as I have before observed, labor is a perfectly natural process, and the majority of cases would terminate favorably with none present but an ordinary nurse, yet events *may* occur which would call for prompt interference, and such interference as none but a well-educated and qualified medical man could afford.

PREPARATIONS FOR THE BIRTH.

Immediately on your arrival you should take care that every necessary preparation is made for the occasion : the room should be put in perfect order, the clothing for both mother and child be placed in readiness, arranged in the order in which they will be required, and placed in such a convenient position that they can be obtained without trouble. You should also have convenient a pair of sharp scissors and a couple of short pieces of strong cotton cord.

As soon as possible you should " make the bed," that is, place a square of oiled silk or rubber sheet over the mattress, to protect it and the bedclothes from the " discharges." Over this place the under-blanket and sheet, and upon them two or three sheets folded square. These will absorb the greater portion of the discharges, leaving the dry bed-linen beneath.

When you have put the patient to bed, draw up her night-gown above the hips, to escape soiling. Assist her to the best of your ability in promoting the expulsive pains and mitigating her sufferings.

Place your patient in the most convenient position for the delivery, — on the left side, and near the edge of the bed, the knees drawn up, and a pillow between them. When the last pain, which expels the head of the child, comes, receive the child upon your extended hands, taking care that the umbilical cord is not wound round the child's neck ; wait patiently until the entire body is expelled, and then convey the body to a sufficient distance to avoid the discharges ; place it in such a position that it shall rest easy and breathe freely.

As soon as respiration is fully established, the umbilical cord should be tied at about two inches from the navel, and again a few inches further on, cutting the cord between the two ligatures with the scissors.

As soon as the cord is tied and cut, the child should be wrapped up closely in a soft flannel blanket which has previously been well warmed, and then be removed.

Immediately after the birth the binder should be applied to the mother. This may consist of a folded towel, or other broad bandage, placed around the whole abdomen and extending down over the hips. It should be pinned firmly, but not too tight. Be careful to have it smooth, so as to give an even support to the whole abdomen. This is especially serviceable for the first few weeks, particularly in the case of feeble women, and also when the patient suffers faintness immediately after delivery. If properly applied at first, it is very useful in maintaining a certain degree of contraction of the uterus, and giving support to the abdominal walls. It also assists in promoting a return to the natural condition of the abdomen, preventing that loose, flabby state of the abdominal walls which so frequently follows confinement. I recommend that it should be worn some time after getting up, as it has a happy effect in preserving the natural form and dimensions, especially of women who have many children in the course of a few years.

After the placenta or after-birth has been taken away, a warm napkin should be applied to the external parts, and the binder tightened, if necessary. The soiled bed-linen should then be removed, the night-dress drawn down, and the patient be induced to sleep. After a few hours' rest, the napkin should be removed, the parts washed with soft warm water, to which a few drops of the tincture of arnica should be added, and another napkin applied. This operation should be repeated twice or thrice per day for the first few days.

The room should be kept neat and clean, well lighted, and well ventilated, and of an equal temperature, between 67° and 73°. The patient should be kept free from the excitement of company and conversation, for the first few days at least, all visitors or children being excluded but the nurse and attendants. Take care that the patient has appropriate and nourishing, but not exciting food, rigidly following out the physician's orders, who will direct the changes in diet according to the patient's condition.

The directions as to the care of the infant will be found in a subsequent chapter.

CHAPTER VIII.

TREATMENT AFTER DELIVERY.

MANY persons labor under the preposterous impression that it is necessary to keep the chamber constantly darkened, alleging many reasons therefor, the principal of which is, that the infant's eyes are too delicate and sensitive to bear the ordinary light of day. It would be just as rational to close all the doors, windows, and other apertures to exclude the air, on the supposition that the atmosphere was too strong for the infantile lungs. It is well enough, for the first two or three days, that the light should be so modified that its rays should not be *too strong* for both mother and infant; but light, equable temperature, and thorough ventilation are as necessary for all in the apartment as *food* and *rest.* Let the room be *neat and clean, light and airy.*

Ventilation is far too frequently neglected. Not many years since, the doors of the chamber would be barricaded with sand-bags, and the windows closely encased in weather-strips or some similar material, to preclude all ingress and egress of air, the inmates being thereby obliged to breathe over and over again the same vitiated atmosphere, and thus made painfully susceptible of puerperal and other diseases, which common-sense treatment would have prevented.

The temperature of the room should always be kept at a given standard, from 67° to 75°. It should not be increased, the air vitiated, or your own health and life endangered, by the excitement of visitors or children, for the first few days at least. All *friendly calls* should be positively forbidden during the first week, the only persons admitted being your physician and your attendants.

AFTER-PAINS.

These seldom occur as the result of a *first confinement*, though they are very frequently met with in second and subsequent labors. They are the direct result of uterine contractions. As a general rule, they commence within half an hour after delivery, and ordinarily cease within thirty or forty hours, though they may continue longer. They vary much in their frequency, severity, and duration, but are usually unaccompanied by any sense of bearing-down. Within certain limits, their operation is undoubtedly salutary, as they prevent flooding, diminish the size of the uterus, and expel its contents. But when they occur in an aggravated form, and are unduly protracted, — an occurrence not at all uncommon in females of excitable, nervous sensibility, — they should be subdued as speedily as possible.

TREATMENT.

In the event of soreness, local pain, or nervous excitement supervening, arnica, coffea, and aconitum may be given in doses of 10 globules every two or three hours. They may also be used in the liquid form externally as a lotion. (For other remedies, see Nos. 16, 17, 25, or 26, according to circumstances.)

FLOODING AFTER DELIVERY.

Of course in all cases it proceeds from the mouths of the vessels which have failed to contract after the separation of the after-birth.

One of the most frequent causes of hemorrhage after delivery is mental excitement, caused by too much company, the worry arising from children's noise, depression of spirits from disappointment, and indeed excitement of any kind. It is therefore necessary that all excitement should be religiously avoided, and sleep, that great restorer of health and strength, be courted.

TREATMENT.

A drop of the *tincture of cinnamon* in a tumbler half full of water — a teaspoonful every few minutes — has produced happy results in exciting contraction of the womb when all other remedies have failed. Cold water is a valuable auxiliary, and in all severe cases should be freely used. Cloths dipped in the coldest water should be applied to the abdomen and genitals and renewed every few minutes ; or pounded ice, if necessary, may be put in bags and applied in the same manner. Cold drinks are also of great service. (Nos. 24, 219, 220, 221.)

DURATION OF CONFINEMENT.

It will be advisable for you to lie quietly in bed for six or eight days after delivery. The length of time, however, will, in a great measure, depend upon circumstances ; many women are better able to stand upon their feet within six days than others are within three weeks. Should your general health be poor, your strength exhausted, or the discharge profuse, amounting to hemorrhage, and producing great debility, you will be compelled to occupy your bed or couch for as long a period as the symptoms continue. For the first nine days, as a rule, the greater part of your time should be spent in bed, if even your good health should appear to warrant your getting about earlier, as by indiscretion you might bring on local displacements or other serious uterine diseases, which would take many years to recover from, if even they did not become permanent. After this period, you may get up, resting in an easy-chair for a short time every day. When twelve or fifteen days have elapsed, if you feel pretty strong, it would be advisable to take gentle exercise about your room; but you should not resume your ordinary household duties or go up and down stairs until the close of the third week, nor are you, in reality, thoroughly recovered until the expiration of the sixth week.

DIET AND REGIMEN DURING CONFINEMENT.

By a strict and well-regulated regimen during confinement, you will be able to ward off a great many accidents. Great care must be taken that the utmost cleanliness is observed; in washing the body, warm water should be used, being gradually reduced in temperature, from time to time, until it is nearly but not quite cold. The linen should be changed at least once in twenty-four hours. The food must be easy of digestion, moderate in quantity, and not stimulating. For the first three days it should consist of gruel, light custards, toast, bread, weak black tea, broths, and other similar articles. After the third day, or when the supply of milk is fully established, a little soup, light, nourishing meats, such as chicken, lamb, etc., can be partaken of, until, gradually, the ordinary diet may be resumed without danger.

Ales, wines, coffee, and stimulating drinks generally, which are commonly used to promote the secretion of milk, should be studiously avoided as being peculiarly injurious. Most of these preparations predispose to fevers, and not unfrequently to night-sweats. Coffee especially deranges the nervous systems of both mother and child, and produces numerous diseases of the digestive organs. For drinks, weak tea, claret, and cold water, either pure or flavored with sirups, are excellent for women in confinement, as is also broma, which is a specially nutritious preparation of cocoa, and can be procured at any respectable grocery.

DISEASES FOLLOWING PARTURITION.

The Lochia. — The discharge of blood which accompanies delivery continues for several days afterwards, doubtless from the mouths of the vessels exposed by the separation of the after-birth. After three or four days the character of this discharge changes, and, instead of continuing a mere escape of blood, it takes on the character of a secretion. This discharge is called the "lochia." For the first three or four days it continues of a red color, but much thinner and more watery than blood; it then sometimes becomes thick and yellow, but more frequently maintains its watery consistence, and changes its color successively to greenish, yellowish, and lastly that of soiled water.

The duration of the lochial discharge varies greatly in different women; in some it is thin and scanty, and ceases in a few days; while in others it continues for several weeks, and is sometimes so profuse as to almost amount to hemorrhage. As this secretion is necessary to health, its *sudden* suppression is generally attended with evil results. Frequent washings with soft, warm water should be practised as long as it continues.

Suppression of the Lochia. — This may be caused by exposure to cold, errors in diet, or sudden mental emotions. The symptoms are generally chilliness, fever, thirst, headaches, and occasionally delirium, pain in the back, limbs, etc.

TREATMENT.

Warm compresses around the abdomen, and warm hip and foot baths will prove excellent remedial agents. For internal remedies, see Nos. 219, 220.

Excessive or Protracted Lochia. — When the lochial discharge is too profuse or continues too long, tepid hip-baths are valuable auxiliaries, and, in all severe or obstinate cases, should be freely used; complete rest and good nourishment are indispensable to the correction of this derangement. In all affections during or subsequent to parturition, where the symptoms do not at once succumb to the temporary treatment, skilful medical aid should at once be obtained.

MILK-FEVER.

About the third or fourth day after confinement you may expect your breasts to become distended with milk; and at the same time you may experience a chill, followed, more or less, by fever and headache. This is called milk-fever. It is but seldom, however, that this disturbance becomes sufficiently serious to call for medical interference, especially if you nurse your own infant, when the milk can, of course, be drawn off as soon as it commences to flow. If, however, you do not nurse your child, this fever may become complicated with other ailments, which it is necessary to prevent. External applications are of little use during milk-fever (excepting the arnica lotion). The milk should be drawn out as soon as possible, either by child or nurse; during the continuance of this fever none but the lightest articles of diet should be partaken of, such as gruel, boiled rice, toast, toasted crackers, weak tea, broma, or other equally light food. (*See Nos.* 210, 211, 212.)

Suppressed Secretion of Milk. — The secretion of milk being a natural function, its sudden suppression not unfrequently produces sudden disorders, such as internal or local congestion and inflammation, determination of blood to the head, chest, or abdomen, and the usual train of symptoms constituting childbed fever. The evil effects arising from the suppression of the milk are frequently of so serious a nature that the slightest diminution in the supply should excite your apprehension and place you upon your guard; for, in the great majority of cases, at the outset of this difficulty, the flow of milk may be restored by the administration of Nos. 210, 211, 212.

Excessive Secretion of Milk is generally accompanied by painful distention of the breasts, emaciation, debility, and not unfrequently initiates nervous and inflammatory disorders. In all cases wrap the breasts in cotton batting; it will reduce the swelling and mitigate the pain.

CONSTIPATION AFTER CONFINEMENT.

It is somewhat common to find the bowels inactive for some few days after delivery, the secretion from the intestinal tube being partially or wholly suspended; and this is not to be wondered at, when we take into account the great changes going on at this time within the female organism, whereby a great quantity of liquid is discharged from the womb and breasts. This, together with the vicarious action of the skin, demonstrating itself by the increased perspiration, amply compensates for the temporary inactivity of the alimentary canal; and, by this provision of nature, the balance of the system is kept up.

TREATMENT.

We cannot too strongly condemn the use of aperients in such cases; as they only tend to promote irritation, which is indeed but the stepping-stone to inflammation. And, besides, the relaxation thus produced always interferes with the proper secretion of the milk. It was, and still is, to a great extent, among "old-school" physicians, the practice to give a mild cathartic on the second or third day after delivery. What reason or utility there can possibly be in such an unwarrantable interference, is totally beyond our comprehension. Nature purposely provides other means of relieving the system, in order that the patient shall be undisturbed until the uterine organs have been enabled to recuperate their energies and resume their normal form and position. We have frequently known the use of cathartics at this time to produce the most serious results; but we have never yet seen an instance where any trouble has arisen from this temporary inactivity of the bowels. As a general thing, the bowels will move spontaneously about the fifth or sixth day. In very obstinate cases, which seldom occur, an injection of lukewarm water, with linseed oil, will produce the desired result. (Nos. 11, 27, 179, 180, 181, 182, 183.)

DIARRHŒA AFTER CONFINEMENT.

This should always be looked upon as a highly dangerous condition, and prompt means should at once be taken for its speedy removal. It is generally caused by cold, errors in diet, or the abuse of aperient medicines.

TREATMENT.

For appropriate treatment see Nos. 12, 190, 191, 192.

RETENTION OF URINE, OR PAINFUL URINATION, DURING CONFINEMENT.

It not unfrequently happens, especially after severe labor, that the neck of the bladder, and the whole tract of the urethra becomes extremely sensitive, causing painful emissions, and sometimes even entire retention of urine. This sensitiveness arises from the great amount and long-continued pressure to which the parts have been subjected.

Retention of the urine, when it lasts for any considerable length of time, is an extremely dangerous affection, because if relief is not obtained, and the pressure on the inner surface of the bladder is not relieved by the removal of the accumulated water, inflammation must necessarily follow. Fortunately, complete retention is seldom met with, and the painful and difficult emissions of urine, which are frequent, as a general thing, yield readily to treatment. The application of warm fomentations to the parts will sometimes be of great benefit, or sitting over a pan containing warm water will often have the desired effect.

SORE NIPPLES.

This frequent and exceedingly annoying complaint may, in a large majority of cases, be prevented if proper care of the breasts is taken previous to confinement. Of this we have spoken at large in the article on " Preparation of the Breasts." There appears to be a constitutional tenderness of the skin in some females, which predisposes it, upon the slightest occasion, to the development of cracks and sores of a most distressing nature, which at times prove most obstinate to heal. Wherever a tendency of this kind exists, the utmost care should be taken to avoid the least irritation or abrasion of the skin, either by your clothing, by a shield, if you use one, or by the breast-pump. When a shield is made use of, it should be frequently and carefully removed, and the parts bathed with a weak solution of tincture of arnica, or brandy and cold water. This will obviate the otherwise certain result of tenderness and consequent excoriation.

There is no doubt that many cases of broken breasts owe

their origin to the reluctance of the mother to encounter the pangs of suckling her infant while these cracks and fissures remain unhealed. The most frequent form of sore nipples consists of a long, narrow ulcer, about as wide as a horse-hair, and varying in length from the sixteenth of an inch to the whole circumference of the nipple.

The chief difficulty in healing sores of this nature, you will readily observe, arises from their being constantly torn open afresh by the efforts of the child in nursing. It is, therefore, very important, especially where the fissures are deep and gape open, that some means should be devised to keep the edges pressed together. This can be accomplished with a narrow bit of adhesive plaster, or you can spread some adhesive salve upon a narrow piece of ribbon; the latter, on account of its pliability, I have found to answer the purpose better than the common adhesive plaster. I have also used arnicated collodion in the same manner with great success. This, as well as the other application, will admit of the child's nursing without tearing the fissures open afresh.

In all cases, as soon as the child has left the breast, the nipple should be washed with *cold* water, to which a few drops of tincture of arnica have been added, and should then be thoroughly dried. Then, taking the nipple between the thumb and the first two fingers, gently compress it. This is done for the purpose of disgorging the small vessels that have become distended by the suction of the child. As soon as you have rendered the nipple soft and flexible, cover it over thickly with powdered wheaten starch or gum arabic. Pulverized white sugar, according to Dr. Hering, makes an excellent application. Should this precautionary treatment prove inefficient, and the fissure in the nipple become sore, and refuse to heal in spite of all your care and attention, you will then have to resort to the use of internal remedies, to counteract or remove the constitutional taint to which the disease generally owes its origin.

TREATMENT.

In the majority of cases, sulphur would seem to be specially indicated, especially when the nipples are sore and chapped with deep fissures around the base, which bleed and burn like fire. When these fissures are large, bleed easily, and prove obstinate to heal, you will generally find them to contain little granulations of proud flesh. In all such cases, apply burnt alum, or pulverized tobacco ashes and burnt alum in combination.

All cases of sore nipples, however, do not present themselves in the form above described; sometimes the nipple becomes abraded or excoriated, and even suppuration occasionally takes place. A very important point in the successful treatment of those cases is, to keep the parts perfectly dry. This, I have already remarked, can best be accomplished by wrapping the nipple in pulverized starch or gum arabic. There are numerous domestic remedies, in the form of powders, salves, and lotions, which have been used with various results. Borax, dissolved in mucilage of slippery elm, makes a pleasant and serviceable wash; powdered potter's clay, sprinkled upon the parts, frequently effects a cure. Reliance, however, cannot be placed upon any form of treatment, especially in severe cases, except the internal administration of appropriate remedies. In all cases where external applications of any description have been made use of, the nipple should be carefully cleansed with a little warm milk and water before presenting it to the child.

GATHERED OR BROKEN BREASTS.

To make the nature and importance of this disorder perfectly plain and intelligible, we will give a brief anatomical description of the female breast. Beneath the skin on the front of the chest, there lies — one on each side — a large secretory organ, called the mammary gland. It is composed of milk-tubes, nerves, arteries, veins, and lymphatics, the whole being inclosed by a fibrous investment, which also sends out prolongations through the glands, dividing it into numerous lobes. Between these frequent membranous divisions, especially near the skin, exist numerous small cells in which fat is deposited, giving to the surface its beautiful, soft, smooth, hemispherical form.

THE MAMMARY GLANDS.

The nipple is only a bundle of milk-tubes, nerves and blood-vessels, gathered together and covered with a thin derm, or skin.

The milk-ducts or tubes, resembling little canals, vary from ten to fifteen in number. When distended they are about the size of a small goose-quill. Starting from the extremity of the nipple they enter the breast, soon become divided and subdivided, becoming finer and finer as they go inward, until each minute tube terminates in a small hollow globule or granule about the size of a mustard-seed, from the inner surface of which the milk is secreted. The number of these little granules it would be impossible to count. If you should take a small syringe and inject each of these ten or fifteen distinct milk-tubes from the nipple with different colored substances, thus filling one canal with yellow, another with green, a third with violet, and so on, until the whole breast was completely distended, you would see no amalgamation of colors, no uniting or coalescing of tubes, but each injection would follow its own canal, through all its divisions and subdivisions, to its granular termination. Thus, you observe, we can trace the course of each milk-tube from its exit at the nipple, through all its divisions and divergences, to the actual minute milk-producing granule, just as we can trace a river on the map from the broad Atlantic, where it empties, to the very springlets among the mountains, where it has its origin.

The quantity of milk that a given gland will produce at one time does not so much depend upon the size of the organ as upon its secretory power. With a breast-pump some women can draw out a half-pint from one breast at one sitting; not that it was actually all present in the breast when she began, but was secreted, as it were, upon demand, the flow of milk only ceasing when the secretory power of the gland becomes exhausted, and then a period of rest is demanded. To carry on this process of milk-secretion, it is necessary that the organ should be supplied with a large amount of blood and nerve-power. Accordingly, we find numerous branches from large arteries distributed throughout the breast, while by a great number of nerve fibres it is intimately connected with the two great nervous systems.

CAUSES OF DISEASE.

During lactation the breasts are in a high state of activity, which, together with their intimate connection with the rest of the system, renders them exceedingly liable to partake of any disorder, either physical or mental, which happens to affect a woman while nursing. Thus we shall find ague in the breast, as it is called, arising from a cold, a chill, fright, anger, fear, grief, etc.

Gathered breasts not unfrequently arise from a too tardy application of the child to the breasts, or from sudden cessation of suckling, occasioned either by the death of the child or an unwillingness on the part of the mother to encounter the pangs of nursing the infant, consequent upon sore nipples.

When the breasts become distended with milk, and all the little milk-tubes are filled and crowded against one another, you will often find it incompressible, and its sensibility so greatly increased that the least handling produces great pain. Now, unless this tension is speedily reduced, as a natural consequence, inflammation must follow, or fever soon arises, ushered in by rigors or severe chills.

A chill acts in the same manner, or at least is productive of the same results ; the breast increases in size from congestion of its blood-vessels and consequent obstruction of the milk-tubes ; and the result, if not prevented by prompt interference, as before, will be inflammation and suppuration.

TREATMENT.

The treatment is, of course, to take away the milk, when the breasts soon become cool and flaccid, and the freest handling produces no pain.

Do not let the breasts become distended ; apply the child often, — as often as necessary to keep the breasts in proper order.

Where hard lumps or cakes are felt deep down in the breast you must, by some means or other, soften them, and extract the milk. These lumps, or cakes, as they are commonly called, are caused by the milk-tubes becoming clogged up ; or rather they become distended, and crowd against each other, until they are so compressed that the flow of milk is obstructed ; and thus one division of the gland becomes caked, while the rest remain open.

Nurses make use of all sorts of embrocations and hot applications to scatter the cakes, which simply means to soften and relax these particular tubes so that the milk can flow. And this *must* be done, or inflammation, followed by suppuration, will be the result.

When the breasts become swollen and very tender, the following receipts may be successfully used: Nos. 29, 136, 137, 138.

Should the swelling and tenderness subside, but there still remain lumps or cakes in the breast, you will find relief from applying a plaster made of beeswax and sweet oil. The great art in preventing gathered breasts is to keep the breasts well drawn; if the child is unable to do it, then you must resort to nipple-glasses, the breast-pump, or, what is better than either, the lips of the nurse or some other adult person.

You will seldom find a nurse who will acknowledge that ever such a thing as a broken breast did occur to a patient of whom she had the entire charge; but all such assertions it is well to take with a few grains of allowance, for in spite of all precautions the breast will sometimes gather and break.

In the early stages of this disorder, it is best to abstain from applying warm poultices, as they have a tendency to involve a still larger part of the breast within the suppurative sphere. But as soon as the gathering points, or when it becomes evident that it must soon break, it should be hurried along as fast as possible; and if you employ a physician, he will at this period undoubtedly lance it. Ground flax-seed makes the best poultice: it should be applied warm, and changed once in three hours.

When the abscess has opened and the matter has been discharged, the breast should be compressed either by strips of adhesive plaster or a bandage. This you will find will facilitate the process of healing.

Should the above remedies fail to produce a cure, you can have recourse to Nos. 29, 236, 137, 138.

During all the time that the breasts have been gathering, and still after the abscess has broken, the infant should be permitted to nurse; for you must recollect that milk is secreted by that portion of the gland which is not involved in the abscess, and it must be withdrawn. If the infant cannot, or refuse to do it, you must resort to artificial means.

Diet. — The diet should be plain and nourishing, but not stimulating.

CHILDBED FEVER, OR PUERPERAL PERITONITIS.

I shall not enter into any detailed description of this disease, because I do not deem it safe for any but an experienced physician to attempt its treatment. I shall, therefore, briefly give its nature and characteristic symptoms, together with such remedial measures as will be adapted to the premonitory symptoms and first stages of an attack.

Definition. — Childbed fever, or puerperal peritonitis, as it is technically called by physicians, is an inflammation of the peritoneum, or serous membrane lining the abdomen and covering the bowels. It is not unfrequently complicated with inflammation of the womb and its appendages.

Causes. — Among the exciting causes of this disease, may be enumerated, violence during delivery, taking cold, diarrhœa, irritation of the bowels induced by cathartic medicines, severe mental emotions, suppressed secretion of milk, and so on.

Symptoms. — Childbed fever is generally preceded or attended by shivering, and sickness or vomiting, and is marked by pain in the abdomen, which is sometimes very much distended, though in other cases it is at first confined to one small spot. The abdomen soon becomes swelled and tense, and the tension rapidly increases. The pulse is frequent, small, and sharp; the skin hot, the tongue either clean or white and dry; the patient thirsty; she vomits frequently, and the milk and lochia are usually obstructed. These symptoms often come on very acutely, but they may also approach insidiously. But whether the early symptoms come rapidly or slowly, they soon increase, the abdomen becoming as large as previous to delivery, and often so tender that the weight of the bed-clothes can scarcely be endured; the patient also feels much pain when she turns; the respiration becomes difficult, and sometimes a cough comes on, which aggravates the distress; or it appears from the first to be attended with pain in the side, as a prominent symptom. Sometimes the patient has a great inclination to belch, which always gives pain. The bowels are either costive or the patient purges bilious or dark-colored fæces. These symptoms are more or less acute, according to the extent to which the peritoneum is affected. They are, at first, milder and more protracted in those cases where the inflammation begins in the uterus, and in such the pain is not very great or very extensive for some time. In fatal cases, the swelling and tension increase, the vomiting continues, the pulse becomes very frequent and irregular, the extremities become cold, and the pain ceases rather suddenly. The patient has unrefreshing slumber, and sometimes delirium, but she may remain sensible to the last.

TREATMENT.

In all such cases, but little can be done until the arrival of a physician; but temporary alleviation can be secured by administering No. 30, 134, or 135.

MILK-LEG, OR CRURAL PHLEBITIS.

Definition. — Milk-leg is the common name given to a peculiar form of disease which sometimes affects women during confinement. As the name implies, it was once supposed that the milk had fallen into the woman's leg. I cannot say that physicians ever took this view of the disorder, but certainly the people did, and it is no uncommon occurrence to meet with persons who still insist that the milk has gone into the leg, because the limb is swollen and looks white; and besides, the milk has partially or entirely left the breast. All the reasoning in the world will not make them believe differently. But it is the sheerest nonsense to say that the milk has fallen into the woman's leg, for such a thing is impossible.

Physicians, now, who know anything about the disease, call it crural phlebitis, which name signifies inflammation of the veins of the leg; and this is the true seat and nature of the affection. The swelling of the limb is due to the effusion of lymph and serum from the blood into the cellular tissue.

Causes. — The origin of the affection may generally be found in exposure to draughts, severe cold, or sudden alternations of temperature.

Symptoms. — The ordinary premonitory symptoms of an attack of this disease often resemble and are not unfrequently mistaken for after-pains. There is uneasiness or pain in the lower part of the abdomen, extending along the brim of the pelvis through the hips. The patient is irritable, depressed, and complains of great weakness. Often, however, there will be no precursory symptoms, the patient being suddenly seized with pain in the groin or calf of the leg, and not unfrequently she will complain of pain in the hip-joint, calling it neuralgia or rheumatism. As soon as the inflammation is fairly set in, the region about the groin becomes tumefied; and in a short time, — twenty-four or forty-eight hours, — the thigh becomes swollen, tense, white, and shiny. The swelling, which sometimes increases the limb to the size of a man's body or an elephant's leg, may be confined to the thigh, or it may extend down to the foot. When the pain commences in the calf of the leg, the swelling is first observed there, and gradually extends itself up the leg and thigh. The temperature of the limb is generally increased, although in some cases it falls below the natural standard. Along the course of the inflamed vein, although there is great tenderness, there is neither redness nor other discoloration. In most cases, the vein may be traced from the groin down the thigh, feeling hard, and rolling under the finger like a cord. Either leg may be affected, although the left appears to be more frequently attacked, and it not unfrequently happens that the sound leg participates in the disease before the affection is entirely removed, and then it runs a similar course the second time.

TREATMENT.

The treatment of this disease should be undertaken only by an experienced physician. A few remedies, which may be used at the commencement of the attack, will be found at No. 31.

NURSING SORE MOUTH.

In this disease the soft part, and sometimes the whole interior of the mouth, becomes very red, and so sensitive and tender as to render it almost impossible for the patient to partake of any solid food whatever. This is quite a different disease from what is generally called canker sore mouth. In some females it appears to be constitutional. As I have before remarked, the breasts are intimately connected with the whole nervous system; you will not be surprised, therefore, to learn that this form of sore mouth arises from the peculiar irritation which the act of nursing produces upon the digestive organs. If not properly treated, it sometimes becomes so severe, and is attended with so much suffering and debility, that the weaning of the child becomes absolutely necessary, and has a magical effect upon this disease, — the whole of it vanishing as soon as nursing is discontinued.   (Nos. 32, 285, will be found serviceable remedies.)

Diet and Regimen. — The diet of a woman suffering from nursing sore mouth should be generous and nourishing, but not flatulent.   Whatever articles of food are found to disagree should be strictly avoided.   Exercise in the open air would be found beneficial.

PERSPIRATION AFTER DELIVERY.

The increased perspiration which takes place immediately after delivery, and continues for several days, acts, as I have before remarked, as a substitute for the suspended mucous secretion and consequent inactivity of the alimentary canal.   Therefore, its sudden suppression from exposure to cold, or a sudden chill, is unavoidably followed by some injurious result, not unfrequently gathered breasts, diarrhœa, or childbed fever.   No. 33 will meet the ordinary emergencies of the case.

EXCESSIVE PERSPIRATION AFTER DELIVERY.

Excessive perspiration, besides causing great debility, predisposes to other disorders, by the high susceptibility of taking cold which it occasions.   It is sometimes occasioned by the too high temperature at which your room is kept, in which case the remedy is obvious.   When it still remains, after the proper regulation of the temperature of your room, and the removal of all superfluous clothing, or when the perspiration is profuse while lying still, but diminished by moving about, remedies No. 34 will prove efficacious.

CHAPTER IX.

THE CARE OF THE INFANT.

Let us now return to the infant, which, you will remember, we left wrapped in a warm flannel blanket, and laid on one side while the bandage was being applied, and the mother otherwise cared for or attended to. If the infant appears feeble, and its respiration not well established, the skin having a leaden hue instead of the healthy pink or rose color, it should be permitted to remain undisturbed for some little time, until it is better able to undergo the fatigue of being washed and dressed. But if it appears strong, and cries lustily, it may be washed and dressed as soon as convenient. Some people use cold water to wash the child with, even for the first ablution, under the absurd impression that this early introduction to the vicissitudes of temperature will invigorate and harden the child, and thus make it less liable to the injurious effects of sudden atmospheric changes. I hope Providence has endowed you with more sense than to imagine that any such happy results would follow this barbarous practice.

WASHING THE INFANT.

For the whole period of its uterine existence the infant has experienced a uniform temperature of 98°; now to wash it with or put it into a basin of cold water must give it a shock, which cannot fail to be highly injurious. I would about as soon think of putting the child into a kettle of boiling hot water. In my estimation, the temperature of the water in which the child is first washed should be as

high as 90° at least ; and this, you will observe, is still eight
degrees below the temperature to which, till within a short
time, it has been accustomed.   It is not necessary that you
should stand with a thermometer in one hand, and a kettle
of hot water in the other, and thus temper your bath to the
fraction of a degree.   All that is necessary is to be certain
that the water is *warm* and *soft*, instead of hard and cold.

The white caseous substance which, to a greater or less
extent, covers the body of every new-born infant, and
which sometimes adheres with great tenacity, can best be
removed by rubbing those parts to which it adheres freely
with hog's lard or sweet oil, until the two substances become
thoroughly mixed, and then wash with soap and water.

Owing to the extreme sensibility of the infant's skin, you
should use none but the finest quality of white soap and a
soft flannel wash-cloth.   This is important, for a slight
abrasion of the cuticle, or even the least irritation, may
cause troublesome sores.   After the child has been well
washed, it should be wiped *perfectly* dry with a fine, soft
napkin.

THE USE OF POWDERS.

It is customary, as soon as the child is washed and dried,
to dust it over with some kind of powder, especially about
the neck, armpits, and joints, or wherever the skin is folded
upon itself.   I would advise you to get along without this,
if you possibly can, because the powders that are usually
sold for this purpose are most of them highly injurious ; and
if your child is properly washed and dried, you will have
but little, if any, call for them.   If, however, you think you
must use something of the kind, pulverized starch is the
best.   Both the washing and dressing of infants should be
done as expeditiously as possible, and with the greatest
care, so as neither to hurt or fatigue them.   During this

daily process of washing, which should not be done languidly, but briskly and expeditiously, the mind of the infant should be amused and excited. In this manner, the time of washing and dressing, instead of being dreaded as a period of daily suffering, instead of being painful and one continued fit of crying, will become a recreation and an amusement. In this, treat your infant, from the first, as a sensitive and intelligent creature. Let everything which *must* be done be made a source, not of pain, but of pleasure, and it will then become a source of health, and that both to body and mind, — source of exercise to the one and of early discipline to the other. Even at this tender age the little creature may be taught to be patient, and even cheerful, under suffering. Let it be remembered, that every act of the nurse toward the little infant is productive of good or evil upon its character as well as its health. Even the acts of washing and clothing may be made to discipline and improve the temper, or to try and improve it, and may, therefore, be very influential on its happiness in future life. For thus it may be taught to endure affliction with patience and cheerfulness, instead of crying and fretting at every operation necessary to its well-being. The parent and the nurse should, therefore, endeavor to throw their whole mind into their duties toward the tender being. And in their intention of controlling the infant's temper, let them not forget that the first step is to control their own. How often have I observed that an unhappy mother is the parent of unhappy children.

DRESSING THE NAVEL.

Most nurses and many physicians have fanciful notions in regard to dressing the navel. Some think nothing will do but a piece of scorched linen ; others want a flannel, either scorched or well besmeared with grease. I am acquainted with one old nurse who always keeps a box of powdered cob-web, a little of which she sprinkles over the navel before doing it up with a piece of scorched linen. Now this is all useless; the simplest way is the best, and that is, to take a folded piece of soft, plain cotton or linen cloth, about six inches long and three wide; cut a hole in the centre, and pass the cord through. The cord should then be laid up toward the child's breast, and the lower end of the linen or muslin folded up over it. Over this place a compress, made of several thicknesses of soft muslin, about the size of a silver dollar, or perhaps a little larger. The whole is to be kept in place by the belly-band, which should always be made of a strip of fine flannel of four or six inches in width. This band should be applied smoothly, so as to give equal strength to the whole abdomen; pin it just tight enough to keep it in place. For the first few days the condition of the navel-cord should be carefully examined, to see that the child's movements have not disturbed it, nor caused it to bleed. In the course of six or seven days it will become separated from the child, when you can remove it. The parts are now to be carefully washed, and the compress reapplied. If the parts around the navel are not properly washed and dried, and perhaps dusted with a little starch-powder once or twice a day, they are apt to become red and sore. In case of soreness, or inflammation of the umbilicus or navel, after the falling off of the ligature, or even before, you had better give an occasional dose of sulphur.

In case there is an evident tendency to rupture of the navel, after the ligature has dropped off, great care should be taken to apply a proper bandage, and this bandage should be worn some time after the cure, as a precautionary measure against its return.

CLOTHING OF INFANTS.

I presume it will be entirely useless for me to say one word in regard to the infant's dress. Fashion dictates here, as well as almost everywhere else, frequently to the detriment of the child, and always to the great inconvenience of the mother. But this has ever been the case, and I presume always will be. However, I would have you remember that the power of generating heat at this early period is very feeble indeed, and the child up to this time has been confined in a temperature of 98°, and at the same time most perfectly protected from the possibility of atmospheric changes. You will therefore see the necessity of clothing the infant warmly. Flannel should always be worn next the skin, for various reasons. First, it is warmer, being a bad conductor of heat; and, what is very important, it is much lighter than cotton goods; besides, it is a bad conductor of electricity. The flannel should of course be light, soft, and of the finest texture. In my opinion, if your child's clothing were all made of this material, it would be far preferable to any other; you would then have a *warm*, *light* dress; whereas, should you use cotton, it will require a much greater weight of it than of flannel to obtain the same amount of warmth. Besides, cotton or linen goods do not produce upon the skin that healthy degree of friction which flannel does. No doubt you will object to flannel frocks, and say they do not look as nice and pretty as tucked or ruffled muslin ones do. Well, I do not think they do; but you will acknowledge that health and comfort should always be studied in preference to appearance.

Another important item in infants' dress is looseness; the clothes should be so adjusted as to admit of the freest motion of the chest and limbs. The imperfectly developed organization of the child, you will bear in mind, is liable to compressions and distortions from the most trivial causes; many of the bones are as yet but mere ligaments, and as easily bent as the twig of a tree; the ribs, from the slightest pressure, may become crowded from their natural position, making the child pigeon-breasted, or deformed in other ways.

THE ESSENTIALS OF INFANTS' CLOTHING.

The essentials of the clothing of children are *lightness, simplicity,* and *warmth.* By its being as light as is consistent with warmth, it will neither encumber the child, nor cause any waste of its powers; in consequence of its *simplicity,* it will be readily and easily put on, so as to prevent many cries and tears; while, by its *looseness,* it will leave full room for the growth and due and regular expansion of the entire form, a matter of *infinite* importance for the securing of health and comfort in after-life. *Short sleeves and low-necked dresses are never suitable, under any circumstances,* for children or young persons, much less a delicate infant. To leave the neck, shoulders, and arms of a child nearly or quite bare, however warmly the rest of the body may be clad, is a *sure* means of endangering its comfort and health; violent attacks of croup, bronchitis, or even inflammation of the lungs, pneumonia, angina, catarrh, general fevers in cold seasons of the year, and bowel complaints in summer, and the seeds of pulmonary consumption, are often induced by this irrational custom; and it is not improbable that the foundation of pulmonary consumption is often thus laid during childhood. It is an important precaution, therefore, to have the dress worn by children so constructed as to protect the neck, breast, and shoulders, and with sleeves long enough to reach the wrist. The fact is, that vanity is, in many persons, a stronger passion than parental love; but it should never be forgotten, in reference to the dress of infants, that *the power of generating animal heat is lowest at the time of birth, and gradually increases with the advancing age of the individual till past the period of childhood.*

APPARENT DEATH, OR ASPHYXIA.

It sometimes happens, after severe or protracted labor, that the new-born infant presents all the appearance of being dead; it does not breathe, the blood does not seem to circulate, and there is no apparent motion. This may be termed the first danger to which the infant is subject on its entrance into this world of trouble and vexation. Cases of this kind demand the energetic and immediate attention of the physician and nurse; for, if means are not speedily taken to revive it, the child will not probably recover from this suspension of vitality.

The first thing to be done is to place the child in such a position that there will be no impediment to the circulation through the umbilical cord; then wrap the body and limbs in warm flannel cloths, and rub the hands and feet with soft, warm flannel, or with what perhaps is better, the warm, naked hand. Ordinarily, this will be sufficient to reëstablish the circulation; the pulsation in the cord will soon manifest itself, the action of the heart will become apparent, breathing will soon follow, and nothing more will be required. When the infant has fully recovered, the cord may be tied and divided. Now and then, however, cases do occur which do not yield so readily, but we must not be easily discouraged in our efforts, for infants have been restored after laboring with them three or four hours; we should, therefore, persevere, as our efforts may ultimately prove successful.

If, after rubbing the infant with warm flannels, the naked hand, or some stimulant, for five or ten minutes, still no pulsation shall be felt in the cord, it should be tied and cut, and the infant be immersed in a warm bath. While in the bath, the friction of the skin should be continued, and the chest pressed and rubbed; also dip your hand in *cold* water or alcohol, and rub the breast. Some physicians have directed a stream of *cold* water to be poured upon the chest from the spout of a teapot placed some two or three feet above the infant, and have found the action very efficacious.

If, in the course of ten or fifteen minutes, there is no sign of returning animation, or if there is but feeble pulsation of the cord, and limbs relaxed, or if the face is purple and swollen, *tartar emetic* should be administered; and if this fail, *opium* should be tried.

All other means having proved unsuccessful, artificial inflation of the lungs should be attempted. This may be done by placing your mouth over the child's mouth and blowing gently, so as to inflate the lungs, at the same time pressing the child's nostrils between the finger and thumb, so as to prevent the air from passing out through the nose. After the lungs are filled, the chest should be compressed gently with the hand. Care must be taken not to force too much air into the child's lungs, lest you injure them. An excellent internal remedy will be found in Prescription No. 35; and a mild current of electricity carefully applied to the spinal column, nerve centres, and chest, will oftentimes prove beneficial.

SWELLING AND ELONGATION OF THE HEAD.

It is quite common for the head of the infant to be swollen and elongated immediately after birth, and especially when the labor has been difficult or protracted; sometimes the head is so drawn out or swollen as to be shockingly deformed; and to the uninitiated its appearance not unfrequently causes great alarm. In most cases this is but a trifling affection, and generally disappears of its own accord.

In case the swelling be extensive, or does not disappear in a day or two, repeated washings with cold water or a weak solution of tincture of arnica will prove beneficial.

SWELLING OF THE INFANT'S BREASTS.

Sometimes at birth, or immediately after, the breasts of infants are found inflamed and swollen. It is only a simple inflammation of the gland, and should be treated as such. Our first endeavor should be to reduce the swelling; and to accomplish this, we generally cover the breast with a piece of lint or soft linen, dipped in sweet-oil. This is all the application I have ever found it necessary to make. Sometimes, when the inflammation has been excessive, I have deemed it advisable to apply a poultice of chamomile flowers steeped in warm water.

Some authors speak of a propensity on the part of nurses to squeeze the breasts, under the absurd impression that there is milk, or some like matter, in them which should be pressed out. I never have had the misfortune to meet with such ignoramuses; but, nevertheless, I can easily conceive how they might do a considerable amount of injury by exciting an inflammation which would end in the suppuration and disorganization of the whole breast, and thereby, in females, destroy its usefulness forever.

THE MECONIUM, OR FIRST DISCHARGE FROM THE BOWELS.

The first evacuation from the infant's bowels consists of a dark, bottle-green substance, called the meconium. Nurses are never content until the infant has had a free evacuation of the bowels; and, to make sure of an early movement, they, upon its first arrival, give the little stranger a good dose of some laxative trash. I have often wondered if an infant had the use of its reasoning faculties, what would be its first impression of the inhabitants of this world, where the ladies in attendance, without even saying, "By your leave, sir," just open its mouth and force down a teaspoonful of molasses, or perhaps the same quantity of some nauseous compound. It must think it had come into a strange land.

Now, this does seem to me the most absurd thing in all the world. Suppose the large intestines are full of meconium: have they not been in the same condition for a long time? What is the great haste to get rid of it? Will it kill the child if it remains there a few hours longer? Nature, who is wise in all her dealings, will take just as good care of the bowels as of the brain or lungs. In fact, she has already made provision for the expulsion of this bugbear in the kind and quality of the milk secreted in the mother's breast. But it is a fact that some people, in their self-conceit, imagine themselves wiser than their Creator, and, at the very threshold of life, commence marring the truly beautiful frame of God's image.

Although it may seem perfectly rational that the early contents of the bowels, called the meconium, should be purged off, you should never forget that nature has made wise provisions for this very want.

As soon as the mother feels herself sufficiently recovered to permit it, the infant should be placed at the breast, where it will obtain just the quality and quantity of *medicine* necessary for its welfare.

THE USES OF COLOSTRUM.

The generally received opinion is, I am well aware, that at this early period there is no milk secreted, and this is true; but every physician knows, and it is high time that mothers and nurses were aware of the fact, also, that there is secreted within the mother's breast, long before the birth of the infant, a fluid technically called *colostrum*, exactly fitted for, and containing the properties to produce just the necessary amount of mechanical action in the alimentary canal to assist in the expulsion of the meconium.

If the mother is able to nurse her child, absolutely nothing should be allowed to enter its mouth, for the first few days at least, but what it gets from her, except perhaps a little cool water, which all children should have. The nurse should always be careful to wash the infant's mouth out well with cool water *every morning*.

The colostrum furnished by the breast does not act like physic, producing a succession of stools, but more slowly, so that it may take two or three days for all the meconium to pass away; but when the work is thus once done, it is well done.

Mothers need be under no apprehension should a temporary delay occur in the passing of the meconium; far greater evil results from the violent method taken for its expulsion than could possibly occur from its continuance in the alimentary canal for a longer period than natural.

Should, however, an unusually long period elapse, and the child appear costive, uneasy, and restless, a few teaspoonfuls of warm sugar and water may be given to it, which will generally have the desired effect.

NURSING.

Every healthy and well-organized woman should support her child from the natural secretion of her own bosom, which is the dicta of both nature and reason. The mortality among infants fed wholly on artificial nourishment is far greater than among those which are nourished from the maternal source. It is extremely difficult to estimate the injury sustained by the infant being deprived of its natural food; as farinaceous and other artificial substitutes, however carefully prepared, cannot possibly supply its place. No animal refuses to nurse its young ; it is only among the human species that we find mothers cruel enough to deprive a new-born infant of its natural food. If this is done from wilful neglect or indifference, mothers often pay dearly for such violations of nature's laws. If the baby is allowed to nurse as soon as it seems hungry, and the mother has obtained rest, there will be no need of giving any other laxative or cathartic, such as molasses, castor-oil, etc., for nature has made all the provision in this direction which is necessary. For the last fifteen or twenty years I have not given, in a single instance, any form of laxative medicine to new-born infants, aside from that nourishment provided in the mother's breast ; and I am satisfied that children do much better without than with such articles as are frequently given to them to move their bowels. The nearer we follow nature the better. If the infant is fed a few times before nursing, it often loses the faculty of nursing, and it is, in such cases, exceedingly difficult to induce it to nurse.

NURSING NECESSARY TO HEALTH.

Nor does the child alone suffer from its not being allowed to nurse. It is exceedingly rare for a woman's constitution to suffer from the secretion of milk ; but, on the contrary, their health is, very generally, materially improved by the performance of the duties of nursing. Parental affection, and occasional self-denial, would be abundantly recompensed by blooming and vigorous children. By this practice, too, the patient is generally preserved from fever, inflamed or broken breasts, and other maladies.

Where the supply of milk is not sufficiently copious, or the mother is not sufficiently strong and vigorous to maintain the infant's demand for sustenance, both mother and child may be materially benefited by feeding the infant with nicely-made panada, gruel, or farina, in the intervals of nursing, thereby averting the undue drain on the mother's strength, and aiding the infant in its approaching period of dentition.

REGIMEN DURING NURSING.

It is of the utmost importance that nothing should occur to the nursing mother that may interfere with or arrest the secretion of milk, or alter and diminish its nutritive qualities.

Nature always provides for her new-born, and the fountain of life which she has opened within the mother's bosom would ever give forth a bounteous supply of pure and healthy nourishment were it not for our follies, sins, and fashionable dissipations.

Mental and moral emotions, improper diet, and irregular habits have a decidedly injurious effect upon both the quantity and the quality of the milk. This is a point which it seems almost superfluous to discuss, but, nevertheless, in the face of all the proofs which can be brought in support of this fact, there are still in existence those persons who wholly ignore the idea that mental emotions or changes in diet in any way affect the lacteal secretion, and who very much doubt that errors in diet ever produce any very marked changes in the quality of the milk.

CONSEQUENCES OF IMPROPER DIET.

Now, who has not seen children suffer from indigestion, vomiting, colic, and diarrhœa, in consequence of the mother having indulged in a very rich diet? Some parents cannot even partake of fruit or vegetables, or make the slightest change in their food, without its having an immediate effect upon the nursling. I would not, for a moment, counsel entire abstinence from fruits, vegetables, or any other ordinary article of food, but only emphatically enjoin the greatest and most vigilant watchfulness and care in selecting the articles and regulating the quantity to be eaten.

We are all aware that butter made from the milk of a cow fed upon Swede turnips, or garlic, or strong-smelling oils, herbs, or plants, will contain the flavor and odor of that plant or substance, to a greater or less extent. The same principle obtains, most markedly, in the maternal secretion. The worst case of colic, I think, that I ever saw in an infant was produced by the nurse eating unripe fruit. I am acquainted with a lady who cannot eat the least thing that is at all sour or acid but that her nursing infant is sure to have an attack of colic. It therefore follows that a nursing mother should be specially careful in the choice of her nourishment, in order to impart to the milk such properties only as will make it a wholesome and nutritive agent. Plain, wholesome food, as a general thing, will produce wholesome milk, while a diet of highly-seasoned and fancifully-cooked dishes, served perhaps at irregular hours, and accompanied with tea or coffee, is almost certain to impart something to the milk which will prove injurious to the child.

If, after a proper regulation of the diet, the milk still proves unwholesome, you may rest assured that there is some constitutional difficulty resting with the mother, which will have to be removed by internal medication.

DIETETIC REGULATIONS.

The diet should be simple and nourishing; not too rich nor too stimulating; bread, fruit, and vegetables may be freely used; while meats should be partaken of in modera-

tion. The mother's own wishes will generally point out
what kind of food is most wholesome for herself and child.
A little experience will soon teach her what does and what
does not agree with her infant; and if she be a true mother
she will be willing to sacrifice some of her choice dishes,
her coffee and tea, and any other little luxuries which she
finds to disagree with her child. Regularity in eating is of
the utmost importance. As I have already observed that a
stimulating diet is, under no circumstances, advisable, it
may be well here to make a few remarks upon the popular
beverages, such as ale, porter, and the like, so extensively
made use of for the purpose of increasing the flow of milk.

It has been asserted that " no idea can be more erroneous
than that women, during the nursing period, stand in need
of stimulants to support their strength and increase the
flow of milk." When you come to look into the subject a
little, you will find that this is true.

DETERIORATION OF MILK.

A great ado was made, not many years ago, by the citizens of New York, because the dairymen from the country
and suburbs of the city insisted upon supplying them with
swill-milk, or milk secreted by a cow constantly fed upon
swill. Now, if people are so opposed to using swill-milk
themselves, why will they insist upon manufacturing it for
their children, by introducing the alcoholic element into the
lacteal secretion? No one doubts the fact that swill-milk
is unwholesome. In the first place, the milk contains more
or less of the properties of the substance from which it is
manufactured. Now, if you manufacture milk by passing
swill through a cow, — the udder acting simply as a filter, —
you of course get more or less of the properties of the swill,
whatever they may be. In the second place, a cow fed
upon swill soon becomes diseased, and of course gives dis-

eased milk. You will now readily observe that the milk which you get, in addition to containing more or less of its original properties, as affected by swill, is still further contaminated by being drawn from a sick cow.

ALCOHOLIC LIQUORS INJURIOUS.

Now it is just the same with a nursing woman fed upon ale and porter; not to so great an extent, it is true, because her diet is not exclusively confined to one unwholesome article, but the milk which she produces is unhealthy, and therefore not a proper nourishment for the infant. Drugs enter largely into the composition of all malt liquors, wines, and brandies, and to a far greater extent, too, than is generally supposed. Milk, impregnated with either of these drugged articles, can scarcely fail to engender obstinate and formidable chronic diseases both in mother and child. The regular administration of alcohol, with the professed object of supporting the system under the demand occasioned by the flow of milk, is "a mockery, a delusion, and a snare," for alcohol affords no single element of the secretion, and is much more likely to impair than to improve the quality of the milk. If a woman cannot afford the necessary supply without these indulgences, she should give over the infant to some one who can, and drop nursing altogether. The only cases in which a moderate portion of malt liquor is justifiable are when the milk is deficient, and the nurse is averse or unable to put another in her place. Here of two evils we choose the least, and rather give the infant milk of an inferior quality than endanger its health by weaning it prematurely, or stinting it of its accustomed nourishment. But, as a general rule, a judicious system of feeding, gradually introduced from a very early period in the life of a child, is infinitely preferable to an imperfect supply of poor milk from the mother; and, if the mother is so foolish as to persist in nursing her infant after nature has repeatedly warned her of her incapacity to do so, it is the duty of the medical man to set before her, as strongly as possible, the risk — the absolute certainty — of future prejudice to herself. The evils which proceed from lactation, protracted beyond the ability of the system to sustain it, may be to a certain degree kept in check by the use of alcoholic stimulants; but we are convinced, from experience and observation, that the arrestation of these evils are only temporary, and that they will, sooner or later, manifest themselves a hundred-fold intensified. Under no circumstances is the habitual or even occasional use of alcoholic liquors during lactation either necessary or beneficial. By the use of alcoholic stimulants the constitution of both mother and infant is stimulated far beyond the limit set by nature. The laws which govern the animal economy are positively infringed, and it is impossible that mother or infant should escape the penalty of that infringement. Both will suffer to a certainty in some shape or other, if not immediately, at some future period. Thousands of infants are annually cut off by convulsions, etc., from the effects of these beverages acting on them through the mother. We wish it to be clearly understood, then, that when the mother does not furnish a sufficient supply of milk for the wants of the child, a wet-nurse should be obtained, or the child should be weaned immediately.

MENTAL EMOTIONS AFFECTING THE MILK.

It is just as important that a nursing mother should pay strict attention to the state of her mind as to her diet and general health. No other secretion so evidently exhibits the influence of the depressing emotions as that of the breast.

The infant's stomach is a very delicate apparatus for testing the quality of the milk, far exceeding anything which the chemist can devise. How a mental emotion can affect the quality of the milk, perhaps it would be difficult to demonstrate, and what that change in the character of the milk consists in, no examination of its physical properties by the chemist can detect; but, nevertheless, we are well aware that, after severe fits of anger, some change takes place in the milk, which alters it from a healthy, nutritive agent to an irritating substance, producing griping in the infant, and a diarrhœa of green stools. Inasmuch, therefore, as the quality of the milk is very liable to be injuriously affected by any sudden or unpleasant excitement of the feelings, or other causes producing a constant and continued state of unhappiness, it is desirable that the most assiduous care should be taken to keep the mind in as quiet and happy a state as possible. It may not be practicable for nursing mothers to avoid all occasions of getting angry or sad, but it certainly is possible to avoid all violent and artificial excitement.

Grief, of course, is an emotion which we cannot entirely control, and it is not an uncommon occurrence for the loss of a relative or friend to have such a depressing effect upon a nursing mother as to cause an almost total suppression of milk.

EFFECT OF EMOTION UPON THE INFANT.

It is not unfrequent, either, for a child to suffer from griping pains and green, frothy stools while sick with some other disease, and yet there be no connection between the two complaints. We, as physicians, can readily understand it, but the mother little apprehends that it is all owing to her own anxiety.

Terror which is sudden, and great fear, instantly stop the secretion of milk.

Sir Astley Cooper remarks: "The secretion of milk proceeds best in a tranquil state of mind, and with a cheerful temper; then the milk is regularly abundant, and agrees well with the child. On the contrary, a fretful temper lessens the quantity of milk, makes it thin and serous, and causes it to disturb the child's bowels, producing intestinal fever and much griping." It is absolutely necessary, therefore, that, if you would have healthy, quiet, and good-natured children, you should always yourself be calm, cheerful, and happy.

It is not well for a woman to nurse her child soon after having recovered from fright, passion, etc. She should wait until she is perfectly composed, and perhaps it would be as well to draw off a portion of the milk before the child is again applied to the breast.

WEANING.

There is a great variety of opinion among the members of the medical fraternity, and a still greater difference in the sentiments and actions of mothers and nurses, as to the period when the infant should be " weaned." There is no doubt that many thousands of infants, from the unskilfulness, ignorance, and recklessness of their mothers and nurses regarding the withdrawal of nature's nutriment and the substitution of unsuitable and badly prepared food, lay the foundation of many serious chronic, inflammatory, and fatal infantile diseases. I do not hesitate to assert most emphatically that nature herself has plainly and unmistakably demonstrated the *proper time* for the *commencement* of this process by the appearance of the *temporary* teeth.

THE FIRST DENTITION.

The operation of dentition generally commences between the fifth and the seventh month of infantile existence, and the first ten teeth have usually cut themselves through by the close of the ninth month, an unmistakable intimation of

Dame Nature that the digestive apparatus is so far prepared for functional action that it can assimilate and utilize food of a different character to that which it has hitherto obtained from the maternal organism. As the office of the teeth is to masticate and divide the solid portions of our food, one may very naturally suppose that their appearance and growth is a fair index of the development of the infant's digestive organs, and of the capabilities and powers of the stomach, as well as the demands of the general system, in regard to nutriment. If we take the protrusion and growth of the teeth as a guide by which we are to regulate the diet of the infant, we shall undoubtedly come to the common-sense conclusion that some children may be weaned far earlier than others ; so that it is impossible to name a definite age at which all children may be entirely deprived of the breast.

As a rule, if the mother be ordinarily robust and healthy, the child should be fed exclusively from the breast until the first two teeth have made their appearance (say the sixth or seventh month), when farinaceous food, the juice of light, delicate meats, gruel, etc., may be alternated with the periods of nursing, so as to accustom the organs to their new aliment. As the other teeth appear successively, the quality of the maternal secretion administered may be lessened, and the variety and quantity of the artificial food be gradually increased, until the nursing is altogether discontinued. There are, as we have said, many exceptions to this rule, which will be considered as they occur. One thing must be specially borne in mind, that the child should never be allowed to swallow solid animal food until the first dentition (that is, all excepting the upper molars and canine teeth) is completed.

We will now proceed to notice some of the reasons why it is not always possible to follow the rule we have laid down. Owing to fever, or some acute or chronic disease, the milk may spontaneously "dry up," in spite of the utmost care; or it may be, that, during the whole life of the mother, there has been a latent tendency toward consumption, scrofula, or even cancer, which the excitement during pregnancy, or the nervous shock of confinement, may have brought into activity, and either of which diseases would so contaminate her milk as to render it highly injurious to the child's health if she continue to nourish it at the breast.

Again, some mothers are unable to support this constant drain upon their system more than six months without becoming pale, weak, and emaciated, their milk thin and watery, and so deficient in nutriment as to be totally inadequate to the support of the child. In such a case the child should be immediately weaned, or recourse had to a healthy wet-nurse.

The return of the menses during the period of nursing sometimes, though not always, has a decidedly prejudicial effect upon the mother's milk; but, as a general rule, it does not render the weaning of the child necessary, especially while the milk continues to agree with it. The same is true if pregnancy should occur while the child is too young to wean, particularly if the mother is strong and healthy; but it is not advisable to continue that nursing longer than three, or, at the most, four months after the commencement of pregnancy.

On the other hand, there are many circumstances which may render it advisable to protract the term of nursing beyond the ordinary period. The child may be delicate and weak, with feeble digestive powers; it may be suffering from some disease incident to teething, or incipient infantile complaint. You would, therefore, naturally wait until the sickness had passed off before you changed its food. Again, it would be hardly prudent to wean a child during the hot months of summer. The months of March, April, May, September, October, and November may, all other things being equal, be regarded as the most favorable for weaning children. Some persons are very particular that weaning should take place during a certain phase of the moon ; but this is all *moonshine.* It would be hardly advisable to wean a child during the prevalence of an epidemic among children ; because the morbific influence prevailing produces a strong disposition to disease. Caution upon the points which we have here briefly glanced at may be the means of preventing a severe fit of sickness, or even of saving the life of your infant.

SUPPLEMENTARY DIET OF INFANTS.

It is universally acknowledged as a fact that the mortality among children " brought up by hand," as it is called, is immeasurably greater than among those who are not deprived of the maternal source of nourishment. Dr. Merriman, an eminent London physician, in remarking upon the infantile mortality of that populous city, says : " I am convinced that the attempt to bring children up by hand proves fatal in London to at least seven out of eight of these miserable sufferers ; and this happens whether the child has never taken the breast, or, having been suckled for three or four weeks only, is then weaned. In the country, the mortality among dry-nursed children is not quite so great as in cities ; but it is abundantly greater than is generally imagined."

COW'S MILK.

It is of course primarily necessary that the special food which the child partakes of in addition to or as a substitute for the mother's milk, should approach the quality of that milk as nearly as possible ; and from chemical analysis we find that by adding a portion of loaf-sugar and water to *pure* and *good cow's milk*, we obtain a substitute closely resembling breast-milk. The great requisite, of course, is *purity;* and there can be no doubt but that perfectly pure, fresh milk from *one* cow, sweetened with loaf-sugar, and diluted at first with water, and gradually reducing the dilution and giving it pure, is the best diet for young infants; but the diabolical concoction which is peddled through the streets of all our large cities in wagons, and which is obtained by filtering distillery swill through sick cows, is utterly unfit either for man or beast, and still more for tender infants and feeble children. What, then, remains to be done? The poor little innocent cannot be fed on such vile stuff, nor can it subsist forever on gruel and the numerous farinaceous mixtures and abominations with which nurseries generally are constantly deluged. The only course that can be taken is to partially avert these disadvantages by taking special care that the child is not *overfed*. No one will doubt that loading the child's stomach with farinaceous or any other description of food will produce gastric derangement. I am thoroughly convinced, from close observation, that it is not so much the article given, as it is the *state* or *quantity* in which it is given, that produces the trouble. For instance, you will find that the gruel prepared for children is made from meal ground very coarse and containing a great deal of feculent matter, as is also the case with panada, crackers, or bread and water, etc. Now, at best, this substance is unfit for the delicate stomach of a tender infant; but how much more so is it when you come to feed it after it has been prepared two, three, or more hours; and, though not actually sour to your sense of perception, it has undergone some change which renders it unwholesome to the infant, occasioning the colic and the gastric derangement which writers attribute to the *kind* instead of the *quality* of the food. Experience also teaches us that we as frequently injure children by overfeeding them as we do by feeding them with unwholesome food. We ourselves are not unfrequently reminded by fits of indigestion that we have indulged our appetites to too great an extent. Some mothers look upon every cry of their offspring as an indication of hunger, and every time the child worries or frets a little it must be fed. By this means the stomach is kept constantly distended with food, and the inevitable result of such a course — *indigestion* — will speedily follow.

As a general rule, a healthy child from one to three weeks old requires a *pint of breast-milk*, or other food equally nutritious, during the twenty-four hours. At the end of the first month, and in the course of the second, the quantity usually taken by the child increases gradually to about a pint and a half or a quart.

FARINACEOUS FOOD.

After a thorough investigation of the subject, I am convinced that, in cases where the maternal supply is deficient, unwholesome, or entirely absent, the best supplementary diet is that made from finely-ground rice or barley-flour. This flour is frequently sold in fancy packages weighing a pound, and in that form, is often impregnated with pepper, cloves, cinnamon, or some other spice, from contact with those articles upon the grocer's shelves; it is much the best, therefore, to purchase it loose, like ordinary meal.

The following is the method by which these articles should be prepared for children's diet: For an infant, take one tablespoonful of the flour — more, of course, for an older child — and moisten it with *cold water*, being careful to have it well stirred, so that it shall contain no lumps; then add a little salt and a sufficient quantity of *hot water*, and boil it for *ten minutes*, during which time it should be constantly stirred to keep it from burning. After it has been removed from the fire, you should add a sufficient quantity of loaf-sugar to make it about as sweet as breast-milk. The quantity of water which you should put to a spoonful of flour will, of course, depend altogether upon the consistency you wish to give it. If it is to be fed through a nursing-bottle, it will have to be quite thin (but, from my own personal experience, I have reason to condemn the use of feeding-bottles, as for several reasons they are decidedly injurious to the infant, and are the primary cause of many infantile ailments) ; if from a spoon, which is by far the most advisable method, it can be made quite thick, — almost as thick as an ordinary farina pudding.

For those children whose bowels are habitually inclined towards constipation, you will find the barley-flour better adapted, as it has a slight loosening tendency. On the contrary, for those whose bowels are inclined to be lax, or tend in that direction, you will find the rice flour preferable. You will observe that I advise that the flour should be cooked with water, *not* with milk. I do this, not specially on account of the difficulty of procuring pure milk in the city, but because I have observed that when a child is taking breast-milk, it should not be fed with cow's milk, as it will not assimilate with the breast-milk, and consequently produces disturbance in the digestive organs.

When the mother does not supply any nourishment for the child from her breast, I would recommend you to add a portion of pure milk to the flour and water. These two articles of diet, with the exception of a small quantity of the gravy of underdone meat occasionally, in addition to the milk furnished by the mother, are all the child will need or ought to have ; and a *strict* adherence to this simple diet, with as few variations as possible, except in case of sickness, until after the temporary teeth have made their appearance, you will find more conducive to the general health, comfort, and happiness of the child than any other you can adopt.

NECESSITY FOR REGULARITY OF DIET.

If it is perfectly true that whatever is taken into the system and digested is assimilated by the vital forces, and goes to make up the tissues of which the body is composed, is it not important that we, who have the selection of the food, should be extremely particular regarding the material from which the thread of life is spun? Experience has taught observing mothers, as well as physicians and nurses, after having made a proper selection of food for the infant, the importance of adhering to one plain, simple course of diet, and not to be constantly flying from one thing to another, giving it cracker and water one day, farina another, and gruel another.

I have chosen rice-flour and barley because I have found them to agree with the infant's digestive apparatus better than anything else; and I recommend them as a constant diet, with the exceptions already mentioned, until after the period of first dentition.

After the fifth or sixth month, the food may be of a more substantial nature. When the child is taking its food it should be supported in an easy, semi-recumbent position upon the arm or lap of the person feeding it, and should be kept quiet for at least thirty or forty minutes after having received its nourishment. Rest is peculiarly favorable to digestion, because the digestive organs require a concentration of the vital energies upon themselves in order to enable them to perform this important function with due rapidity and ease. Both experience and experiments upon the lower animals have shown that the process of digestion is particularly liable to be impeded by strong mental or corporeal exercise or agitation after a full meal. The practice, therefore, of dandling or jolting infants soon after they have taken nourishment is decidedly improper. You will notice that all the lower animals, as well as your babe, manifest a disposition to this quietness and repose after eating.

We have several times spoken of the impurity of the cow-milk procured in large cities. For the information and benefit of mothers and nurses, we will append a few remarks on the precautions necessary to be taken in its selection and use.

METHOD OF TESTING MILK.

In the first place, it is of the highest importance that the milk should be taken from a single cow, and not be a mixture of that of several. Then it is essential that, in the case of a very young infant, the cow shall have been in full milk for at least three or four weeks, and not more than four months. Cow's milk should be slightly alkaline; but it sometimes occurs that it is slightly acid, in which case it is very apt to disagree with children. Hence, in selecting a cow from which to obtain milk for an infant, it is always well to test the milk by means of blue litmus-paper. Hold the end of a strip of this paper in fresh milk for a short time, and if it changes to a red color, the milk is acid, and not suitable for a young child; another cow should therefore be selected. Good milk will change red litmus-paper to blue after some minutes' contact. Litmus-paper can be obtained at the druggists'. If milk which is being used disagree with a child, or cause disturbance of the stomach and bowels, it should be rejected, and the milk from another cow tried; but test the milk as above directed before using it. For an infant, it is important to use the milk which is first drawn, as it is much weaker than the last which is obtained, and will not require diluting with water, which may impair its quality. The first drawn milk need not be diluted, but should be sweetened with a little white sugar. Milk which has been boiled is not so easily digested as unboiled milk, and it is generally better only to heat it to the right temperature for drinking; and it is best that this should be done by setting the dish containing the milk into a vessel of boiling water.

WET-NURSES.

That the nurse's milk is the best substitute for the mother's milk, we presume will not be questioned. Should any, however, be sceptical enough to doubt it, we have only to refer them to those children who have " been brought up by hand," in comparison with those who have had a nurse. The healthy appearance of the one beside the emaciated condition of the other offers proofs stronger than any argument that we can adduce.

Inasmuch as the child will undoubtedly be influenced, to a greater or less extent, both by the moral and physical condition of the nurse, it is highly important that we should use great discrimination and care in selecting the person to whom we give the entire charge of the infant. It is true we are seldom left much margin for a choice; often, we consider ourselves fortunate indeed if we are able to find a female with a breast of milk who is willing to give her whole time to the care and nursing of another's infant. But, in your eagerness to secure the object of your search, you should not accept the first that offers, irrespective of her general health or moral character; or else, in after years, when perchance your child develops a cross and sour disposition, or is afflicted with some ugly eruption, you may have the unpleasant recollection that, in all probability, it took it from its nurse, and then forever blame yourself for what you can never, though you would gladly, remove.

We have already seen that errors in diet, mental and moral emotions, etc., have a decidedly deleterious effect upon the milk, changing it from a source of nourishment to a substance which seems to act like poison on the infant. If, then, the delicate organism of the infant is so sensibly affected by these changes in the milk, — changes which the most delicate chemical tests are unable to detect, — perhaps we can catch an inkling of the manner in which the whole constitution of the infant might become radically changed; the whole moral and physical disposition, as inherited from the mother, become supplanted, or at least obscured and superseded by the peculiarities of the moral and physical organization of the nurse in whose hands the infant has been placed.

INFLUENCE OF THE NURSE ON THE INFANT.

Humanity, in the first flush of its tender existence, both in its moral and physical aspect, is not unlike the potter's clay; and, like the potters, they who have the handling of it can fashion it into *almost* any form they please. A child of a kind and loving disposition, confiding, and easily led, is very apt to be led astray by an unprincipled or careless nurse; while a child who is perverse and shows a preternatural disposition to wrong, would, in such hands, be ruined beyond all hope of redemption. At no period of life is a child so susceptible of being influenced by the unamiable qualities of a companion as during the early months of infancy.

The impressions made upon an infant at this early period are not simply transient, as most persons are apt to think, but they sink deeply into the mind, and do seriously affect, either for good or evil, the whole future character of the subject of them. And, therefore, I would earnestly impress upon the minds of all parents the importance of early attention to the moral education of children. If there is any-thing in this world that a child does inherit from its parent or nurse, it is fretfulness, ill-humor, vicious propensities, and tendencies to physical derangement. Now, if the nurse, in whose society the child is constantly kept, pos-sesses a genial disposition, the prominent points of which are cheerfulness, contentment, gratitude, hope, joy, and love, don't you suppose that, as the child becomes developed, as each mental petal of that mind unfolds to the influence of surrounding objects, the impressions it receives are quite different from what they would have been had the nurse possessed all of those little satanic embellishments which we call moroseness, ill-humor, selfishness, envy, jealousy, hatred, revenge, and the like? I am pretty certain you will see in whose hands it is best to place the infant, especially when you come to remember, or, if you do not already know, to learn, that " *the feelings constitute an ever-acting source of bodily health or disease,* and also a principal source of enjoyment, as well as of suffering; and that upon their proper regulation most of the happiness and true value of human life depend."

As yet I have said scarcely anything in regard to the *physical* diseases a child may inherit from its nurse. These are *legion*, — acute, chronic, hereditary, and blood-diseases of all kinds. This part of the subject is so palpable to the perceptions of all, that elucidation is unnecessary.

QUALIFICATIONS OF A NURSE.

I would advise you never to engage a wet-nurse, however favorably you may be impressed with her appearance, until your family physician, in whom you put implicit confidence, has first seen her. In fact, the question should rest upon his decision, especially as to her physical condition. If there is any disease about her, he will be able to detect it. The best nurses are those who possess all the evidences of good health : the tongue clean, teeth and gums sound, indicating healthy digestion, breath free from unpleasant odor, the surface of the body free from eruptions, and the insensible perspiration inoffensive, the breast smooth, firm and prominent, the nipples well-developed, rosy-colored, and easily swelling when excited. The milk should flow easily, be thin, bland, of a bluish tint, and of a sweet taste, and, when allowed to remain in a cup or other vessel, be covered with a considerable amount of cream. She should be thoroughly healthy, free from any discoverable tendency to chronic diarrhœa, about the same age or even younger than the mother, and delivered at least within a few months of the same time ; let her complexion be clear, skin smooth and healthy, eyes and eyelids free from any redness or swelling. She should be of an amiable disposition, not irritable, nor prone to anger or passion ; of regular habits, not indulging in any of the forms of dissipation ; naturally kind and fond of children.

The nurse should make it her duty to guard the child as much as possible against diseases. This she will be best able to do by paying strict attention to her diet and her general mode of living. A nurse who loves children will cheerfully deny herself the pleasure of eating or drinking any articles whatever which injuriously affect her milk. She should, by all means, avoid all heating or spirituous beverages, spices, flatulent food, or food that is very salt. In a word, her diet should be simple and easily digested, consisting of a proper proportion of animal and vegetable food. As little change as possible should be made from her former mode of living, lest the change should affect her health, and thus disturb the child, causing flatulence, colic, diarrhœa, constipation, or some other of children's many ailments.

CHAPTER X.

INFANTILE AFFECTIONS AND DISEASES OF CHILDHOOD.

CRYING, WAKEFULNESS, AND RESTLESSNESS OF INFANTS.

IT may be taken for granted that infants do not cry — that is, have frequent and long-continued fits of crying — without there being some occasion for it. What that occasion is can usually be ascertained upon careful examination. A fit of crying is not unfrequently caused by some mechanical irritation; the child's dress may be wrinkled, or so adjusted as to be uncomfortable, or a pin may be misplaced or pricking into the flesh. Perhaps the most frequent cause of crying in infants is derangement of the stomach and intestines, such as cramps, colic, griping pains, and so forth. These are indicated by writhing of the body, drawing up of the legs, and diarrhœa.

Occasional crying of infants should cause no uneasiness in the mind of the mother, because this is the only method by which the child can manifest its wants. It may cry or worry from hunger, or from lying too long in one position; but when attention to these and other particulars, which will suggest themselves to every thoughtful parent, has been given, and the infant still refuses to be pacified, *chamomilla,* *belladonna,* *rhubarb,* or some other remedy mentioned in the Appendix, according to the abdominal and other symptoms, will be found beneficial.

Restlessness and wakefulness, like crying, are not diseases, but simply symptoms of some derangement of the system. It is not always possible to say with exactitude what causes the child to worry and prevents it from sleeping. We can often trace it to flatulence, and not unfrequently to an overloaded stomach, but we are quite as often in the dark as to its cause.

TREATMENT. — Nos. 260, 261.

As the difficulty is sometimes occasioned by the condition of the mother's milk, it being in some way unwholesome, it will be occasionally necessary to prescribe for the mother, as well as to make some restrictions or regulations regarding her diet.

DISEASES OF THE AIR-PASSAGES AND LUNGS.

When taking into consideration the alarming prevalence of disease of the air-passages and lungs, especially among young persons and children, half-grown maidens and tiny infants, together with the large percentage of deaths caused thereby, one would naturally suppose that if those who had given this subject its due attention could devise any method whereby these numerous affections could be warded off or prevented, their advice would be eagerly sought for and implicitly followed. But no, it is not till grim disease, in the shape of some appalling epidemic, wrapped in a malarious robe, mounts his chariot and comes sweeping over fair sections of our country, spreading dismay and desolation on every side, snatching from circles here and there a bud, a blossom, or possibly a full-blown rose, that the oft-repeated advice of the family physician, though listened to with marked attention, is actually heeded.

Every time a physician is called upon to prescribe for a patient he is reminded of the necessity of administering a short lecture upon the general laws of health, including dress, diet, and the like. It is a noticeable fact that sick persons are very penitent, sorry for past transgressions, willing observants *now* of the Decalogue, anxious beyond measure to obey implicitly every wish of the physician. But no sooner does the first glimmer of health irradiate their sickly forms than their self-reliance and independence return : —

God and the doctor they alike adore,
But only when in danger; not before:
The danger o'er, both are alike requited;
God is forgotten and the doctor slighted.

CHILDREN'S DRESSES.

The subject of dress, in connection with this class of diseases, is a very important one. Nearly if not quite all the diseases of the air-passages are caused by the sudden chilling of the body. Our climate, with its sudden vicissitudes of heat and cold, together with the exquisite method of our American mothers of dressing, or rather I would say, of *undressing* their children : the low neck, to show the beautiful contour of shoulders and of bust; the half pants, exposing the knees of small boys — yet what beauty there is in a *boy's* knee I never could ascertain, but I presume they must be charming, or certainly they would not be left bare ; all these add their quota to the full development of throat and lung affections.

The universal, deplorable ignorance or inattention, or both, in regard to the subject of dress, is astonishing, and cannot be too frequently brought before the minds of those who have the special care of young children.

Prevention is in all cases better than cure ; and certain it is that by careful and wise attention to the physical education of young children you can ward off such diseases as croup, bronchitis, laryngitis, pneumonia, and the like, even in those who have shown a predisposition or a liability to

them. Undoubtedly one of the most important means to
be made use of is the adoption of a proper dress ; and this,
in cold weather, should be one that will cover the *whole*
body.

LOW-NECKED DRESSES.

You can see, at any time, ladies wearing warm and comfortable dresses, with high necks and long sleeves, sitting in
the same room with their children, who are almost naked.
The dear little creatures, their arms and necks must not be
covered up, they look " so cunning" and " so sweet."
Their dresses are made so low and loose about the neck
that the whole chest, even down to the waist, is virtually
exposed. Yet, mark you, as soon as the children grow
older, and therefore become stronger, and better able to
bear exposure, they are dressed warmer. What inconsist-
ency ! Is it any wonder that children are more liable to
diseases of the air-passages and lungs than adults? O
Fashion ! thy potent sway fills many an infant grave !

I do not wish to dictate to any parent how she should
dress her children ; at least any further than is necessary to
preserve their health, by protecting them against the evil
effects of sudden transitions of temperature. Children
should never be dressed with low neck and short sleeves,
except in the heat of summer, and in the New England
climate not at all. I am well aware that it is the custom
so to dress them even in midwinter ; but you yourself
would be uncomfortable, to say the least, clothed in this
manner ; and how much more so must they be, with their
extreme sensibility of skin.

WHAT NURSERIES SHOULD BE.

But, you may argue, the child, especially the infant, is
never exposed ; the nursery is always warm, and it seldom
goes out of it : why be so particular to cover the neck

and arms? That is true; the rooms are always warm, and, in the vast majority of cases, too warm; but the doors are continually being opened and shut, subjecting the child to a constant fanning.

Now, the nursery, or room where the children are kept, should be large, airy, and well ventilated. Plenty of cool, fresh, and pure air should be constantly admitted, for the purpose of respiration. The temperature, while the children are well, should never exceed seventy-two degrees; and, generally speaking, from sixty-seven to seventy degrees will be sufficient to be comfortable, provided the children are properly clad.

I am aware that you will frequently be told, and that, too, by those who ought to know better, that early exposures harden the children, and make them robust. Would you expect to harden a tender plant by exposing it to chilling winds, or to the cold and biting frosts of a winter's night? Would you expect your flowers to grow, your roses to bud and blossom, without the genial warmth of a summer's sun? No, indeed! Neither can you harden your children by allowing their little shoulders, arms, legs, or feet to be cold; and you will often see them so cold that they are fairly blue.

It is cruel; and you may rest assured, that if these children do not suffer in infancy they will, as they grow up, be more liable to diseases of the air-passages and of the lungs than those who have been properly cared for.

Croup is a rare disease among the Germans; they are very particular in regard to children's dresses, taking great care to have the throat and chest well protected.

WHAT CHILDREN SHOULD WEAR.

Delicate children should *invariably* wear a flannel under-shirt, or a shirt made of some woollen material, next the skin, made high up about the neck, and with sleeves to come below the elbows. Then put on the accustomed underclothes, and even those had better be made of woollen, not only on account of its warmth, but because it is lighter than other goods; and over all a stout muslin or a light woollen dress.

The stockings should also be of wool, and come high up, always above the knees. The old way of tying a garter around the leg, to keep the stocking up, is open to many objections. In the first place, it spoils the beauty of the leg, by preventing a full development of the calf, by cutting off, or at least retarding the circulation. This alone would be sufficient reason to condemn it; but, what is of more consequence, it also produces cold feet, and causes congestion of the veins, making them knotty and uneven. An elastic strap, going from a button upon the outside of the top of the stocking to a button upon the waistband of the drawers, will answer every purpose and be quite as convenient.

As I have before stated, all children should be accustomed to cold (or very slightly tepid) bathing. For puny, weak, and delicate children, subject to croup, catarrh, and cough, in fact, taking cold upon the slighest exposure, I have found bathing, always in conjunction with warm clothing, of valuable assistance in strengthening the child, giving a good healthy tone to the system, and thus protecting it from many diseases to which it would otherwise have fallen a prey. Our city houses are generally warmed — no, *heated,* that is the word — with furnaces, another prolific source of disease. The children are virtually parboiled, or rather baked, while indoors, and, consequently, when they are taken out, the first draught of air that strikes the tender little hot-house plant produces a shock, drives the blood from the surface to the delicate membrane lining the throat or lungs, and thus produces some one of the innumerable diseases of the air-passages so prevalent in our midst.

CORYZA, SNUFFLES, COLD IN THE HEAD.

This disorder, which consists of an inflammation and consequent thickening-up of the mucous membrane lining the nasal passages, occurs as a distinct disease; but it is also frequently connected with inflammation of the lungs, with measles, but more frequently with scarlet fever.

It attacks all, indiscriminately, both old and young. In the older children it is but of little account, never injuring the general health by its own action; but in the infant it is quite a different thing, and becomes a serious, even a dangerous disease. In these little sufferers, who are unable or unwilling to breathe otherwise than through the nose, it is quite an impediment to respiration, especially after the first few days, when the head and nose become completely filled with a thick, tenacious secretion, which it is impossible to remove. Being prevented from breathing through the nose, the child, when nursing, is obliged to frequently relinquish the nipple in order to obtain breath, which makes it cross and fretful. When coryza exists in connection with other diseases, it of course adds to their severity.

Causes. — As a general thing, cold is the exciting cause. Children, when put to sleep, should never lie with their head toward or near a window, or in any other position where there is the least liability of a draught of air, however slight, blowing upon them. A person takes cold much more readily while asleep than awake.

Nurses are in the habit of covering the child's face with a little blanket after it has been put to sleep. This, by confining the breath, invariably produces perspiration. Children covered in this way always waken with their head dripping with sweat, and, when taken up in this condition, are very liable to become chilled, — and snuffles is the result. Do not cover the face.

Symptoms. — All are acquainted with the symptoms of an ordinary cold in the head. It usually commences with shivering, some little fever, sneezing, obstruction, and dryness of the nose. This dryness is soon followed by a discharge, more or less profuse, with watering of the eyes, pain through forehead and temples, as well as about the root of the nose. Of course the little infant does not complain of this pain, but the older children do; therefore we are led to infer that all suffer more or less from it.

The secretion from the nose interferes with respiration, and when the passage from the head is completely filled, the patient is compelled to breathe through the mouth; and this soon causes dryness and stiffness of the tongue and throat.

TREATMENT.

For the premonitory symptoms of coryza, with shivering and headache, camphor is the best remedy; and, if administered promptly, a few doses will, in the vast majority of cases, be sufficient to effect a cure. In case you have nothing but the ordinary spirits of camphor convenient, you may put one or two drops upon a lump of sugar, and dissolve the whole in water.

It is sometimes advisable, when the secretion becomes suppressed, or before it has commenced, when the nose is hot and dry, to apply with a feather or camel-hair pencil a little *almond-oil* or cold cream to the interior of the nose, or let the vapor of hot water pass up the nostrils. Goose-grease rubbed upon the bridge of the nose in any quantity is of no earthly use. (Nos. 36, 245.)

COUGH, OR TUSSIS.

A cough is not a disease in itself, but rather a symptom denoting an abnormal condition of the lungs or throat.

Cough is a violent and sonorous expulsion of air from the lungs, preceded by, rapidly followed by, or alternating with, quick inspirations.

This, in fact, is but an effort on the part of nature to remove some obstruction or to throw off some accumulation which disease has created. During the course of an inflammation of the lungs, there is always more or less mucus secreted ; and, were it not for these forcible and violent expirations, the air-passages would become clogged up, and respiration materially interfered with. This is but one of nature's ways to rid herself of an offending substance : she has many. You will see an illustration of this parental care exhibited in the young infant ; the child, not knowing how to eject air violently through the nose for the purpose of clearing that organ, has been provided with a " sneeze."

Cough is often combined with a cold in the head, both originating from the same cause, namely, exposure. In the majority of cases cough is but a slight inflammation or irritation of the throat or upper part of the windpipe, accompanied with more or less fever.

Sometimes, where cough originates from a high state of inflammation, the soreness in the throat, the fever, in fact, all the acute inflammatory symptoms will have passed away, and the cough, though diminished, still remains. Such a cough should not be neglected, or it will become chronic, prove troublesome, and not easily be gotten rid of.

Causes. — Like every other disease of the air-passages, cough usually originates from exposure. But, then, there are a great many *indirect* causes which produce coughs ; that is, it may be sympathetic, depending, as it not unfrequently does, upon some derangement of the digestive apparatus. A very troublesome kind frequently met with is one occasioned by an elongated palate ; this keeps up a constant tickling, which is very provoking, and the cause being overlooked, it not unfrequently proves intractable. It would be impossible to enumerate all the causes, direct and indirect, which give rise to cough; in selecting a remedy you must not look upon it as an isolated symptom, you must take into account all the attendant circumstances, the source, and the peculiar condition of the system at the time. If much fever, chilliness, headache, sore throat, pain in the windpipe upon pressure, inflammation, elongation of the palate, enlargement of the tonsils, congestion, irritation, or presence of a foreign body,—all these concurrent and coincident symptoms will require special treatment, and the cough be treated and looked upon as a result or relative effect of that peculiar condition of things. Again, the cough may be entirely sympathetic, and originating in a derangement of some other important viscera besides the lungs.

TREATMENT.

Now, looking upon cough in this light, you will readily see the folly, the utter absurdity, of *cough panaceas.* I would therefore advise you never to have recourse to them. Their effect, to say the least, is uncertain, and not unfrequently they do a great deal of mischief. You should study each particular case thoroughly and on its own merits; ascertain, if possible, from whence comes the difficulty, and endeavor to select a remedy that will give temporary relief. Then, if it does not yield to that palliative treatment, at once seek the aid of a skilful, intelligent physician. In all ordinary cases you will have no trouble whatever in making prompt and perfect cures; but occasionally you will meet with chronic, obstinate cases, which can only be successfully treated by the physician. (Nos. 37, 169.)

Diet. — Patients suffering from cough, particularly if it is chronic, should live upon a good, plain, substantial diet, avoiding all articles of food which are found to disagree with them. Avoid all rich, high-seasoned food, fat meats, new bread, and all articles of a stimulating nature, or having a strong, pungent taste or smell, strong drinks, acids, beer, and so forth; also spices of every description.

Regimen. — Free exercise in the open air is highly beneficial; a morning walk, exercise with the dumb-bells; drawing large quantities of air into the lungs, then beating upon the chest with the hand; all this will not only expand and strengthen the lungs, but the whole bodily frame. Children should be encouraged in lively out-of-door play; it makes them active; let them run, skip, and jump; let them play at any and all games calculated to develop their physical and expand their mental faculties. During a portion of the day children should be permitted uncontrolled liberty of action. The daily bathing with cold or slightly tepid water is the most effectual method of overcoming a predisposition to coughs and colds. A sponge, sitz, or shower bath should be taken every morning, and the skin should afterwards be rapidly dried and rubbed to a glow, either with the hand or a coarse towel, after which the child should be warmly dressed.

BRONCHITIS.

This disease has several appellations; by some it is called catarrhal fever, or catarrh on the chest, by others, cold on the chest, etc. It is simply an inflammation of the mucous membrane lining the bronchial tubes, those formed by the division of the windpipe, and leading directly to the lungs, their office being to convey air into the lungs. In mild cases, ordinary bronchitis, or cold on the chest, the inflammation, which is slight, is confined only to the larger tubes; there is little or no difficulty of breathing, moderate cough, and slight fever; while in the severer forms, the inflammation extends down into the most minute bronchial ramifications, and all the symptoms from the outset are of a severe nature.

Causes. — The chief causes are transitions from warm to cold temperatures, or *vice versa*, and inadequate or unsuitable clothing, especially in children, whose absurd styles of dress originate fully three-fourths of the bronchial troubles. There is many a long row of little white stones in Forest Hills, Mount Auburn, and all our suburban cemeteries, that would never have been erected but for the weekly holocausts of innocent little victims which are offered at the shrine of Fashion. Every physician, as soon as he commences to treat a case of bronchitis, orders the child to be warmly dressed about the chest and arms, and to be kept from the cold air. He knows that without this precaution, in the large majority of cases, his remedies would be prescribed in vain.

Symptoms. — For convenience' sake we divide this disease into three forms: 1st, Simple Acute Bronchitis; 2d, Acute Suffocative or Capillary Bronchitis; 3d, Chronic Bronchitis. The first form, *simple acute bronchitis*, is a very frequent disease among children of all ages. It seldom sets in suddenly as an inflammatory affection, but gradually develops itself from an ordinary catarrh or cold in the head. The breathing becomes somewhat accelerated, there is more or less cough, stuffing of the chest, some fever, and skin a little hotter than natural. On applying your ear to the chest you will hear a wheezing sound, or a rattling of mucus in the air-tubes; sometimes, after a severe coughing spell, vomiting will take place. As a general thing, toward night the patient is more restless and uneasy, fever higher, and cough more troublesome. Remarkable remissions at times take place in the course of this disease, the child appearing quite well for hours at a time, or it may wake up quite bright in the morning, but, as the day wears on, the fever rises, the skin again becomes hot and dry, respiration hurried and anxious, cough frequent, with a sensation or an appearance of tightness across the chest, so that during the day and forepart of the night it appears to be quite ill, but, as morning approaches, the fever diminishes, the skin becomes moist, the cough less frequent, and the child gets a quiet nap, which so much refreshes it that during the next forenoon it appears quite like itself. These symptoms may run along for four or five days, when the difficulty of breathing, with the fever and the restlessness, disappears; the cough grows less, gradually diminishes, and the child soon regains its accustomed health.

In cases rather more severe than this, the cough is a prominent symptom from the beginning; at first dry and violent, very frequent and harassing as well as painful, the paroxysms of coughing sometimes lasting a quarter of an hour, during which the child cries, throws its arms up or its head back, thus evincing its anxiety and pain. The cough is excited by crying and sucking.

As the disease progresses, the cough becomes loose; small children vomit up quantities of phlegm, while larger children expectorate quite freely. The mucous rattle may now be heard over almost every part of the lung, the fever is high, breathing quick and oppressed, skin hot and dry, pulse frequent, child fretful and restless. Older children complain of pain when coughing, and the infant evinces it by its wincing as well as by its endeavor to suppress the cough. The expectoration, at first scanty and viscid, later becomes copious and streaked with blood. There is an entire loss of appetite, foul tongue, great weakness, paleness of the lips, countenance anxious or dull, and the child drowsy.

Symptoms of improvement, which generally take place in three or four days, are diminution of the fever; the skin, instead of continuing hot and dry, becomes moist, and feels more natural to the touch; respiration becomes less frequent, soreness and pain diminished; the cough becomes loose; the appetite returns, and the child rests better. Ordinary bronchitis is a very frequent disease among children, and often follows in the wake of whooping-cough, scarlet fever, or measles. This form of disease is rarely fatal. During convalescence, there is profuse secretion of mucus, which can be heard rattling in the chest, from a contractive tightening of the muscles.

Capillary Bronchitis is so named from the fact that the inflammation extends down into the capillaries or small subdivisions of the bronchial tubes. It may appear as an idiopathic or primary affection, but, as a general thing, it succeeds the form just described, particularly when that form has been neglected or improperly treated.

Chronic Bronchitis usually follows an acute attack, either on account of improper treatment, or the presence of some hereditary taint, predisposing the child to scrofula or consumption. The cough from the acute form never entirely ceases; it becomes loose, and the expectoration may be considerable; the difficulty of breathing, though diminished, never entirely disappears; every night, or perhaps only every other night, fever arises, and is followed by more or less perspiration; the lips crack and become ulcerated, sores break out around the nostrils, the skin looks blanched, eyes are sunken, appetite lost, strength diminished, thirst is excessive. The neighbors and friends remark that the child is going into a " decline." These symptoms may last for weeks, months, or even years; but at any time a colliquative or watery diarrhœa may set in, and this will soon put out the last ray of its glimmering existence, and the little sufferer die of marasmus.

TREATMENT.

It needs considerable skill, nicety of discrimination, and practical experience in the selection and administration of remedies for this disease ; and a great many things in regard to the general health and constitution of the patient have to be taken into consideration. For these reasons we have refrained from giving any general treatment, as, in the great majority of instances, the advice and experience of a skilful physician will alone meet the requirements of the case, and carry it through to a successful issue. (For Temporary Remedies, see Nos. 38, 139, 140, 141, 142.)

PLEURISY, OR PLEURITIS.

The lungs are enclosed and their structure maintained by a serous membrane called the *pleura.* This membrane forms a shut sac, as in fact do all the serous membranes ; and the lungs fit into it, as a boy's head would into a tippet when it is inverted or partially folded within itself. You will observe, therefore, that the lungs, though enclosed by this membrane, are still upon the outside of it. After covering the lungs as far as their roots, the pleura is reflected over the inner surface of the chest. Pleurisy, or, as physicians call it, pleuritis, consists of an inflammation of this membrane ; at every act of respiration, every time the lungs expand and contract, the opposing surfaces of this membrane must glide upon each other, and, when in a healthy state, they do this freely, for the parts are well lubricated with serum, just as a piece of machinery is with oil, and for the same purpose ; but when inflamed, the pleura becomes hot and dry, the supply of serum is diminished or entirely suppressed, and the *friction* thus inevitably produced causes the pain or stitches in the side and chest. Pleurisy may either terminate by an adhesion or a gluing together of the opposed surfaces of the empty sac, or its walls may be widely separated by a pouring forth of serum; this latter effect constitutes dropsy of the chest. This disease seldom attacks infants and young children; it is not as frequent, neither is it as dangerous a disease, as inflammation of the lungs, with which, however, it is often connected.

Causes. — The exciting cause, as a general thing, is exposure to cold or damp. It may also arise from severe injuries to the chest, as from a blow or a fall.

Symptoms. — Pleurisy, from the onset, is marked by a sharp, stabbing pain, on a level with, or just beneath one or the other of, the breasts, preceded or accompanied by chilliness or shivering; a dry, ineffectual cough is usually present with no expectoration, or, if any, very little, and of a frothy, whitish look; some difficulty of respiration, high fever, pulse quick and hard, great thirst, hot, dry skin, loss of appetite, headache, and sometimes bilious vomiting. The pain beneath the breast may diffuse itself throughout the chest, but usually it is confined to a small space, and is of a sharp, stabbing nature, seemingly as though a knife were thrust into the side, which prevents the patient from taking a long breath, and produces great suffering; when coughing or sneezing, the child endeavors to suppress the cough. The pain is always aggravated by deep inspirations, change of position, or by pressing upon the parts; it usually lasts three or four days, and then subsides. In some cases — but these are few indeed — there is little or no pain.

The patient cannot lie upon the affected side, at least during the first stages of the disease; that position increases the pain; however, as the pain subsides, and effusion takes place, she is unable to lie on either side, on account of the pressure made upon the sound lungs by the effused serum, which produces great difficulty of breathing. The patient is, therefore, compelled to lie upon her back, or nearly so.

This effusion into the pleural sac, sometimes amounting to several pints, causes the affected side to bulge out and become evidently larger than the other.

TREATMENT.

As in pneumonia, the application of cold bandages is often of great service. The diet is substantially the same. (Palliative remedies may be found in Nos. 39, 290, 291, 292, 293.)

PNEUMONIA, OR INFLAMMATION OF THE LUNGS.

Pneumonia is an inflammation of the *substance* of the lungs; but the majority of the cases of pneumonia are attended with more or less inflammation of the serous membrane lining the interior of the chest, and inverting over the lungs; that is, there is some pleurisy. Bronchitis is also a frequent accompaniment. Pneumonia may be either single or double; one lung may be affected or both. It is more common upon the right side than upon the left, and generally commences in the lower lobes. Why it does so is not known, but such is the fact.

Causes. — Inflammation of the lungs, or lung-fever, as some persons call it, is a very important, because frequent, disease of childhood. As a general thing it does not occur as a primary affection, but supervenes as a complication either in scarlet fever, measles, whooping-cough, inflammation of the bowels, or bilious remittent fever. As cold is an active exciting cause, you will find pneumonia much more frequent during the winter than during the summer months. A severe blow or fall upon the chest, the inhalation of noxious or irritating gases may, and often do, produce it. I have known children to inhale hot steam from the spout of a coffee-pot or tea-kettle, and thereby excite an inflammation of the lungs. Children of all ages are liable to its invasion; but, from statistical reports, we are forced to believe that it is more frequent from the third to the fourth year; nursing infants and children under two years of age being less liable to it than those older.

Symptoms. — Pneumonia, in the majority of cases, commences, as do all inflammatory or febrile diseases, with a chill or shivering, followed by heat and an increased frequency of the pulse. Cough is always present, at first dry and deep, or quick and spontaneous. The respiration is accelerated, the breathing from 50 to 60, sometimes even 60 to 80, in a minute. Pain, or, more properly speaking, a stitch in the side, usually the right, on taking a long breath or deep inspiration. If you will now, in this the *first* stage of the disease, place your ear to the patient's chest you will hear a peculiar crackling sound, similar to that produced by throwing salt upon hot coals, or like the sound produced by rubbing between your finger and thumb a lock of one's own hair, close to the ear. This is an important symptom; it gives an early and sure intimation that engorgement or congestion, the forerunner of inflammation, has taken place. The expectoration, which, however, is seldom present in children under four or five years of age, is at first tough and sticky, but soon changes to a bloody mucus; sometimes, especially in older children, the sputa is of a rusty color. The face is flushed, and wears an anxious look; it is, in severe cases, blanched, and the features pinched; the skin is hot and dry, and of a shiny or glazed appearance; thirst is excesssive; the pulse ranges from 130 to 140; in young children it may run as high as 160, or even 180; the tongue may be hot and parched; but, as a general thing, you will find it moist, and covered with a yellowish or whitish fur. The patient does not wish to be disturbed, would much rather be let alone, usually lies upon his back, and desires nothing but plenty of cold water.

Now, the train of symptoms presented in a young infant — a babe at the breast — differs in some respects. Of course the child cannot tell you that it has a pain in the side; it cannot express its sufferings in words. How, then, are you going to ascertain what is going on within that little chest? In fact, *how* are you to know what is the difficulty, and where located? Children are not deceitful; and if you are attentive, and at all discriminating, you will have but little trouble in interpreting their look of anguish or their cry of pain.

The child will be peevish, restless, and uneasy; cries and frets all the time; does not care to nurse; skin hot and dry, respiration short and hurried. You will observe that the chest does not rise and fall regularly with each inspiration, but the movements are short, uneven, or jerking.

Respiration is carried on chiefly through the action of the abdominal muscles. From the onset, cough is present; at first dry, short, and hacking, but it soon becomes loose; vomiting is frequently present; sometimes a spell of coughing will end in vomiting, and, thereby, the expulsion of a quantity of glutinous mucus, or mucus tinged with blood. That the child suffers from pain when coughing is evident from the expression of its face; the grimaces and twistings of the features are always marked; and then, as you will observe, when the cough comes on, the little sufferer attempts to smother it, instead of taking a full inspiration, as it would if its chest were not sore; it tries to make it short and sudden; it tries to suppress it. Each spell of coughing is *accompanied* or *instantly* followed by a scream of pain, or a fit of crying. The cry, also, is peculiar. It is not a healthy cry, but a kind of a suppressed cry, more of a sobbing nature, but still sharp and shrill, indicative of real suffering.

When the inflammation has reached its height, which it does generally by the fifth or sixth day, the symptoms, not invariably, but usually, remain stationary for one or two days, and then begin to subside. The fever diminishes, the skin loses its hot and harsh feel, becomes soft and moist, the cough becomes quite loose, less frequent, and ceases to be painful, the child can take a deep inspiration, or even cry aloud, without suffering pain. The flashing of the cheeks passes away, the expression of the face becomes more natural, the child looks around, and notices all that is going on. At this period of the disease children are apt to be quite cross and fretful, wanting everything, and throwing all away as soon as gotten. Mothers say that this is a good symptom.

When pneumonia ends unfavorably, the patient lingers along for a great while; the disease runs the course we have described; but instead of taking a favorable turn, the fever continues, the breathing becomes less frequent, but more laborious and irregular, the child gradually fails, the strength diminishes, the face looks blanched and sunken, low muttering delirium may be present, but usually intelligence is retained to the last.

TREATMENT.

The temporary treatment, which will have a palliative effect until medical aid can be obtained, will be found at Nos. 40, 294, 295, 296.

Diet and Regimen. — The diet should be plain, consisting of light, easily digested substances, panadas, gruels, etc. Cocoa makes an excellent drink. Cold water may be allowed, when desired. The breast, of course, is the diet for infants. While suffering under a short attack of pneumonia, or more particularly when recovering from it, great care should be taken that the child is not exposed by taking it from one room to another, through cold halls or passages, or into damp basements.

HOARSENESS, OR RAUCITUS.

This affection, like croup, does not, in itself, constitute a disease, but is dependent on some morbid condition of the throat or larynx, such as irritation or a congested condition of the parts. The causes are the same as in almost all chest difficulties, and is frequently the sequence of a common cold. It ought never to be neglected, for it is one of the premonitory symptoms of membranous croup. No. 41 is an excellent temporary remedy.

CROUP

is one of the most frequent diseases to which childhood is exposed. It is almost peculiar to children, and occurs, as a general thing, during the period of first dentition, that is, about the second year, though children from one to twelve years of age are more or less liable to it. It occurs indifferently in the weak and strong, in boys and girls. Though not *contagious*, as some people suppose, there is strong argument in favor of its hereditary character, a predisposition to this disease being very frequently traceable for three or four generations. From whatever cause it originates, it consists in a simple ordinary inflammation of the upper part of the windpipe — the larynx — with a violent spasmodic action of that organ. Its attacks are usually abrupt, without any premonitory signs, awakening the child from his slumber with a paroxysm of spasmodic coughing.

Causes. — Croup is more common in cold, damp climates than in warm, dry ones. Rapid and frequent changes of season, weather, and temperature have considerable influence in producing it. Certain conditions of weather specially predispose children to its attacks, without doubt; hence the popular idea that it is epidemic in character, and that it is a *contagious* disease, which we do not believe.

Symptoms. — The symptoms of croup are well marked, and need never be mistaken for those of any other disease. In the evening, or before midnight, the child will be aroused by a paroxysm of spasmodic coughing. The cough is rough, barking, and is accompanied by a shrill, sharp sound; during the paroxysms of cough, the breathing is spasmodically oppressed, at times seemingly almost to suffocation. The face and neck are at first highly flushed, but, as the paroxysms become more violent, assume a dark, livid red, which afterward passes into a deadly paleness, if the fit is of long duration. The veins swell, and beads of perspiration stand out upon the forehead; sometimes the whole head is wet with sweat. The disease seems to threaten immediate suffocation, the countenance presenting a picture of the utmost anxiety. The patient may remain in this condition for fifteen or twenty minutes, or from half an hour to even an hour. As soon as the violent symptoms abate, the child falls asleep; and, on awakening, only a slight hoarseness, a loose cough, and slight fever will remain.

If the disease be improperly treated, or neglected, these frightful attacks will continue and increase in intensity, successive days and evenings; and before the third day has passed the inflammation will have extended through the trachea or lower part of the windpipe, toward, and sometimes even into, the bronchial tubes; and then, with all its attendant horrors, you will have a case of true membranous croup.

TREATMENT.

An ordinary case of croup no mother need fear, if she only have the proper remedies at hand. A warm bath, about 96° temperature, and kept up to that standing or raised three or four degrees higher, is an invaluable auxiliary, lessening the agitation and subduing the patient's symptoms.

Poultices of flaxseed meal should be applied to the throat and chest, and a moderate emetic, such as a teaspoonful of ipecacuanha wine, a spoonful of dry mustard in water, etc., should be immediately administered.

MEMBRANOUS CROUP

is an aggravated or exaggerated form of the disease we have just described, and consists in inflammation of the larynx (the upper part of the windpipe) of a highly acute character, terminating, in the majority of cases, in the exudation of false membrane, more or less abundantly, upon the affected surface. The inflammation usually begins high up, near that part which contains the vocal cord, or what physicians call the larynx. Perhaps you would better understand me if I should say that it commences in the region of that projecting cartilage called " Adam's Apple," and extends down into the bronchial tubes. This form differs from ordinary croup in the formation of a false membrane upon the inflamed surface, which obstructs the air-passages, and, in severe cases, completely closes them up, so that the patient dies from actual suffocation. This membrane, when coughed up, or when taken from the dead body, looks like a stick of boiled macaroni, is commonly of a yellowish color, and from one-sixteenth to one-twelfth of an inch in thickness. Its general symptoms are very similar to those of catarrh, being attended by slight fever, drowsiness, watering of the eyes, and running from the nose. In the last stage, and in very severe cases, the child is wholly unable to speak, even in a whisper, or cry; the only noise it is able to make is the peculiar violent, short, shrill, barking cough. Between the paroxysms of coughing, the wheezing is heard in the air-passages at every inspiration. As the disease grows worse, the voice becomes more hoarse, the accumulation of false membrane and the mucus increases, till at last the tube is entirely filled up or completely lined, so as to preclude the possibility of respiration — and death is the inevitable result. The duration of the disease is from three to twelve or fourteen days, though many cases have occurred in which death has taken place on the first day.

TREATMENT.

In all cases, the treatment must necessarily be left to some skilful and experienced physician, the life of the patient being in too great danger to permit of its being left to nurses or relatives, however careful, intelligent, or affectionate they may be. (Remedies at Nos. 42, 43, 174, 175, 176, 177.)

WHOOPING-COUGH, OR PERTUSSIS.

This is one of that peculiar class of diseases that seldom, if ever, attacks the same individual but once in a lifetime. It is essentially a disease of childhood. Not but that adults would be just as liable to it as children were it not for the fact that they had already had it as children. Let it once enter a family of children, and the whole group is pretty certain to have an attack. It is undoubtedly contagious, and usually appears in the spring and fall. When it occurs in the fall, it is generally more severe, from its frequent complication with catarrhs, lung diseases, and other ailments. Particular care should be taken that feeble children, as well as young and delicate infants, are not exposed to whooping-cough during the fall months. Its duration extends from six weeks to six months.

Symptoms. — Whooping-cough may be divided into three stages: 1. The Catarrhal; 2. The Spasmodic; and 3. The stage of Decline. The catarrhal period commences with the ordinary symptoms of a common cold. For ten or twelve days, the child will generally evince all the characteristics of catarrh, though occasionally this stage will be entirely absent. The spasmodic stage is marked by violent spasmodic paroxysms of coughing, which occur at longer or shorter intervals, lasting from a quarter to three-quarters of a minute. These fits may succeed each other so rapidly as to make one continued paroxysm of fifteen minutes' duration. They are made up of a succession of *expirations* without any intervening *inspirations*, until the little sufferer gets almost black in the face, and appears upon the point of suffocation. This is followed by one long-drawn act of inspiration, which produces that peculiar shrill sound, or whoop, from which the disease derives its name. This operation is repeated time after time, until all the air is expelled from the lungs, the long inspiration again filling them; and the paroxysm usually terminates in the expulsion of a quantity of thick, ropy mucus, or else in vomiting. In some very severe cases, during a fit of coughing, blood will fly from the nose and mouth, and occasionally from the eyes and ears. The eyes, bloodshot and sunken, will fairly start from their sockets, presenting a horrid spectacle of suffering. The stage of decline consists in an amelioration of the severe symptoms; the paroxysms become less frequent and of shorter duration; the child's appetite returns, and he again resumes his natural habits and disposition. In this stage of improvement, when all is going on smoothly, a slight cold may reproduce all the distinct characteristics of this peculiar cough. (For remedies, *see Nos.* 44, 170, 171.)

Complications. — Simple whooping-cough, when unconnected with any other disease, is seldom or never attended with much danger. But its complications are many, and of various forms; therefore it is highly important that all the accidents apt to occur should receive a careful consideration. BRONCHITIS is a frequent complication, and may be recognized by a greater amount of fever, increased difficulty of breathing, and an incessant cough during the first stage. The expectoration will be more difficult, less profuse, and have a frothy or yellowish look. A marked expression of pain will cross the child's face in every fit of coughing. Sometimes the whooping-cough will be entirely superseded by the bronchial affection. CONVULSIONS are by no means rare as a complication, and are by far the most dangerous ; and consequently require the most vigilant care and watchfulness. Convulsions and head troubles are usually found in connection with whooping-cough at about the second year, or during dentition, and may be considered serious and especially liable to a fatal termination. PNEUMONIA is another complication often met with, the symptoms of which will be found indicated under that head.

Diet and Regimen. — This is a matter of the greatest importance. Anything stimulating should be especially avoided. It should be particularly plain and nutritious. Light and easily digested food is the best. Spices and hot stimulating drinks should never be permitted. Cold water, oatmeal gruel, barley-water, rice-water, toast-water, etc., are the only suitable beverages. Exposure to cold will very much aggravate the cough and even reproduce all the severe symptoms when the child is in a fair way of recovery. The dress should be so regulated as to guard against sudden atmospheric changes, and the body be kept at an even temperature.

INFANTILE ASTHMA.

This disease has frequently been confounded with spasmodic croup, and treated as such. It is purely a nervous affection, and is chiefly manifested in children of a strumous or scrofulous habit, occurring between the first and third year of life, and frequently connected with dentition or a deranged state of the digestive system. Many of the symptoms are similar to those of croup, but asthma may be distinguished by the absence of premonitory symptoms, its

occurrence in the daytime as well as at night, the absence
of fever, and its leaving no cough or hoarseness behind it.
No. 45 will be found an efficient remedy.

LARYNGITIS, OR INFLAMMATION OF THE LARYNX.

Laryngitis is simply an inflammation of the mucous
membrane lining the larynx, just as bronchitis is an inflammation of the mucous membrane lining the bronchial tubes.
It occurs at all times of life, but as a general thing is confined to children of from three to six years of age. It very
much resembles an ordinary case of croup, the only point
of difference being the absence of that peculiar spasmodic
cough. On looking into the throat, you will observe more
or less inflammation about the tonsils and palate, which may
be diffused over the whole surface, or be in patches varying
in color from a mere blush, as in mild cases, to a deep-rose
or even a violet-red. The first symptom which should excite your alarm and prompt you to call in a physician, is
the *difficulty of swallowing*, for which you can find no *adequate* cause. To this will be added difficulty of breathing.
The respiration is peculiar, being attended with a throttling
noise, each inspiration producing a wheezing sound, as if
the air were drawn through a narrow reed. The larynx is
painful upon external pressure, the face flushed, skin hot
and dry, pulse more frequent than in health, rising to 120°
or 130° per minute, and the child is thirsty, restless and
uneasy. The disease is, happily, not very frequent, at least
the severer form of it. It is easily distinguished from a
common sore throat by the difficulty of breathing and swallowing. It is true, extreme enlargement of the tonsils obstructs respiration, but then on inspection this swelling will
be *visible*. In laryngitis, there is but slight inflammation and
swelling, at least little can be *seen* of it. (*See Remedy
No. 46.*)

Diet and Regimen. — A child suffering from this disease should be confined to a warm room, and not be allowed to roam all over the house, through cold rooms and in draughts of air. A slight reduction of diet is advisable, forbidding all condiments or anything of a stimulating nature. A farinaceous diet is the best. The application of cold water is always advisable, often affording great relief. Rubbing the throat with camphorated oil, goose-grease, or Ready Relief, or, in fact, the application of embrocations or stimulating lotions of any kind, is highly objectionable.

COLDS.

The term "cold" is a relative one, used to express a certain condition or sensation produced by the abstraction of heat from the system by any substance of a lower temperature than that of the body. This may not always be occasioned by the same degree of temperature. For instance, a temperature that, to a healthy, vigorous, active man would seem warm or comfortable, would, to one enfeebled by disease, appear quite the reverse. A man or child, though not physically strong, but full of energy, courage, and excitement, would resist a greater amount of cold than one who is faint-hearted, nervously depressed, or despondent. Children are more susceptible to atmospheric depressions than adults, and simply because the power of generating heat within themselves is weak, undeveloped. Cold does not always cause disease in the exact part to which it has been applied; that is to say, because a person sits through a tedious concert with a draft of air continuously playing a disease-tattoo on his back, he must not necessarily have rheumatic pains or some other trouble in the part, although this may be the case. As a general rule, the cold, by diminishing vital actions in the parts on which it acts, so determines and increases the same in distant parts, as to give rise to congestions and inflammations, or to a train of diseased action, more or less definite, which, by common consent, is usually termed a cold, such as chills, general soreness and lameness, pains and aches in the head and limbs, followed, as soon as reaction comes on, by accelerated respiration and circulation, as well as other symptoms which constitute fever.

Cold does not affect all persons alike. Two ladies, exposed to the same current of air, may, as the result, suffer from diseases quite dissimilar. This depends upon peculiarities in temperament, predisposition, and habits of the individual. As a general rule, however, those organs or parts of the system are first affected which are the weakest. If the lungs are predisposed to disease, cold will develop some difficulty in these organs. Should a person be subject to catarrh, cold will act as an exciting cause to bring it into action. Children subject to croup, glandular enlargements, or gatherings in the head, need only a cold to set the disease in motion. The same principle is true with other organs and structures of the system. The extent and severity of the disease thus excited will depend upon the amount of exposure and the delicacy of the part affected. The most common results of taking cold are catarrh and croup, sometimes fever, colic, dysentery, diarrhœa, neuralgia, sore throat, pains in the teeth, ears, or general pain and soreness throughout the whole system. (*See Remedy* No. 47.)

DISEASES OF THE STOMACH AND INTESTINES.

The particular age at which children are most liable to these affections is from birth to the termination of the first dentition; this, of course, includes the second summer. From this period onward, as the child increases in years, it becomes less liable to their invasion.

Causes. — By far the most frequent exciting cause of

all gastric diseases during infancy, is an improper or unwholesome diet. They are not unfrequently, in nursing
infants, dependent upon an unhealthy condition of the
mother's milk ; but it appears to me that the chief source of
difficulty is the too early resort to an artificial diet, or an
artificial diet badly chosen. Of course, the natural aliment
of an infant, for the first nine months, is the mother's milk,
which, during the first few months, is very thin, and pos-
sesses properties peculiar to itself. Now contrast this with
the various articles of food prepared for children, consisting
mainly of pap, or thick bread and milk, or crackers moistened with milk and water, to which sugar is added ; gruels
of all kinds ; coarse preparations of rice, barley, etc. As
before intimated, the stomach of an infant is only intended
to receive the milk provided by its parent ; and it is entirely
incapable of digesting the thick or coarse, and often too
rich, food, which is so frequently substituted for that which
nature has provided. But it is not only the quality of the
food that is at fault, but the quantity. The *overfeeding*
of children is a constant and never-failing source of mischief. Children artificially fed scarcely ever escape these
intestinal derangements. Diarrhœa frequently sets in immediately after weaning a child, and this is the necessary
result of irritation of the mucous membrane lining the intestinal canal, produced by the change of food.

Indigestion, or loss of digestive power, and the consequent enervation and wasting away of the system from imperfect assimilation, is but the direct effect of an improper
diet, or the over-taxation of the digestive apparatus from
excessive feeding.

The heats of summer and sudden atmospheric changes
are undoubtedly powerful predisposing causes to infantile
bowel complaints ; in fact, we seldom have these diseases to
any great extent, presenting all their characteristic features,

except during the hot months of summer. To the heats of summer we have usually to add impure air and badly-ventilated houses. As you pass through some of the streets of our cities, and inhale the effluvia from the dirty gutters, you wonder, not that so many are taken sick, but that all do not die; and then, when you come to enter the damp basements, and find huddled together whole families of ten or a dozen persons, occupying one room, in which they cook, eat, and sleep, you are actually bewildered and in amazement, wondering how any mortal can draw the breath of *life* from such a vitiated atmosphere. To people living thus, and all those who reside in narrow, crowded streets and alleys, these diseases are as scourges. But these disorders are not confined exclusively to the poor and to those living beneath the ground, and away from the light and air which God has given us. No, they are only too common among all classes of the inhabitants of our large cities.

Dentition being a natural physiological process, we should not expect it to be productive of any evil results; nevertheless, it is a well-established fact that the cutting of teeth is a powerful predisposing cause to intestinal irritation, and it frequently impairs or diminishes the tone of the digestive function, so that the infant is often unable, during the period of dentition, to digest food, which at other times agrees with it perfectly well. Gastric derangements of children, from the completion of the first dentition to the age of eight or ten years, may, in the majority of cases, be traced directly to the persistent inattention on the part of mothers and nurses to the general laws of health. It is the strangest thing in all the world to me, that poor human nature, plain and simple in all its requirements, should be so wholly disregarded. It is either from ignorance or thoughtlessness, or most likely both. It is quite a common thing, in fact the general custom, to allow children of from two to three years to sit at the table and partake of the same food that is prepared for the adult members of the family : hot rolls and butter, hot buckwheat cakes, sausages, salt fish, radishes, cucumbers, candies, meats, and indigestible dishes of all kinds.   These are partaken of at all hours, and in excess, from breakfast-time to just before going to bed ; and the wonder is, not that we are a pale, thin, dyspeptic, and anxious looking race of people compared with Europeans, but that we have any health at all, especially when our children are allowed to make use of the indiscriminate and unwholesome diet we have described.

Now, one would think this alone would be sufficient to exterminate the whole race in a short time ; but to all this is yet to be added that vile, pernicious habit of drugging children with medicines.   Most mothers and nurses have each their little collection of remedies.   For every little ailment that may overtake a child, brought about by some error in diet, a dose of medicine, usually a cathartic, must be given.   And what is the result?   Why, a slight indisposition, which a little care or judicious restriction in diet would have speedily removed, is transformed into some serious disorder.   The medicine given is so repulsive to nature that the whole system is thrown into commotion in the effort to reject it ; the child is vomited, physicked, — thoroughly " cleaned out."   If the child recovers, it is *in spite* of the treatment ; but usually the digestive apparatus and the entire nervous system are shattered, the child becomes irritable, cross, morose, a burden to itself and to all around it.   I never pass through our cemeteries, and contemplate the rows of small white stones, without the thought crossing my mind, —

> Sleep on, sweet child, thy trouble's past;
> *Physic* has freed thy soul at last.

THRUSH, OR APHTHÆ.

The term *thrush* or *aphthæ* is applied to an ulcerated sore mouth peculiar to infants, which makes its appearance during the first year, as a general thing within the first fortnight. Nurses, and women of experience generally, anticipate and avert its arrival by frequently washing the infant's mouth with a soft linen rag dipped in cold water. The most prolific cause is the trash which is forced down the poor infant's throat during the first few weeks of its existence. It consists of a series of vesicles or pimples, capped with small white spots, which break out into ulcers, occupying the internal surface of the under lip and cheeks, the edges of the tongue and gums, sometimes extending over the entire mouth. There is little or no fever, though the mouth is hot, and the quantity of saliva is largely increased. When the ulceration extends far back into the mouth, the stomach, and the alimentary canal, the act of swallowing is materially interfered with and performed with great pain. In such cases diarrhœa often sets in, and the affection assumes a serious aspect.

TREATMENT.

Borax is an excellent remedy, and may either be given dry, in the form of pills, or in solution, and is very frequently used as a gargle. (No. 48.)

CANKER OF THE MOUTH.

This form of sore mouth is usually found in children of from five to ten years of age; by many it is considered contagious, but upon this point physicians are divided, though all agree in considering it epidemical. It is an inflammation of the mucous membrane, with an exudation upon the surface of a yellowish, plastic lymph, with erosion or ulceration, which occasionally, particularly if improperly treated, becomes very destructive in character, running into dark, deep, sloughing sores. It is also known as *cancrum oris*, scurvy of the mouth, or canker-sores.

Symptoms. — The peculiar characteristics of this disease are : first, pain and uneasy sensations in the gums, which soon become hot, red, and very sensitive ; they also swell, become spongy, and bleed when touched. The gums and internal surface of the cheeks are covered, or rather spotted over, with patches of false membrane, which adhere with considerable force to the tissue beneath. Under this layer of exudation small ulcers make their appearance on the gums, the inside of the lips and cheeks, the soft palate, and edges of the tongue. Sometimes this false membrane is entirely wanting, when the ulcers are plainly visible, and present a grayish or livid appearance, with swollen, softened, or bleeding edges.

These ulcerated spots may be but few in number, either upon the inner surface of the lips and cheeks, or edges of the gums, or they may be studded over the whole cavity of the cheek. The breath is always more or less fetid, and not unfrequently has a putrescent odor ; and sometimes, especially in severe cases, there is a copious discharge of offensive bloody serum from the mouth. The glands about the throat and neck are swollen and painful, the movements of the under-jaw are stiff ; this, together with the looseness of the teeth, makes mastication very difficult, while swallowing is interfered with from soreness of the tongue and throat. There is generally more or less of a low grade of fever ; the patient loses his strength, and becomes very much prostrated. The course of this disease is short, if under judicious treatment, but not unfrequently sudden, severe, and destructive salivation is set up by the excessive administration of calomel, which, if not ending in gangrene of the mouth, prolongs the difficulty to an indefinite length of time. (For Temporary Remedies, *see Nos.* 49, 145, 146.)

Diet and Regimen. — The diet should be plain, and of either a farinaceous or vegetable form; animal food, either solid or in soups or broths, had better be dispensed with. It is desirable that the mouth should be frequently gargled or rinsed out, and especially after eating, that no offensive matter or particles of food may remain to irritate the parts. A weak solution of brandy and water makes the best wash; lemon-juice and water or a decoction of sage is also frequently used. Decayed teeth, or stumps of teeth remaining in the mouth, are a frequent source of irritation, and should be speedily removed.

GANGRENE OF THE MOUTH.

This term — gangrene — signifies mortification, which is justly the terror of all who have to contend with it. It generally commences with ulceration of the mucous membrane lining the cheek and gums. The mucous tissues and substance of the mouth, gums, lips, etc., are destroyed, turn black, and slough away, the teeth loosened, and the jaw-bone denuded and exposed. The affection is seldom met with in private practice, and is almost exclusively confined to institutions where large numbers of children are gathered promiscuously together. It almost always follows upon some previous acute or chronic disease, such as long-continued fevers, measles, or other acute exanthemata, during which the patient suffered more from the treatment than from the actual disease. Unfavorable hygienic conditions, debilitated constitutions, a scrofulous habit, etc., are conceded on all sides to constitute the predisposing cause of this affection, but the *exciting cause* has been and still is a bone of contention among physicians encountering this disease. It is perfectly plain, however, to those who are disposed to see, that gangrene of the mouth is nothing more nor less than poisoning by mercury. The duration of the disease extends from six to twenty days ; and the aspect of the poor little patient is as sad as it is hideous. This disease is usually terminated by entire prostration, insensibility, and death.

PTYALISM, OR SALIVATION.

This disease consists in an irritation, inflammation, and swelling of the salivary glands of the mouth and throat, with a profuse discharge of saliva or spittle. Most persons are too apt to attribute it to the injudicious use of mercury ; but we often see patients recovering from severe attacks of fever, with all the symptoms of salivation, where there has not been one particle of mercury given. Salivation is produced by administration of copper, antimony, potassium, arsenic, castor-oil, digitalis, and opium, under certain conditions of the system, quite as readily as by mercury. Sometimes it occurs spontaneously, as a result of local irritation, decayed teeth, cold, fever, etc. It occasionally occurs as a critical discharge, by the action of nature, and is then beneficial.

The salivary glands and mucous membrane of the mouth and throat are red, swollen, and considerably inflamed, the glands beneath the under jaw being enlarged and very tender. The saliva, which is discharged in large quantities, is much changed in its character and appearance. Instead of being thin, watery, colorless, inodorous, and tasteless, as in health, it becomes dark, thick, stringy, fetid, and very offensive.

All astringent washes or gargles which directly diminish the salivary discharge are injurious. Mild washes or gargles, such as milk and water, may be used with considerable benefit. (Also Remedy No. 50.)

The diet must be of the mildest kind, — gruels, milk and water, crackers soaked in water, plain puddings, and the like. For a drink, cold water may be used, or cocoa, if the patient likes it.

RANULA, OR SWELLING UNDER THE TONGUE.

This is a swelling of the sublingual glands, caused by some obstruction of the salivary duct — which is the little canal that carries the saliva from the gland to the mouth — from cold, inflammation, or some irritating cause. Tumors of this kind are not generally painful; but when they are of any considerable size, they interfere with the free motion of the tongue, and thus materially impede the power of speech. (*See Remedy No.* 51.)

GUM-BOILS. — ABSCESS IN THE GUMS.

Almost every form of swelling with inflammation affecting the gums (even abscesses and suppurations) are popularly but erroneously classed under the head of gum-boils. This is a very annoying affection, and arises from various causes, such as cold, decaying teeth, or the cutting of new teeth, especially the molars and bicuspids, or from their unskilful abstraction. Sometimes a slight incision by knife or lancet will be found necessary; but outward application of anodynes, warm cloths, fomentations, etc., will usually be found effective. (*See Remedy No.* 52.)

MUMPS, OR PAROTITIS.

The salivary glands are six in number, three upon either side of the throat, and are called the *parotid*, the *submaxillary*, and the *sublingual*. The *parotid* are situate below and in front of the ear; the *submaxillary* below the lower jaw, and the *sublingual* under the tongue. The office of these glands is to furnish the saliva or spittle with which the food during mastication is softened, so that when carried into the throat it passes with ease through the œsophagus into the stomach.

Now, *mumps* is an inflammation of the largest and most important of these glands, the *parotid;* hence the name *parotitis.* It often prevails as an epidemic. When it attacks one child in a family or a school, it is almost certain to affect all the others, *simultaneously* or *in succession.* It is undoubtedly contagious, chiefly attacks children and young persons, and seldom, we may say *never*, attacks a person the second time.

There are no marked symptoms at the commencement of the disease, except the tumefaction and swelling under the ear. Sometimes one side only is affected, and sometimes both at once, but most frequently one side is first affected, and the disease afterwards extends to the other. The swelling is hot, dry, and painful, and very tender to the touch. There is usually some fever; the motion of the under jaw is interfered with from the swelling in the vicinity of the joint. The inflammation reaches its height in about four days, and then begins to decline; its whole duration may be stated, on an average, at eight or ten days.

Mumps is not considered dangerous, unless from imprudent exposure the patient takes cold, or from any other cause the disease "strikes in," that is, becomes thrown back upon the system, so as to involve some of the vital organs. In many cases, under these circumstances, the swelling about the throat and neck subsides quickly on the fifth or seventh day, and shows itself upon the testicles in the male sex, and upon the breast in the female, and these parts become hot, swollen, and painful. Another dangerous transfer of this disease, but particularly rare, is from the testicles to the brain. (*See Remedy No. 53.*)

Diet and Regimen.—The diet must be light.

Toast and black tea, cocoa, custards without spice, bread-puddings, baked apples, and stewed prunes may be allowed. If it is during cold weather, the patient should be kept in a moderately warm room; if there is much fever, he had better lie in bed. No external application need be made, unless it be simply a handkerchief tied around the neck. Should the neck get very tense, hot, and dry, it will be advisable to apply hot flannel cloths. Great care must be taken to prevent the patient from taking cold. Never apply cold water or any of the many lotions; follow simply the directions above given.

INFLAMMATION AND SWELLING OF THE TONGUE.— GLOSSITIS.

Glossitis is an inflammation of the substance of the tongue, characterized by pain, redness, hardness, and swelling, either with dryness of the mouth or profuse discharge of saliva, and accompanied with the usual symptoms of inflammatory fevers. The inflammation may be confined to one side of the tongue, or the whole organ may be implicated.

It usually arises from mechanical injuries, or from contact with chemical agents or acrid substances which excite irritation. This affection is sometimes induced by exposure to cold or to currents of cold air about the head after the use of mercurials, or by the suppression of the salivary discharges. In many cases the attack is very sudden, a severe inflammatory action setting in without any apparent cause. The first symptom complained of is usually an acrid, stinging sense of heat or burning pain in the tongue. The inflammation, as a general thing, sets in very suddenly, and proceeds rapidly; the pain and swelling is very great; the tongue presents a livid or dark red appearance. The inflammation may commence upon one side, or be restricted to a very small portion, but gradually it may extend until the whole organ becomes involved. During the progress of the disease the pain becomes more acute, and of a burning and lancinating character, which is aggravated by the slightest movement; the attempt to talk or swallow causes great suffering. In severe cases, the tongue becomes enormously swollen, filling the entire mouth, speaking and swallowing being prevented while respiration is obstructed, even to threatened suffocation. In other cases, the swollen and inflamed organ is protruded from the mouth, presenting a horrid picture of suffering. The tongue is usually furred over with a thick coating, and a profuse secretion of saliva flows from the mouth. Should you meet with a case where the swelling has become so enormous as to threaten suffocation before a physician can arrive, do not hesitate to take your knife, or any sharp instrument, and make a free longitudinal incision in the tongue. This gives egress to the blood, which removes the congestion and relieves the patient. (*See Remedy No.* 54.)

DENTITION, OR TEETHING.

Dentition being a natural process, we should scarcely expect it to occasion disease or suffering of any sort, and were all children in a perfectly healthy condition at the time of its commencement, they would suffer but little, if any, during this period. But all children are not born healthy, and many that are ordinarily healthy at birth, are, by neglect and mismanagement in dress, diet, and exercise, speedily rendered unhealthy. Under these circumstances, dentition frequently becomes complicated, difficult, and dangerous, the digestive organs and nervous system being the first to feel the baneful influence.

The first, milk, or temporary teeth, are twenty in number, and begin to make their appearance at about the sixth

month, continuing until the end of the second year, those
of the lower jaw preceding the upper. The regular order
and time of teething, however, is subject to considerable variation. Some children get their teeth two or three
weeks after birth, or are even born with them, while others
do not cut any teeth until they are ten or twelve months
old. Teething, in the most favorable cases, is preceded by
slight salivation or drooling, as it is called, by heat and
swelling of the gums, increased thirst, restlessness, or fretfulness, etc. Sometimes there is a rash upon the skin,
called " red gum " or tooth-rash. Connected with teething,
there are often many sympathetic affections, such as determination of blood to the head, convulsions, constipation,
swelling and suppuration of glands, eruptions of various
kinds, both upon the head and body, gatherings and discharges from the ears, cough, and general irritability of the
nervous system, so that trifling ailments, which at other
times would scarcely trouble the child, would, during this
period, excite a train of acute and serious symptoms which
only prompt and judicious treatment could successfully combat. In children of deficient vital power, a cold, an error
in diet, or some undiscoverable cause, may excite a slight
derangement, at first scarcely noticeable, but which, by
improper treatment or neglect, leads to a permanent state
of bad health, ending in tubercular degeneration of the
lungs or digestive apparatus. The necessity, therefore, of
zealously guarding the children from every source of disease,
to which they might be exposed at this time, will be obvious to all. Unfortunately, however, for the children, most
young mothers have aunts, grandmothers, or some well-
meaning but officious female friends who " know all about
these little complaints of teething, and can treat them quite
as well as any doctor." The mantel-piece is accordingly
adorned with lotions, pills, and powders, bottles of all kinds

in warlike array, — ipecac, squills, Godfrey's cordial, paregoric, soothing sirups, castor-oil, sulphur and molasses,
peppermint, goose-grease, catnip-tea, mustard, and onion
draughts, — all of which the poor infant is bound to have
thrust down its throat in regular and constant succession.
This is continued day after day, until the child is "doctored" into some serious disease. Then the physician is
sent for, but, alas! too late to be of any service : the child
dies, either from the disease or the treatment; most frequently from the treatment, for the disease itself seldom
kills.

The Care of the Teeth. — The proper culture and
preservation of children's teeth is a subject demanding the
attention of every thoughtful parent. When taking into
consideration the importance of sound and regular teeth,
alike in regard to health, comfort, and appearance, the little
care and attention requisite to keep them in a proper state
seems almost insignificant. The soundness of the teeth
depends in a great measure upon a healthy state of the
stomach and bowels, so that whatever tends to the derangement of those organs will exert a deleterious influence upon
the teeth. Children are often refused candies because they
are said to rot the teeth. Now, sugar itself never directly
injures the teeth, but the confectioners' preparations, with
the mysteriously-made but pretty-looking fixings of which
children are in the habit of partaking, have a direct and
extremely injurious effect upon the stomach, deranging the
bowels, causing dyspepsia, flatulence, and gassy eructations,
which blacken and corrode the enamel of the teeth, thus
laying the foundation for their decay and speedy destruction. As a general thing, healthy persons have sound
teeth, while sickly, feeble persons have decayed teeth.

A few brief hints for the *preservation* of the teeth will
here, we doubt not, prove acceptable to our readers.

1st. *They must be kept clean;* not by the use of dentifrices, tooth-pastes, powders, etc., but by the use of pure water and powdered charcoal or white Castile soap, applied with a soft brush.

2d. By avoiding the introduction of very hot or very cold substances into the mouth, as all sudden changes of temperature eventually crack the enamel and produce decay.

3d. By peremptorily forbidding the use of metallic toothpicks of any kind.

4th. By removing the temporary teeth as fast as they get loose.

5th. By refraining from the practice of cracking nuts, biting threads, or lifting heavy bodies, etc., with the teeth. (*See Remedy No.* 55.)

TOOTHACHE.

This troublesome and painful affection, for which so little sympathy is felt, either in youth or adult, has its origin in many causes : some are hereditarily predisposed to it ; in others it is induced by exposure ; it may originate from disturbances in other parts of the system, or it may be purely nervous. It is often rheumatic, or may have its origin in carious teeth and the excessive use of coffee or calomel, etc.

TREATMENT.

Do not have the teeth extracted, unless they are decayed or the roots ulcerated. Many of the ordinary remedies for toothache are not only useless, but positively injurious to the general health, as well as to the teeth themselves, such as creosote, laudanum, tincture of cloves, etc. Better remove the *cause* of the diseased condition. (Treatment Nos. 56, 321, 322, 323, 324.)

SORE THROAT, OR QUINSY.

This is known by a variety of names, and consists in an inflammation of the back part of the throat, including the palate and tonsils, frequently terminating in the formation of abscesses in the tonsils or adjacent parts. It is not strictly limited to any particular age, the infant, child, young girl, and adult being alike susceptible to attack.

Symptoms. — Ordinary quinsy, of moderate severity, generally begins with restlessness, irritability, fever, sometimes a slight cough, and more or less soreness or pricking sensation in the throat, especially when swallowing; the older children complain of this pain and refuse all diet except drinks and soft food, while the infant betrays it by refusing to nurse, and wincing its face whenever swallowing is attempted. The face is flushed, respiration accelerated, voice thick, and speaking difficult or painful. The combined symptoms are very similar to, and often mistaken for, inflammation of the lungs; but upon placing your ear to the chest, you will readily mark the difference in the two diseases by the entire absence of all physical signs of pneumonia. To examine the parts well, the head should be thrown back, the mouth widely opened, and the root of the tongue depressed with the handle of a spoon. By this means the whole of the interior of the throat will be exposed to view.

In its severer forms it is quite a serious affection, and if not at once attended to becomes dangerous. Relief cannot reasonably be looked for until the abscess bursts. (*See Remedies Nos.* 57, 312.)

Diet and Regimen. — The diet will have to be regulated according to the degree of inflammation. If extensive, the throat much swollen, and swallowing difficult, solid food cannot be taken. Custards, panadas, gruel, light soups, are all that can possibly be given. In no disease, perhaps, is the beneficial effect of cold water more marked than in sore throat. When going to bed at night put a wet bandage around the throat, and cover it with a dry cloth. If the patient is confined to the house, repeat the same through the day. If it does not yield to this treatment, and the abscess continues its progress, its ripening should be hastened by the external application of warm flaxseed poultices and gargling the throat with warm water. When much pain is present, the inhalation of the vapor from boiling water will afford considerable relief. All medicinal gargles, blisters, leeches, mustard drafts, and kindred remedies, are not only useless, but decidedly injurious.

MALIGNANT OR PUTRID SORE THROAT.

This is usually a symptom of malignant scarlet-fever, but is also an independent form of disease, generally occurring in damp, autumnal seasons, attacking children of vitiated, impoverished, or delicate constitutions, weakened by some previous diseases. It is also more apt to attack children living in low, damp, cold, mouldy, or ill-ventilated dwellings. Under such circumstances, an ordinary sore throat is readily transformed into one of a malignant type. It is an exceedingly dangerous disease whenever and wherever it appears ; the treatment should therefore be prompt and energetic, and should never be attempted by any one but a medical practitioner.

Symptoms. — This disorder commences with a chill, fever, and languor, oppression at the chest, with or without vomiting, more or less inflammation of the throat and tonsils, an acrid discharge from mouth and nostrils, excoriating all the parts with which it comes in contact, weak and rapid, almost imperceptible pulse, swollen throat and glands, face bloated, and general restlessness. Upon examining the throat, numerous small, yellowish ulcers, covered with an ashy-gray crust, will be seen, the surrounding tissue being of a livid or dark-red color. These ulcers are not confined to the throat and tonsils, but extend over the entire mucous membrane of the mouth, and even involve the windpipe. In severe cases they run together, and present a gangrenous appearance; there is excessive prostration; the teeth and tongue are covered with a black crust, similar to that of typhus-fever; there is more or less delirium; the breath is fetid, countenance sunken; vomiting and diarrhœa supervene, the pulse grows feebler, cold and clammy sweats take the place of the previous harsh, dry skin, stupor sets in, and the patient dies. Should the disease yield to treatment, the symptoms we have described gradually subside. (*See Remedies Nos.* 58, 309, 310, 311.)

Diet and Regimen. — The first thing to be done is to place your patient in a dry, airy room; plenty of fresh, pure air is the best adjuvant in the treatment of this or any other disease. The food, as a matter of course, will have to consist of rice, arrow-root, corn-starch, thin flour gruel, broths, and the like. When the mouth is very hot and dry, it is advisable to moisten it with a little warm milk and water. The mouth should be frequently washed out, and this must be done very gently, so as to produce no irritation. As a wash, warm water is the most desirable. During convalescence, great care should be taken that the patient does not overload the stomach, as this would tend to produce a relapse, or at least excite some gastric derangements whereby recovery would be retarded.

TONSILLITIS, OR INFLAMMATION OF THE TONSILS.

As we have already partially considered tonsillitis under the head of "Sore Throat," we will confine our remarks to chronic enlargement of the tonsils. You will frequently hear the ignorant speak of children having *tonsils* in their throat, as though *all* children, and adults, too, were not provided with them by nature. The tonsils are two rounded oblong bodies, placed between the arches of the palate. The use of these glands is to secrete a fluid which makes the passage to the stomach smooth and slippery, so that the food can be easily swallowed. Tonsillitis constitutes the enlargement of these glands from chronic or congenital inflammation, or an inflammatory condition arising from excessive nutrition.

Symptoms. — The first indication is continued snoring, caused by the pressure of the tonsils upon the palate, which partially closes the passage through the nose, the air being forcibly drawn through the narrowed opening. Deafness is another symptom, and originates from the pressure of the elongated tonsils on the small canal leading from the throat to the internal ear, called the "Eustachian tube." But the most serious consequence is the effect it produces on the chest, the obstruction preventing the free entrance of air into the lungs. These organs are but imperfectly developed, and produce the prominence of the breast-bone known as "pigeon-breast."

Treatment. — The application of nitrate of silver and other caustics, or the cutting-out of the tonsils, is barbarous and injurious in the extreme, for in the majority of cases such treatment induces lung-fever, consumption, and a thousand other affections more or less calamitous and fatal. The treatment we have advised in quinsy will also be found beneficial in this disease. (*See Remedies Nos.* 59, 243.)

FALLING OF THE PALATE.

Though so much has been said about this disease, it does not exist; it is purely imaginary. The fact is, that some persons, after a slight cold or attack of indigestion, suffer from a trivial inflammation of the palate, which, from its thickened and elongated state, produces a sensation of looseness or descent. Cold water is very beneficial, used both internally and externally. All stimulating articles of diet, fancy or highly seasoned dishes, should be especially avoided. (*See Remedy No.* 60.)

DIPHTHERIA, OR DIPHTHERITE.

This term is used to designate a specific and peculiar form of inflammation of the throat. Unlike ordinary inflammations of these parts, it is attended with an exudation of false membrane upon the mucous surface, attended usually with a low grade of fever, and is mainly confined to the throat, tonsils, and nasal cavities. It is a constitutional, and by no means a new disease, for we read of its ravages in all parts of the world during more than 2,000 years.

Causes. — Diphtheria is propagated by two causes: epidemic influence, and contagion. Scrofulous children, those subject to glandular enlargements, catarrhal and croupous affections, are usually first affected when the disease rages as an epidemic. It generally spreads through the entire household where it once enters, affecting both adults and children, those most closely in communication being first attacked while those removed from the locality of the contagion at an early period escape. Children are the chief sufferers from this deadful affection.

There is a great diversity of opinion, even in the medical profession, both as to its origin and nature. But in the light of critical investigation and practical experience it may be safely concluded that diphtheria arises from a specific poison taken into the system, which, acting through the blood, produces a true constitutional disease, exhibiting its local manifestations in the formation of false membrane upon mucous and abraded cutaneous surfaces, and becomes capable of transmission from one to another, without any recurrence to the original source of poison. Such being the case, the instant one child in a family is attacked, all the other children should be removed beyond the range of infection.

Symptoms. — There is a strong resemblance in the symptoms of scarlet-fever and diphtheria, so much so that, in regard to the eruption, it would seem to be a sort of cross between that of measles and scarlet-fever, while the other symptoms have a peculiar similarity to scarlet-fever. But, for all that, it is an entirely distinct affection. It generally commences in the same manner as an ordinary cold or influenza : slight chills and fever, general prostration and weariness, occasionally high fever and severe pain in the head, disordered stomach, and loss of appetite, etc. In the course of twenty-four or forty-eight hours there is a decided aggravation of the throat trouble, the glands about the neck becoming sensitive and swollen, with an increased flow of saliva or water into the mouth. In many instances the beginning of the disease is so insidious that its true nature would hardly be suspected were the *patches* of false membrane not seen in the throat. These patches vary from the size of a split pea to half an inch in diameter. When the membrane becomes detached it leaves the surface beneath in appearance not unlike a piece of raw meat. There is stiffness of the neck, more or less fever, unbearable headache, the inflamed surface is bright and glassy, or almost purplish, the breath is offensive, and there is great prostration. Under favorable circumstances, and with judicious treatment, convalescence may usually be established in from eight to ten days, though it may be weeks or even months before the debility and nervous depression is removed and the system restored to its natural elasticity and vigor. Hence the great danger of *relapse*, which is too generally fatal.

It frequently occurs that the patient is suddenly seized with rigors and vomiting of a thin, white, yellowish matter of a very offensive nature, and purging of a similar fluid, followed by prostration, stupor, and more or less violent delirium. The membranous exudation increases to such an extent as to impede respiration and threaten strangulation, the countenance assumes a leaden hue, and the skin becomes cold and shrivelled. In a few hours, at this crisis, if the disease is not effectually arrested, the patient is beyond all hope of recovery, and death closes the scene. If allowed to extend to the windpipe and bronchial tubes, it invariably proves fatal. The physical appearance of the membrane is similar to that thrown out in true inflammatory croup, except that it is soft and is saturated with fluids. It is of a yellowish-white, gray, or light-brown color in the mouth and tonsils, looking like gray velvet or wet chamois, but of rather lighter color in the windpipe and bronchi.

TREATMENT.

In all cases of diphtheria, no matter how mild it may appear, the patient should at once be placed under the care of a skilful physician. (*See Remedies Nos.* 61, 193, 194, 195.)

Diet and Regimen. — The main feature in diphtheria being debility, the most watchful care is necessary.

In the first or fever stage, stimulants would be highly injurious; but so soon as the prostration begins to show itself, a sustaining regimen and good nourishing diet, judiciously combined with stimulants, should be adopted. The amount of both stimulants and diet will, of course, depend entirely upon the nervous and general condition of the patient. In the majority of instances, beef-tea will be found most suitable to their requirements. It should be made palatable by seasoning, and be given by the spoonful every few minutes or half-hour. In cases of extreme prostration, it should be given by enema. Clam-broth, the soft parts of oysters, port wine, champagne, eggs beaten up, brandy in small quantities, are all of great benefit. As a beverage, when the patient is thirsty, barley-water, or toast-water, flavored with lemon-juice, or cold water, with raspberry or strawberry sirup added, will prove grateful.

NAUSEA, VOMITING, AND REGURGITATION OF MILK.

Owing to the imperfect development of the infant's stomach, this affection is common. It is usually a simple act of nature, ridding the stomach of any excess of food received by it. Older children also have these spells, but they are generally the consequences of visits to the apple-orchard, the candy-store, or some place for the sale of cheap and nasty " indigestibilities." This kind of vomiting always affords relief and proves beneficial. Sometimes, however, vomiting arises from other causes, and instead of only a *portion*, the *whole* of the food is thrown up, accompanied by mucus and bile. This is, of course, far from salutary, and needs immediate attention. (*See Remedies Nos.* 63, 331, 332, 333, 334.)

BILIOUSNESS.

This ailment is so vaguely comprehended and generally misunderstood that considerably more permanent mischief is done by the *remedy,* in a great many instances, than by the disease itself. If a child loses its appetite, has a cough, sickness at the stomach, dizziness, or headache, it is declared to be bilious, and is accordingly drenched with powerful emetics, until the whole digestive apparatus is temporarily, if not permanently, injured. A more disgusting and injurious course of procedure cannot possibly be imagined in ordinary cases. There are instances, as in the ejection of any foreign substances from the stomach, in which they are both useful and necessary, but these are, or rather ought to be, exceptions to the rule.

Symptoms. — The patient appears dull and languid; headache, giddiness, great oppression, fulness at the pit of the stomach, nausea, vomiting, offensive belching, smelling like stale meat or rotten eggs; the tongue is covered with a thick, slimy, yellowish coating; there is a disagreeable, bitter, putrid, slimy taste in the mouth, especially in the morning; the bowels are either constipated or quite loose; passages dark, very offensive, and accompanied with much wind; eyes dull and heavy, of a yellowish cast, and the skin, particularly about the mouth and nose, of the same hue. (*See Remedy No. 64.*)

Diet and Regimen. — Meats and soup strictly forbidden; nothing but gruel, oatmeal cakes, dry toast, milk toast, crackers, plain bread, without butter, oranges, and cold water. Even lemon-juice should not be taken if diarrhœa is present.

OFFENSIVE BREATH.

This unpleasant affection arises from one of several causes: decayed teeth, inflammation or other disorder of the gums, ulcers in the mouth, or from want of careful attention to cleanliness, allowing particles of food to collect and remain between and around the roots of the teeth, or the accumulation of tartar. The correction of such matters belongs naturally to the dentist. The mouth and throat should be rinsed with cold water twice or thrice every day, and the teeth thoroughly brushed with a soft brush after *every* meal. When offensive breath arises from a deranged stomach or from other diseases, the proper treatment will be found under the head of such disorders. In other cases, where it is the chief symptom, and its origin can be traced to no apparent or perceptible cause, Remedies Nos. 65, 286, may be employed.

WIND COLIC, OR COLIC OF INFANTS.

All severe pains in the abdomen not dependent upon inflammation are called *colic;* when its principal symptoms are sharp and griping pains, it is called *spasmodic colic;* when accompanied with nausea and vomiting, it is called *bilious colic;* when the abdomen is distended, and relief is afforded by the passage of wind, it is called *wind colic.*

Causes. — It arises from cold, sudden or violent emotion of the mother, improper food, or a confined state of the bowels.

Symptoms. — Disturbed sleep, rolling of the eyes, distortion of the features, drawing up of the knees, abdomen tense and swollen, with rumbling in the bowels. Severe attacks, unless speedily relieved, may end in spasms or convulsions.

TREATMENT. — (*See Remedies Nos.* 66, 293.)

Most cases of colic are attended with constipation ; a free evacuation of the bowels often gives instant relief. It is imperative, therefore, that a movement should be effected as soon as possible. The most efficient way to obtain this is an injection of tepid water with a little salt mixed in it, continuing it until the desired effect is produced. Hot applications to the abdomen should always be made use of.

CHOLERA MORBUS.

This disease is characterized by great anxiety, painful and violent gripings, copious and frequent vomiting and purging, and coldness and cramps in the extremities. The griping pain evidently proceeds from violent spasmodic contraction of the alimentary canal, causing the repeated and frequent ejection of its contents by vomiting and purging.

Causes. — Intense heats of summer, especially when the days are hot and the evenings cool, with heavy dews, sudden atmospheric changes, cold drinks when the body is overheated, and the incautious use of ice, sudden suppression of habitual discharges, diarrhœa, cutaneous eruptions, vexation, fits of anger, errors in diet, partaking of unhealthy food, or of an improper quality or quantity, unripe or indigestible fruits, particularly melons, cucumbers, pineapples, green apples, or poisonous and irritating food of any kind. Large doses of cathartic drugs not unfrequently produce it.

Symptoms. — The attacks are generally sudden, and without premonitory symptoms. The patient has vomiting and purging, severe griping pains in the bowels and stomach, great anxiety and restlessness. The discharges from the bowels, at first fecal, soon become watery, bilious matter,

each evacuation preceded and accompanied by violent burning and cutting, colicky pains, especially in the region of
the navel, extending in severe cases to the arms and hands,
with pinched features, sunken eyes, cold and clammy skin,
and general depression. The substance vomited is the contents of the stomach, largely mixed with bilious matter;
and afterward a watery fluid; the gagging and retching is
continuous.

TREATMENT. — (Remedies Nos. 67, 162, 163, 164.)

CHOLERA INFANTUM, CATARRH OF THE INTESTINES, OR SUMMER COMPLAINT.

The chief seat of this affection is in the large and lower
part of the small intestines, seldom extending to the
stomach. In ordinary and mild cases, it is simple catarrh
or irritation, arising from teething or improper diet. It is
seldom met with at any other time of year than June, July,
and August. It is far more prevalent in cities and in
northern and eastern climates than in country districts, or
in southern and western territory, so that it is evident that
heat alone will not produce it unless allied with close,
unwholesome air, want of cleanliness, and neglect of sanitary precautions. The most prolific causes, however, are,
as we have said, teething and unsuitable diet. It is
astonishing how reckless parents are in reference to the
diet of their children; overfeeding, unripe fruits, rich and
luscious dishes, and indigestible, poisonous messes of all
kinds are placed within their reach, and, consequently, the
children are susceptible to its attacks at all times and seasons. In many cases there would seem to be an hereditary
predisposition to the affection, especially in those families
where the constitutions of the children are feeble and deli-

cate, of a nervous, irritable tendency, or derive scrofulous or consumptive tendencies from their parents.

Symptoms. —This disease is both sudden and gradual in its attacks. A child, apparently in good health, may be suddenly attacked with diarrhœa, vomiting, great exhaustion, anxious and contracted countenance, coldness and paleness of the skin, similar to the cholera of adults. Usually, however, the mode of attack is gradual, commencing with a diarrhœa, which soon proves obstinate and exhausting, nausea and vomiting, preceded by feverish restlessness. The mother lays it to the teeth, and calls the child cross when in reality it is sick. The dejections become more frequent and abundant than natural, spotted and streaked with green, looking like chopped-up greens or spinach, and mixed up with particles of undigested food. Occasionally they contain blood and mucus. Evacuations are accompanied with more or less pain, severe straining and bearing down. The frequency and severity of the vomiting depend upon the violence of the attack. The tongue is coated with a dirty white or yellowish-brown fur, the edges and tip being red. The thirst is intense. There is always more or less fever of a remittent type, the abdomen hot, distended, and tense, the head hot, and the extremities cold. The emaciation is rapid, and if not speedily checked, the child soon has all the appearance of an aged person. The duration of the disease almost entirely depends upon the treatment, its course varying between six weeks and six months, though frequently the patient's fate is decided in twenty-four to forty-eight hours.

TREATMENT. — (Remedies Nos. 68, 160, 161.)

Diet and Regimen. —If the patient be an infant at the breast, and the mother has enough for it, no change should be made. If the child be older, the diet must depend entirely on circumstances. *Overloading* the stomach should be especially avoided, as that would endanger the child's life. The amount should be restricted to the smallest possible quantity. The motto should be, " Little and often." Fresh cow's milk should form the chief ingredient. It should be diluted with about one-third water, boiled for ten or fifteen minutes, and moderately sweetened with loaf-sugar. This may be alternated with rice-flour, arrow-root, sago, tapioca, or wheat-flour. In cases where there is excessive vomiting, a little gum-water or arrow-root or rice-water may be given until it ceases. Fresh air, in the country or at the sea-side, is as important as good diet ; tepid or sponge baths, and a dress suited to the weather, and changed to suit the alterations in the temperature, care being taken not to clothe the child too warmly.

DYSPEPSIA, OR INDIGESTION.

The term Dyspepsia means any condition of the stomach in which the function of digestion is disturbed or suspended, causing want of appetite, distention of the stomach, eructations of various kinds, heartburn, water-brash, pain in the region of the stomach, uneasiness after eating, occasional vomiting, constipation or diarrhœa, with an endless string of nervous symptoms. Indigestion may be, and doubtless is, simply debility, a defect of muscular power in the stomach, or a want of vital power and strength.

Indigestion, though not confined to any period of life, is most common between the ages of twenty and forty-five, and is more frequent in females than in males. The upper and middle classes are most subject to it. The predisposition to it is sometimes hereditary, particularly in persons of a weak, relaxed system, with highly nervous susceptibility and general debility of constitution. Sedentary occupations, indolence, long and intense study, insufficient exercise, breathing impure air, essentially predispose to this complaint. The principal exciting cause of indigestion is imperfect mastication. The fact is, we, as a nation, have not time to eat; business or pleasure is too pressing. From childhood to old age we are in the habit of " bolting" our food, as if our teeth were in our stomach, and we could masticate it at our pleasure, like a cow. Children's stomachs are unable to digest solid lumps or tough masses of food, and whatever passes through them undissolved receives but little digestive aid from the stomach. Another very frequent cause of dyspepsia in this country is the excessive use of cathartic medicines in the shape of pills. In addition to this, great quantities of bitters are used, which exhaust the powers of the stomach and produce numberless functional and structural derangements; so that, were the truth to be told, the epitaph on the tombs of nine-tenths of our pill-takers would be : —

I was well; wished to be better; took physic, and here I am.

Indigestion in infants is frequently caused by the unhealthy condition of the nurse's milk. In all cases, from the infant to the adult, the tongue is generally pale, flabby, or slimy, dry, or loaded with a thick coating, especially on rising in the morning. There is generally headache, languor, and mental depression, nausea and vomiting, the last symptom usually affording relief. One form of pain is usually called *heartburn.* Another frequent symptom is *cramps* or *spasms* in the stomach, accompanied by belching and flatulence, which is occasioned by the generation of gases in the abdominal cavity. It is almost always allied to a sluggish state of the bowels. The evacuations are commonly dry, scanty, and deficient in healthy color and odor. In females, dyspepsia not unfrequently occasions difficult, too frequent, delayed, or irregular menstruation, leucorrhœa, chlorosis, hysteria, and painful affections of the spinal nerves, with tenderness and soreness of the back.

TREATMENT. — (Remedies Nos. 69, 202, 203, 204.)

Diet and Regimen. — Good cooking is a matter of the first importance to the dyspeptic. They should avoid all cured, salt, smoked, or pickled meats, sausages, etc., raw vegetables, salads, cucumbers, pickles, etc. Fresh meats, poultry, fruit, and well-cooked vegetables may be taken in moderation, if thoroughly masticated and partaken of at regular hours. Leave off eating as soon as you are satisfied. At least six hours should elapse between each meal. There should be total abstinence from all astringent or alcoholic drinks. Pure spring water is the best possible drink. *Ice-cold water* is injurious.

CONSTIPATION.

In infants it usually arises from an improper mode of living on the part of nurse or child. Those fed on artificial diet are especially liable to it. In adults and older children it is the consequence of unsuitable diet, the use of stimulating and astringent drinks, too long indulgence in sleep, inattention to desire for evacuation, sedentary habits, impaired condition of the digestive function, and excessive use of aperient medicine.

Symptoms. — The tongue is coated at the root and sides, tip red, urine high-colored, slow pulse, quicker after meals; sallowness of countenance and skin; more or less uneasiness and distention about the lower part of the abdomen, much flatulence, and always more or less headache.

TREATMENT. — (*See Remedies Nos.* 70, 179, 180, 181, 182, 183.)

Diet and Regimen. — The cure of constipation depends more on a proper mode of living than on medicine. Plenty of out-door and manual exercise ; the avoidance of all food of a binding nature, as salt meats, cheese, wheaten flour in any shape, stimulating drinks, high-seasoned dishes ; a liberal allowance of all kinds of fruits and vegetables, soups, coarse bread, etc., a free use of cold water, and special attention to mastication.

DYSENTERY.

This is sometimes called bloody flux, from the fact that the evacuations are scanty, mixed with blood and mucus, and but little fecal matter. It is essentially an inflammation of the mucous lining of the large intestines, accompanied by general constitutional disturbance. The mucous membrane is swollen, thickened, red, and softened, and in severe cases ulcerated. It is not, as many suppose, an aggravated form of diarrhœa, but the very reverse, namely, constipation, with a constant desire to evacuate, caused by the inflammation. The inflamed and congested parts are tender and painful. It most frequently makes its appearance in the autumn, when the days are hot and the evenings cool. It is generally epidemic, and may be excited by cold, exposure to wet, unripe or sour fruit, stale vegetables or meat, drinking cold water when heated, from taking cold, etc.

Symptoms. — In mild cases there is little or no fever ; in severe cases, high fever, hot, dry skin, excessive thirst, etc. It often begins with diarrhœa ; blood passes in considerable quantities, either black or of a dark reddish color, resembling the washings of meat. There is severe pain and burning in the lower bowels, especially just before and after each evacuation, accompanied by a painful constriction of the anus, called tenesmus. Nausea, vomiting, and headache are frequently present. The disease is frequently much aggravated by the administration of cathartics.

TREATMENT. — (*See Remedies Nos.* 72, 200, 201.)

Diet and Regimen. — Care should be specially taken that the patient should not be thinly clad, nor be allowed to sit on cold stone or brick seats. He ought to lie in or upon the bed during the attack. For food, water-toast, arrow-root, sago, gruels, and the like may be taken, and, in convalescence, mutton-broth. The patient should eat little and often. Cold water, toast-water, or barley-water may be drank. All kinds of animal food and wine should be strictly avoided, even during convalescence. The water-closets should be abundantly and frequently disinfected, to prevent the spread of the epidemic.

PROLAPSUS ANI, OR FALLING OF THE BODY.

This is a protrusion or falling down of the lower part of the bowels, and though it may at first cause much unnecessary alarm, there is really nothing dangerous or serious about it. It is very common in infancy, and is frequently met with at all periods of life. It arises from a laxity of the muscles, habitual costiveness, straining at stool, diarrhœa, hemorrhoids, drastic purgatives, worms, and other causes.

TREATMENT.

The first thing to be done is to replace the protruded membrane, which should be accomplished as speedily as possible. If it does not return of its own accord, then, after protecting the protruded parts by laying over them a piece of soft, smooth cloth, wet with warm water or sweet-oil, embrace it with the ends of the fingers, and gently and steadily press it upward, not using a great deal of force, until it slips in, which it will do in a minute or two if the operation is rightly performed. If it has become red, swollen, or inflamed, do not be in a hurry to reduce it, but place upon it rags saturated with a weak solution of arnica-water. As soon as the inflammation subsides, the bowel may be returned. When once returned, great care should be taken to prevent a repetition of the trouble. The child should be accustomed to use the chamber at regular intervals; and should be watched to prevent its overstraining while sitting, or remaining on it too long, particularly if the bowels are in any way constipated. Cold hip-baths or sponging with cold water, and sometimes cold-water injections, are of great service. The temperature of the water should be graduated according to the age and vigor of the child. (*See Remedy No.* 73.)

Diet. — The diet should be the same as that observed in derangements of the digestive organs in general. If possible the diet should be so governed as to prevent either constipation or diarrhœa. The child may be allowed as much cold water as it wants to drink.

RUPTURE, OR HERNIA.

By this we understand a swelling formed by the protrusion or escape of a portion of the intestine from the cavity of the abdomen. The places at which these swellings generally make their appearance are the navel and the region of the groin. The point of egress selected by the hernia gives it a peculiar name to express its position: as umbilical, when it appears at the umbilicus or navel; inguinal, when it appears in the groin. Three descriptions of hernia only are especially met with in children, namely, umbilical, inguinal, and oblique inguinal. The latter variety is where the intestines have intruded into the scrotum.

Hernia is termed *reducible* when it can at any time be returned into the abdomen, and *irreducible* when it cannot be returned to the cavity of the abdomen without inflammation or obstruction to the passages of fæces, either owing to adhesions or entanglements of the intestines; *strangulated*, when the protrusion is not only incapable of being reduced, from constriction of the aperture through which it passed, but the circulation is arrested, the passages of fæces towards the anus cut off; inflammation sets in, the tumor becomes hard and tender to the touch, pain, nausea and vomiting occur, accompanied by other alarming symptoms. These varieties are frequent in children of all ages.

Causes. — Children whose muscular development is not compact, but, on the contrary, relaxed and flabby, leaving the natural outlets of the abdomen unusually large, or capable of easy enlargement, are more prone to accidents of this nature than those who are robust and strong. The weakest parts are those at which the accident most frequently occurs. Crying, coughing, or straining, or great bodily exertion or external injury, may produce hernia.

Symptoms. — Umbilical hernia need not be mistaken for any other tumor. Those appearing at the groin, however, so closely resemble other diseases that mistakes may readily be made by any one else but an experienced physician. It generally shows itself as an indolent tumor upon some part of the abdomen, such as the navel or groin. The tumor appears suddenly, is developed above, and descends gradually. It is subject to changes in size, being smaller upon pressure, or when the patient lies upon his back, and larger when the pressure is removed, or when he stands upright. **Vomiting**, constipation, and colic are frequently the result of the unnatural position of the bowel.

TREATMENT.

In every case of hernia, no matter how slight or trivial it may appear, send at once for your family physician, or some experienced practitioner, and ascertain from him its precise nature and probable termination. It is of the utmost importance that a cure should be effected during childhood, otherwise the individuals will, in after years, suffer great inconvenience, be unfitted for any kind of manual labor, and may any day be in danger of losing their lives. The hernia, or swelling, thus formed varies in size from a hazel-nut to a walnut, always increasing in size under the influence of coughing, straining, or sneezing. It is not often painful, unless it becomes very large. The parts should be bathed with cold water every night and morning, and the child be kept as tranquil as possible. Alarming symptoms sometimes accompany hernial protrusions, such as violent burning in the abdomen, as from a hot coal, with tenderness of the tumor, the least touch giving pain, sickness at the stomach, with bitter bilious vomiting, nervousness, and cold perspiration. (Remedies, Nos. 74, 229.)

WORMS.

Many errors prevail in the popular mind regarding the nature, origin, and consequences of the existence of these animalculæ in the human organism. Worms were never yet the sole or originating *cause* of any disease, either in the child or adult. That worms do exist in the alimentary canal of all children is an indisputable fact, but no experienced physician will assert that any particular disease is *caused* by worms. Beyond all doubt, they serve some useful and necessary purpose in the human economy. Worms, as such, are not injurious. They exist in many children without their presence being suspected. If they were the *cause* of disease, their mere expulsion would be sufficient to remove the symptoms attributed to their presence ; but no such beneficial results follow the administration of vermifuges, although numbers of worms are killed and expelled by their use. In treating these cases it must always be borne in mind that it is not merely the worms you wish to remove, but that habit of body which favors their accumulation in such quantities as we sometimes find them. You cannot possibly get rid of them entirely. There are five different species of worms which infest the alimentary canal, but two of these are peculiar to children. The first and most troublesome is the common seat-worm; *ascaris vermicularis*, thread-worm, pin-worm, or maw-worm, as it is variously called. This is the smallest of the intestinal worms, measuring only from two to five-twelfths of an inch, and resembling a small piece of white cotton thread. They are usually found in the large intestines and rectum, but sometimes crawl into the urethra and vagina, causing a troublesome itching and a mucous discharge. The next species of worms most frequently found in children is the long, round worm, called the *ascarides lumbricoïdes*, which very much resembles the common earth-worm. The small intestines is their favorite locality, but they traverse all parts of the alimentary canal. They are sometimes found in the large intestines, from which they are expelled by stool. They are occasionally found in the stomach, and even in the throat. It is not uncommon for children to eject them by vomiting. The use of much sugar, fat, cheese, butter, fruit, or any other diet which enfeebles or disarranges the digestive system, strongly predisposes to their production. Children of a lymphatic or scrofulous constitution are more liable to them than others, and those living in dark, damp, and unclean dwellings, or in marshy regions, are specially prone to worm affections.

Symptoms. — Sudden and frequent changes in the color of the face, red, pale, or lead-colored, bluish semicircles round the lower eyelid, dilatation of the pupils, itching of the nostrils, bleeding at the nose, headache after meals, excessive flow of saliva in the mouth, dryness of the tongue, pains and enlargement of the abdomen, itching at the anus, abundant fetid stools ; in severe cases, convulsions, delirium, epileptiform attacks, etc.

TREATMENT. — (Remedies, Nos. 77, 336, 337, 338, 339, 340.)

Diet. — Avoid all gross, heavy nourishment, such as too much bread and butter, potatoes, or boiled vegetables of any kind, rich puddings, pies, cakes, and pastry in general. Give the patient meat soups, roasted or broiled meat, plenty of cold water and milk. Exercise in the open air is very essential. Cold water, vinegar and water, or lemon-juice and water injected, will allay the itching.

EPIDEMIC CHOLERA.

This is to a great extent a nervous disorder, and where dread and panic prevail, it will reap a rich harvest. Fear will at once cause the premonitory symptoms. People should understand that cholera is a disease which can generally be warded off if they but pay proper attention to known hygienic laws. It is neither difficult to manage nor is it dangerous, and, ordinarily speaking, is not contagious. The only necessary precaution is to avoid over-taxation, anxiety, long fasting, overeating, damp, ill-ventilated apartments, undue excitement, or exhausting employment, unwholesome food, and irregular or improper diet. It is the necessary accompaniment of filth, negligence, and the collection of noxious, decomposing material of any kind.

Symptoms. — During cholera seasons diarrhœa has a special tendency to run on, if not checked, into the more perilous form of the disease, and nothing is more sure and certain to hasten that catastrophe than purgative medicines. As it has hitherto appeared in the United States it has always been preceded by a well-marked premonitory stage of from one to three days' duration, such as confusion of the head, languor and debility, and derangement of the stomach. It may, however, seize upon the patient suddenly, instantly prostrating and depriving them of almost every element of vitality. A sunken and death-like expression of countenance, feeble pulse, blue, cold, and shrivelled skin, covered with a clammy sweat, cramps throughout the muscular system, with stupidity or extreme anguish, vomiting, and frequent rice-water discharges, may at once set in.

TREATMENT.

Perfect quiet and maintaining a recumbent position are absolutely necessary. Have immediate recourse to Remedies 78, 162, 163, 164.

Diet and Regimen. — As soon as the disease has spent its violence, and the patient begins to mend, he should have a little gruel, toast-bread, meat-broth, etc., frequently, and in small quantities, gradually increasing the diet, both in quality and quantity, until he finally gets back to the accustomed mode of living.

HEMORRHOIDS, OR PILES.

Women, both single and married, are very often subject to piles, and more especially during the period of pregnancy. Many have supposed this disorder to originate from obstructed circulation, but the fact seems to be that the most frequent cause which operates in its production is habitual constipation of the bowels, a common affection among females of all classes. If this is avoided in the way we have pointed out in the article on Constipation, much suffering and inconvenience will be prevented; but if it is permitted to exist, and temporary relief only sought by an occasional cathartic, the disease will become chronic, and possibly remain throughout the remainder of the patient's life. The inexperienced can scarcely imagine the amount of suffering some females undergo from piles; and the pain is constant, day in and day out. Various external applications have been devised for their removal; even the knife has been resorted to. Against all these we would warn you, as they are not only exceedingly painful, but, during pregnancy especially, highly dangerous. It is very important that a pregnant woman, and especially if it be her first pregnancy, should pay strict attention to the state of her bowels, not allowing either constipation or diarrhœa to set in, as early attention to either of these derangements will cause their prompt removal.

TREATMENT.

Remedies Nos. 20, 225, 226, 227, 228. In addition to the internal administration of remedies, much benefit may be obtained from a proper use of cold water. When the piles do not bleed, cold applications, either as sitz-baths, compresses, or injections, are of great benefit. As evil results sometimes follow the sudden suppression of the discharge, it is not advisable to use cold water where there is much if any bleeding. When, however, the bleeding is profuse to such an extent as to cause alarm, cold applications are the best styptic. Warm water or steam is preferable when the tumors do bleed, or when, from any cause, the bleeding has ceased, and there is considerable pain. When, after each evacuation, the bowels, or a small tumor, protrude, causing great pain, relief may be obtained by gently pressing them up again with the ball of the finger. Injections of cold water, when judiciously administered, are of the greatest value ; but more harm than good is so often done from the carelessness of introducing the syringe, that I seldom recommend them.

Diet. — As the use of condiments and stimulants of every description tends to produce gastric and intestinal derangements, it is advisable that, in this disease, they be dispensed with, and the patient confine herself strictly to the hygienic rules of diet. Meat diet should be avoided as much as possible ; some physicians even recommend their patients suffering from this complaint to eat nothing for a few days except bread and water.

CHAPTER XI.

DISEASES OF THE SKIN.

SCARLET-FEVER, OR SCARLATINA.

Considerable mystification exists in the minds of many about this complaint, imagining that scarlatina is a modified form of the disease. But the fact is that they are one and the same; it is epidemic, contagious, and febrile in character, and distinguished by a peculiar rash, which appears upon the first or second day, and by inflammation of the tonsils and mucous membrane of the mouth. The two most important and striking features are the affection of the throat and the skin; and yet either may be entirely absent, or so imperfectly marked as to attract but little attention. It is almost exclusively a disease of childhood, and seldom attacks the same individual more than once in a lifetime. It is less prevalent than measles, with which it is frequently confounded; affects both sexes equally, and usually appears between the ages of one and five. In the majority of cases scarlet-fever is contracted from the epidemic constitution of the atmosphere, and not by direct communication with other individuals or their clothing.

Scarlet-fever commences, like all eruptive diseases, with shivering and lassitude, severe headache, and occasionally delirium, nausea, and vomiting. It is generally sudden in its onset, the child going to bed apparently in its usual health, but awaking with these premonitory symptoms, the eruption showing itself in the course of a few hours over face, shoulders, neck, and breast, and extending rapidly over the entire surface of the body. It first appears in dark-red points, speedily becoming so numerous as to present a universal red blush. It is not usually diffused equally over every part of the body, but is more apparent about the groins, the back, and the flexures of the joints than elsewhere. On the arms and legs the eruption does not present the same appearance as on the trunk; instead of being a uniform smooth redness, it is more spotty and rough. In most cases the fever is attended with a burning irritation of the skin. The redness disappears under slight pressure of the finger, and returns when the pressure is removed. The eruption reaches its height about the fourth day, remains stationary for about twenty-four hours, after which it begins to decline and become indistinct, and usually disappears altogether about the seventh or eighth day. At this time the skin begins to peel off. In some mild cases the whole duration of the eruptive period is not more than two or three days, the skin presenting but a slight blush, and there being but little heat or fever. Sore throat is always present; scarcely perceptible, perhaps, but, on closer inspection, inflammatory action is plainly visible. The fever does not subside on the disappearance of the eruption; the pulse is strong and frequent, running up to 120 or even 160. The tonsils are swollen and red, and the glands of the neck are tumefied and tender to the touch. The appearance of the tongue is also peculiar. At the commencement it is covered with a thick, cream-like fur, the edges and tip sometimes being of a deep-red color. After the first two or three days the tongue clears off and becomes preternaturally red and rough, looking like raw flesh.

DIFFERENCE BETWEEN SCARLET-FEVER AND MEASLES.

Possibly scarlet-fever and measles may be confounded by those unfamiliar with eruptive diseases. The distinguishing marks between the two diseases, therefore, are : —

First. The eruption of measles is always preceded by catarrhal symptoms, such as coughing, sneezing, and running from the nose, while scarlet-fever is *not.*

Second. Scarlet-fever is always accompanied by sore throat ; measles is *not.*

Third. The rash of scarlet-fever appears on the *second* day ; that of measles, at least in its regular form, not until the *fourth.* Generally, the eruption of scarlet-fever is smooth and even to the touch, and of a uniform scarlet color ; in measles, on the contrary, the eruption consists of minute little pimples, which are felt to be slightly elevated, and firm to the touch ; besides, the eruption is not continuous, but cut up in little clusters by portions of healthy skin. In measles, the eruption is of a raspberry hue ; in scarlet-fever, it resembles that of a boiled lobster.

Scarlet-fever in any form is of too critical a character to be treated by any but a medical practitioner, for it not unfrequently happens that for one or two days the case may promise to be mild, but suddenly, and without any ascertainable cause, it may assume the threatening features of the worst form of the disease. The consequences of scarlet-fever are frequently worse than the disease itself. Children who have suffered from it are liable to fall into a state of permanent ill-health, and become a prey to some of the chronic forms of scrofula, boils, ulcers, diseases of the scalp, sores behind the ears, scrofulous swelling of the glands of the neck, chronic inflammation of the eyes and eyelids, etc. The same results sometimes follow measles and other eruptive diseases.

One of the most frequent and important sequels of this disease is *dropsy*. This dropsical effusion attacks the structure or tissue just beneath the skin, or any of the cavities of the body. When it affects the head, dropsy or water on the brain is the result. When the chest becomes the seat of the effusion, we have water on the chest. It is generally the result of cold caught during convalescence. At this time the child needs the most watchful care and attendance, and at no stage of the disease is the patient more apt to be neglected. The mother, thinking the child almost well, leaves it to the care of a friend, or older children, while she goes out, and they, not understanding the necessity of great caution, permit it to stand by an open window or door, or allow the fire to go out or the room to become chilled. The patient, from this exposure, takes cold, becomes drooping, languid, irritable, peevish, and restless, after which swelling about the face soon makes its appearance, at first so slight as to be scarcely perceptible. From the face it extends to the hands and feet, and finally to the whole surface of the body. The patient should never be allowed the free range of the house until at least four weeks have elapsed from the commencement of the disease.

MALIGNANT SCARLET-FEVER.

Scarlet-fever does not always present itself in as mild a form as we have described. In malignant and severe cases, the eruption, if it appears at all, is livid, partial, and fades early, is attended with feeble pulse, cold skin, and by typhoid depression ; sometimes the patient sinks at once, and irretrievably, under the virulence of the poison. Or, where the patient survives the first shock, as the disease progresses, a condition of the throat develops itself which frequently baffles the skill of the physician, and soon destroys the life of the patient.

TREATMENT.

The great fear in this disease, as in many others, is that the child will be too much " doctored." *Simple* scarlet-fever is fatal *only* through the officious and unnecessary administration of drugs. Hygiene, carefulness in sanitary regulation, and unceasing watchfulness are the chief elements necessary to recovery. *Cathartics* are totally out of place. When the skin begins to peel off, a few doses of *sulphur* will place the patient out of reach of danger. The irritation of the skin may be allayed by bathing the child in a weak solution of saleratus, or the application of an ointment composed of one drachm of glycerine and one ounce of ointment of rose-water. The application of cold-water bandages to the throat will usually be found efficient in preventing dropsical affusion. Great benefit may be derived from the judicious use of water, either in the shape of baths or ordinary ablutions. (Remedies, Nos. 79, 236, 237, 238.)

Diet and Regimen. —During the height of the fever the patient seldom cares for anything to eat. When the mouth is dry and parched, small quantities of thin rice-gruel, or gruel made of arrow-root, may be administered; or, if preferred, rice-water, toast and water, or cold water, flavored by raspberry or strawberry syrup. Warm drinks should not be allowed, unless especially craved, and then only sparingly. When the teeth and lips become covered with crusts or scabs, they should be carefully cleansed with tepid milk and water. Great care should be taken to keep the mouth as clean as possible, and this can only be done by constant attention. The return to a nourishing diet should be very gradual, as overtaxing the digestive organs might be productive of the most serious consequences.

The room in which a scarlet-fever patient is confined should be as large and airy as possible, well ventilated, but never fumigated. The bed should be kept sweet and clean; clothes, bandages, in fact, everything about the patient, should be removed as soon as done with. Of course, great care should be taken to guard the patient from cold. A room can be kept thoroughly ventilated without exposing the patient.

SCARLET-RASH.

This, though frequently mistaken for, is quite a different disease from, scarlet-fever. Scarlet-rash consists of small granular elevations, easily felt on passing the finger over the skin. The eruption is of a dark-red color, sometimes almost purple; the pressure of the finger leaves no white imprint, as in scarlet-fever, and there is seldom much, or indeed any, sore throat. Scarlet-rash may easily be confounded with measles, as the eruption in the two diseases is very similar. This malady is most common in summer and autumn, though it does occur at all seasons of the year. It attacks children of all ages. It is not a contagious disease, and is occasioned by gastric derangement, sudden atmospheric changes, violent exercises, the use of cold drinks while heated, and by checked perspiration.

The eruption is generally preceded by chilliness, alternating with heat, accompanied by loss of strength, heaviness and fulness of the head, restlessness, sometimes with vertigo, severe pain in the head, and even mild delirium. There is for the first few days, in connection with the above symptoms, more or less fever, heat and dryness of the skin, loss of appetite, and perhaps some gastric derangement. After these symptoms have continued for an indefinite length of time, the rash appears, sometimes upon the third or fourth day, and in its regularity and appearance nearly resembles measles. There is, however, this difference between the two : measles is attended by catarrhal symptoms, has a definite time and special succession of localities for the appearance of the eruption ; whereas scarlet-rash is not accompanied by catarrhal symptoms, and the eruption is irregular in its appearance, or occurs suddenly over the whole body. It cannot be mistaken for scarlet-fever, because it is not contagious, is not accompanied by sore throat, and is composed of irregular circular patches of a deep rose-red color. (Remedies, Nos. 80, 236, 237, 238.)

MEASLES — RUBEOLA.

This disease is characterized by inflammatory fever, catarrhal symptoms, hoarseness, dry cough, sneezing, drowsiness, and an eruption. The eruption generally appears on the fourth day, in the shape of small red dots, like flea-bites, which, as they multiply, unite together into irregular circles or horseshoe shapes, leaving the intermediate portions of skin of their natural color. These red points are slightly elevated, and can readily be felt by passing the hand over the surface. The *causes* of measles are epidemic influences and contagion by personal contact. The particular period of the disease in which its infectious power is most potent has not yet been clearly ascertained. The average period of incubation, or time required to develop the disease after exposure, is from seven to twenty days. It occurs but once to the same person.

As a general rule, the first symptoms complained of are lassitude, irritability, aching in the back and limbs, and shivering, which is soon followed by fever, thirst, and headache, irritation of the mucous membrane of the eyes, nose, mouth, and larynx. The premonitory symptoms are those of a severe cold in the head : the eyes are bloodshot, the eyelids heavy, turgid, and red, excessive sneezing, watering of the eyes, copious defluxion from the nose, soreness of the throat, and a dry, hoarse, peculiar cough, arising from the irritation and inflammation of the mucous membrane lining the throat and nasal passage. This first stage lasts generally about three days ; upon the fourth day, seldom earlier, frequently later, the eruption makes its appearance ; the rash is two or three days in coming out, beginning upon the chin, cheeks, or some other portion of the face, and extending to the neck, arms, and trunk of the body, and finally to the lower extremities. This stage lasts from twenty-four to forty-eight hours. The fever does not diminish when the eruption makes its appearance. All the symptoms are at their height, but the moment the eruption passes its highest point of intensity, the whole of the symptoms gradually subside. After the eruption has passed away, the parts which it recently occupied are left covered with a dry, small scurf, and small bran-like scales. The skin does not peel off in large flakes, as it sometimes does in scarlet-fever, but it crumbles away like dust or fine powder. This stage of desquamation, as it is called, is more indefinite in its duration than those which precede it ; but, as a general thing, it lasts six or seven days, and during this period the patient ought to receive as much care as when the disease was at its height. There are, however, frequent exceptions to this course, which we will now notice.

The severity of the measles does not depend upon the amount of eruption ; the early and plentiful appearance of the rash is, in itself, no sign that the disease will be more severe or dangerous ; on the contrary, the worst cases are those where the eruption is but partial, does not come out well, appears late, or irregular. In what is called the " black measles," the eruption comes out slowly and imperfectly, and is of a livid, purplish, or even blackish color. This is a very dangerous form of the disease; the patient may die early from exhaustion or congestion of the brain or lungs. A retrocession of the eruption is very apt to be followed by unpleasant, if not alarming, symptoms.

Sometimes measles are complicated with gastric derangements; in such cases the tongue will be found coated; there is some nausea, and, perhaps, sickness at the stomach; the eruption does not stand out as prominent as it should, and the healthy portions of the skin between patches of eruption have a yellowish tinge. Perhaps the most frequent and important complication of measles is inflammation of the lungs. Inflammation of the bowels is also a frequent complication.

Treatment. — In ordinary cases, *aconite* is the only remedy called for; the uncomplicated forms of the disease need scarcely any other treatment than a strict attention to hygiene. In all cases, no matter how mild, the patient should be confined in a large, well-ventilated room. In most cases, the patient is quite willing to lie on his bed during the first part of the disease; but as soon as the eruption begins to disappear and the fever subside, he will want to be dressed; and, when once dressed, he will think it strange that he cannot go out, especially if he feels quite well; however, he should not leave the room, and certainly not the house, until he has regained his accustomed healthful look. It has always been the custom to shut a measles patient in a hot room, and allow him nothing but hot drinks. This is a most pernicious habit, and has no doubt led to a great many serious and even fatal results. The patient should never be allowed *hot drinks*, and especially those which are recommended to throw out the eruption. If he is thirsty, give him *cold water*, as much as he wants. It is the most palatable, and by far the best drink you can procure. I have seen the happiest results brought about by its free use. In those cases where the eruption is backward in coming out, give the patient a glass of cold water and cover him up warm in bed. This is especially advisable where the fever is violent and the heat of the skin very great. (Remedies Nos. 81, 267, 268, 269, 270.)

The diet during the febrile stage should be very light. The patient usually will ask for but little, but that little should consist of thin wheat or rice flour gruel, barley-water, toast-water, milk and water, tapioca, crackers soaked in water, or some similar food. When the fever begins to abate, the allowance may be increased to plain or toast bread, bread-pudding, or some light broth, either animal or vegetable, and even a small quantity of chicken or beefsteak once a day until the strength is regained, when the usual diet can be resumed. By observing these rules strictly all trouble will be avoided.

Disorders consequent upon measles are frequently even more dangerous than the primary affection. Running at the ears, inflammation and swelling of the glands, especially about the neck, are apt to occur. This is frequently the case in scrofulous children.

NETTLE-RASH, HIVES, URTICARIA.

This disease, called by each of these names, is a non-contagious eruptive disease, characterized by little, hard elevations upon the skin, of uncertain size and shape, and generally of a red color with a whitish tinge. Sometimes there is little or no redness, and the elevated parts are even paler than the surface around them; more frequently, however, the elevated spots are partly red and partly white. The eruption, on making its appearance, is attended with intense heat, tingling and itching, a sensation much like that produced by the sting of the nettle, from which it takes its name.

Causes. — Some people have a constitutional predisposition to this disease, and the slightest error in diet, or the most trivial functional derangement of the digestive apparatus, is sufficient to bring on an attack. Children possessing a fine, delicate skin are particularly predisposed to attacks of hives ; in such, slight gastric affection, a warm day, excessive clothing, dentition, or almost any little disturbance, will produce it.

Symptoms. — As a general thing the disorder in children manifests itself without any premonitory symptoms. The eruption is attended with heat, burning and itching, the blotches continually changing from one position to another, or disappearing in a few hours on one part and appearing on another. The most frequent form of the disease which we meet with in small children consists in large inflamed blotches, of an irregular shape, being either round or oblong, appearing suddenly, and preceded by very slight, if any, constitutional symptoms. The blotches are of a bright red color, excepting the slightly elevated centre, which is white. The form of the disease is not dangerous, but very annoying, and occasions great irritability and crying. The eruption most commonly makes its appearance about the face, the upper part of the arms, thighs, and buttocks.

In some cases, especially with older children, the eruption is preceded by headache, bitter taste in the mouth, coated tongue, nausea, vomiting, and fever. This is particularly the case in that form of the rash which is induced by errors in diet and exposure to cold. Another form of the disease which is preceded for a few hours or a few days by feverishness, headache, nausea, chilliness, and languor, is where the blotches become reddish and solid elevations, either round or oblong, often called wheals. They resemble as much as anything the ridges caused by the stroke of a whiplash. This eruption, like the other forms, is attended with violent itching and burning. During the attack the patient is usually more or less feverish, and suffers from headache, languor, loss of appetite, and other signs of gastric derangement.

TREATMENT.

Aconite, internally, and *myro-petroleum album*, externally, in the form of a soap, will be found especially valuable and effectual in subduing it and accomplishing a thorough cure *of this and all other* skin diseases. Every effort should be made to promote perspiration by covering the patient well and giving him plenty of cold water to drink. The same diet and regimen should be adopted as for measles. (Remedies Nos. 82, 236, 237, 238.)

ERYSIPELAS, OR ST. ANTHONY'S FIRE.

Erysipelas is a non-contagious disease, characterized by a deep, red rash, or superficial inflammation of the skin, which has the peculiarity of spreading from place to place, the part first attacked recovering while the neighboring parts are becoming affected. Erysipelas is rarely experienced during childhood. The few cases that I have seen arose indirectly from vaccination, the vaccine virus and the local irritation produced by it bringing into activity a disease the seeds of which already existed in the system.

The causes of erysipelas are obscure; slight points of irritation upon the skin may form a nucleus from which the erysipelatous inflammation may spread, but these certainly cannot be the *real cause*. There must be a general epidemic constitution of the air at times, in certain localities or districts, which predisposes to the disease, or else there is an hereditary taint in the system.

Symptoms. — Generally there are but few, if any, marked premonitory constitutional symptoms, the appearance of the eruption being the first indication of the disease, after which we soon have fever, heat, dryness of the skin, and thirst. The inflamed surface is at first of a bright red and shining appearance, but it soon assumes a purplish hue ; and as this change takes place, the parts become tense, hard to the touch, and more or less swollen and painful. The color disappears under pressure of the finger, but returns as soon as the pressure is removed. When the inflammation once begins, if not soon arrested there is no knowing where it will end. When it starts upon the face, it may extend to the scalp and cover its whole surface ; or when commencing upon the arm, it may extend down to the fingers or up to the shoulder, and from there over the whole trunk of the body. For this disease, it is always best, when possible, to consult a good, skilful physician.

TREATMENT. — Remedies Nos. 83, 206, 207, 208, 209.

Diet and Regimen. — The same as for any other febrile disease, *measles*, or *scarlet fever*. To allay the itching, which is sometimes intolerable, dust the parts over with powdered starch, or, which is better, wash with a solution of myro-petroleum soap. Wet or greasy applications of every description should be specially avoided, as they *always* aggravate the disease.

ITCH, PSORA, OR SCABIES.

This is a contagious eruptive disease, characterized by more or less numerous distinct pointed vesicles, transparent at the summit, and filled with a viscid, serous fluid, while, from the base of each vesicle small red lines usually run off. It is comparatively rare in America, though it is prevalent among the poorer classes in Europe ; and is contracted only by actual contact. It is generally the result of want of cleanliness. These little vesicles which rise upon the skin are caused by the presence of a small insect called " acarus scabæi." The zigzag track which the mite makes in burrowing beneath the scarf-skin to deposit its eggs can readily be seen, but not so the mite itself, for it is very small, and only discoverable by a powerful microscope.

As a general thing the eruption first appears upon the wrists and between the fingers, and extends more or less rapidly over the whole body, except the face. It is frequently, however, confined to the hands, fingers, and the joints. The number of these vesicles is variable ; in some cases they are very abundant, while in others they are few, and confined to the flexures of the joints. At first they are of a pinkish color, and contain a drop of sticky transparent serum ; these soon becomes broken by the clothes or fingers, or burst spontaneously, and form their scabs. The disease is always attended by severe itching, the most prominent and distressing feature of the affection. It is most troublesome at night, being increased by the warmth of the bedclothes.

TREATMENT.

Sulphur ointment, or, what is much better, paraffine soap in solution, well rubbed into the skin before a fire, night and morning, for two days, will eradicate the disease. During this treatment, the patient should wear a flannel gown and keep his bed. On the third day the skin should be washed off with soap and water. Should the first attempt not succeed in removing the trouble, repeat it. The disease scarcely requires any constitutional treatment. (Remedies Nos. 84, 247, 248, 249, 250.)

ITCHING OF THE SKIN.

Simple itching of the skin is scarcely a disease of itself, but rather a symptom of some disease; and, indefinite though it is, it may direct us in the selection of a remedy for the morbid condition which gives rise to the irritation. For itching produced by mosquito bites, camphor is a specific, applied externally. (Remedies Nos. 85, 247, 248, 249, 250.)

HERPES, OR TETTER; ZOSTER, OR SHINGLES; CIR-CINATUS, OR RINGWORM.

Herpes is a contagious, non-eruptive disease, characterized by an assemblage of numerous little vesicles or watery pimples in clusters. These patches are surrounded by more or less inflammation, or rather the vesicles are situated on an inflamed surface, and are separated from each other by portions of perfectly healthy skin. The fluid in each vesicle, at first transparent and colorless, soon becomes milky and opaque, and in the course of eight or ten days is entirely absorbed, or concretes into furfuraceous, bran-like scales. The most common varieties among children are *shingles* and *ringworm*.

The causes of skin diseases are obscure and uncertain; but decidedly the most frequent and appreciable are want of cleanliness, disturbance of the digestive function, bilious disorders, sudden transitions of temperature, suppressed perspiration, irregularity in diet, and local irritants.

The characteristic feature of ringworm is the peculiar arrangement of the vesicles in small circular rings. The first indication of its presence is the more or less vivid redness of the skin at the point affected. This inflammation is rapidly filled in with vesicles. The circular patches vary considerably in size, from that of a ten-cent piece to two or three inches in diameter. When small, the whole surface of the patch is inflamed, the centre being of a lighter shade than the circumference. When larger the circumference alone is red, the centre retaining the natural color of the skin. These eruptive patches or rings may appear upon any part of the body, but are most frequent upon the upper extremities and neck.

Shingles is an uncommon variety of the disease, the eruption appearing in the form of a half-zone or belt surrounding the body. Old ladies will tell you that if the two ends of this belt should meet, that is, extend clear round the body, the child will die ; but as this *never* happens, it need not alarm you. The most frequent seat of shingles is at the waist, the belt seldom extending more than half-way around the body. It is preceded by constitutional symptoms, more or less severe, such as languor, loss of appetite, rigors, headache, sickness, and fever. The local symptoms are pungent and burning pain at the points where the eruption makes its appearance. It is variable in duration, is an acute disease, and seldom lasts over eight or ten days. Sometimes the rings appear, and in a short time fade away, only to reappear in some other part of the body ; and thus, by the formation of successive rings or patches, the disease is continued for three or four weeks.

TREATMENT.

Ringworm usually yields readily to the action of sepia, and, even in the severest cases, will succumb to the administration of some of the preparations of refined petroleum. The only external application called for is a solution of the remedy which you are giving internally. (Remedies Nos. 86, 248, 249, 250, 251.)

Diet and Regimen. — As the complaint often arises from gastric derangement, particular care should be taken as to the patient's diet. Avoid all highly-seasoned food, all rich dishes, all irritating substances ; in a word, place the child upon a plain, farinaceous diet. The skin should be kept perfectly clean ; avoid all irritating or scented soaps, and be careful to have the clothes so adjusted that they will not rub and irritate the eruptive patches.

PRICKLY HEAT.

During the heat of summer, adults, infants and young children are frequently much annoyed with an eruption consisting of small papulæ, or pimples, few of them being larger than a pin's head, scattered more or less thickly over the affected surface. The pimples are about the size of a pin's head, and are of a red color, more or less bright, according to the intensity of the eruption. As a general thing, the skin between the papulæ retains its natural appearance. The eruption is most abundant on those parts covered by the dress ; its development is undoubtedly favored by warm rooms and excess of clothing. It will usually be found more copious about the neck, the upper part of the chest, and on the arms and legs. More or less fever usually accompanies the affection, and the intolerable itching of the parts causes much fretfulness and a desire to scratch. In the infant, there is considerable restlessness, worrying, and disturbance of the sleep.

TREATMENT.

In most cases, scarcely any treatment is called for. The eruption is rather beneficial than otherwise, so far as the health is concerned. It is a very bad practice to apply anything having a tendency to repel it; it is only when the heat and itching is intolerable, that any attention should be paid to it, and even then nothing beyond a cooling and cleansing wash or lotion should be applied. Great comfort and benefit will be obtained by frequent bathing. Sponging the child off two or three times a day with bran-water, slippery-elm water, or other mucilaginous water, will often allay the irritation and afford considerable relief. (Remedies Nos. 87, 247, 248, 249, 250.)

STROPHULUS—RED GUM, WHITE GUM, TOOTH-RASH.

These eruptions are most common during dentition. They are caused by disturbances of the digestive apparatus, are never attended with danger; and as they are about the only *pimply* eruption to which young infants are subject, there is no difficulty in distinguishing them. *Red Gum.* — The papulæ or pimples in this variety rise sensibly above the level of the skin, are of a vivid red color, and scattered here and there over different parts of the body, but more generally over the cheeks, forearms, and back of the hands. Red gum occurs chiefly within the first two months of lactation. The eruption remains on the skin for one or two weeks, the pimples disappearing and reappearing in successive crops. It usually terminates in the peeling off of the skin. *White Gum* runs the same course, only differing in the color of the pimples. *Tooth-rash.* — In this variety, the pimples are much smaller, more numerous, and set more closely together than in the others; their color is not so vivid, but they are generally more prominent, and constitute a more severe disorder. The eruption appearing generally during dentition, has, for this reason, been called "tooth-rash."

TREATMENT.

As a general rule, it is hardly worth while to prescribe for either of these complaints. A gentle *aperient* is all that is needed. Cleanliness and attention to the dress is the chief necessity in all such affections. (Nos. 88, 247, 248.)

CHICKEN-POX.

Chicken-pox, or Varicella, as it is technically called, is a contagious, eruptive disease, febrile, and characterized by more or less numerous transparent vesicles or little bladders, which appear first as a small red dot, and gradually change into a bladder about the size of a small pea, containing a watery or milky fluid. Chicken-pox was at one time considered a modification or variety of small-pox, but experience has proved that it is not in the remotest degree related to varioloid or small-pox. It is propagated by contagion, and by epidemic influences.

Symptoms. — The constitutional symptoms are only trifling. The preliminary indications are chills followed by heat, hurried pulse, loss of appetite, nausea, and sometimes vomiting; after which the eruption makes its appearance, but without that regularity which marks variola. It is first observed upon the back or face more frequently than on other parts of the body, though it may appear on any part. The eruption appears in the form of small red pimples, which, in the course of a few hours, show small transparent vesicles in their centre. About the second day they change into globular bladders, the size of a small pea, filled with a transparent orange-colored or colorless fluid. Generally they are not numerous, and all scattered over the body. Sometimes we find them crowded together, even running into each other. On the fourth day they begin to shrink and turn into a thin, brownish, horny scurf, which falls off in two or three days, leaving only a faint red spot behind, which soon disappears. The eruption is usually accompanied by a sensation of heat and itching, which is the occasion of a great deal of uneasiness. The child rubs and scratches those vesicles that are within reach, thereby breaking and preventing them from running the regular course above described.

TREATMENT.

Unless complicated, this disease requires but little treatment beyond attention to diet and the avoidance of cold during convalescence. Poor people let their children, during the whole course of the disease, run about the streets the same as ever, and they recover. The diet and regimen should be the same as in measles. (Remedies Nos. 89, 185.)

VARIOLA AND VARIOLOID.

Small-pox is an epidemic and contagious eruptive febrile disease, characterized by an initial fever, which, upon the third or fourth day, is followed by an eruption of red pimples. In the course of two or three days these pimples are gradually changed into small vesicles, which contain a drop of transparent fluid. From the fourth to the sixth day these again change into pustules, for the suppurative process now commences, converting the serum or transparent fluid contained in these vesicles into pus or matter, after which the pustules dry up and are converted into scabs, which fall off between the fifteenth and twentieth day.

Owing to the attention now everywhere given to vaccination, small-pox is comparatively a rare disease in children belonging to the upper and middle classes of society; but as, among the careless and the poorer clases, vaccination is sometimes neglected, the disease will occasionally break out, and one case is enough to alarm the whole neighborhood. It is as well that all should understand its nature and appropriate treatment. The principal cause of the disease is *contagion*, the propagation by epidemic influence being a matter of very considerable doubt. At what particular period of its course the disease acquires its power of infection has not been precisely ascertained; and as it is always best to err, if err we must, on the safe side, it is advisable to avoid the patient and his house from the moment the real nature of the disease becomes apparent. The period of incubation, after exposure to the disease, before the first symptoms manifest themselves, varies from nine to twelve or fourteen days. Like scarlet-fever, one attack protects the constitution, in the majority of cases, against subsequent contagion.

Symptoms and Treatment. — The disease has been divided into four stages, which we will proceed to describe, and give the treatment appropriate to each as we go along. The *first* or *febrile stage* commences, as we have said, from nine to twelve or fourteen days after exposure to the contagion. The patient first complains of pains in the bones and loins, similar to, and indeed often mistaken for, those of a common cold, or he may be taken with a more or less severe chill, accompanied with headache and fever, dry, hot skin, and great thirst. Nausea and vomiting often exist from the beginning of the attack; there are at the same time loss of appetite, oppression in the stomach, and constipation, more or less obstinate; tongue red and dry. The principal symptom during this stage of the disease is the pain in the loins, which, though varying much in degree, is always severe. In some cases, the head symptoms are especially marked, consisting of restlessness and irritability ; light hurts the eyes ; there is swimming in the head ; the mind wanders ; the patient is flighty, and occasionally there are convulsions. These symptoms continue up to the time the eruption makes its appearance, which is usually in from forty-eight to seventy-two hours. (*See Remedies Nos.* 90, 302, 303, 304, 305.)

Second or *eruptive stage.* — Some time in the course of the third day, after the patient is first stricken with fever, the eruption begins to make its appearance in the shape of small red pimples, of the size of pin-heads ; as the eruption comes out, the fever subsides. This pimply eruption first shows itself upon the face, and then extends to the neck, trunk, and limbs. This stage of the disease lasts about three days, during which time the papulæ or pimples gradually increase in size, and are changed into vesicles, or little pouches, filled with a transparent fluid. At the same time the eruption appears upon the skin we have something corresponding to it affecting the mucous membrane of the mouth, throat, and nose. Sometimes there is severe inflammation of the throat, with tenderness and swelling of the glands about the neck.

Third or *suppurative stage.* — At this stage, the eruption changes from vesicular to pustular — the fluid changes from serum to pus or matter. This change takes place from the fourth to the sixth day of the eruption, or the eighth or ninth day of the disease. During this stage the pustule completes its development, the pock becoming distended, and as large as a split bean. During the filling up of the pock the face swells, often to such a degree that the eyes are completely closed. As the eruption occupies about three days in coming out, those pustules which appeared first upon the face are quite in advance of those which appeared last upon the extremities. In fact, while those upon the face are in the third stage, those upon the breast are only in the second stage, and those upon the extremities are in the first, or just making their appearance. Without this division of the burden, the disease certainly would be unbearable. The *treatment* adapted to this disease depends greatly upon the condition of the patient at the time of its arrival. If there are no alarming symptoms, if the fever which is reproduced during this time is not severe, if the color of the skin between the pustules is not of a livid hue, the remedies which have already been given to the patient may be continued.

Fourth, or *stage of desquamation.* — This is the stage of decline. At about the eighth day of the eruption, a small dark spot makes its appearance on the top of each distended pustule. At this point the pock bursts, a portion of the matter oozes out, and the pustule dries up into a scab. This, however, is not always the case ; sometimes the dark point formed upon the apex extends itself until the whole pustule is converted into a hard crust. The formation of crusts begins upon the face, and extends thence to the trunk and extremities. When, at length, these crusts fall off, the appearance of the skin beneath is peculiar : there is left a purplish-red stain, which gradually fades away, or else, in severe cases, where there has been true ulceration of the skin beneath, there is a depressed scar, or, as it is said, the patient is " pitted." Desquamation, or the falling off of the crusts, does not reach the limbs until about three or four days after it has commenced on the face.

The above description refers only to the regular and favorable course of the disease ; where the pustules are not so numerous as to run together, it is called the *distinct*, in contradistinction to the *confluent*, or that severe form where the pustules are numerous, come in contact, and, running together, form one immense scab, covering the whole surface, the latter being necessarily more severe and dangerous than the former. The *treatment* for this stage is very simple, scarcely anything is called for except cleanliness. Simple ablution with tepid water will generally be all that is required. At the beginning of this stage it is as well to give an occasional dose of sulphur.

Diet and Regimen. —The room in which the patient is confined should be as large and airy as possible ; it should be kept at a moderate temperature, well ventilated, and almost dark. A straw-bed or mattress is preferable to a feather-bed. The diet should be cooling, such, for instance, as water, ice-cream, lemonade, oranges, roasted apples, stewed prunes, strawberries, gruel, toast, etc. Avoid the fruits and acids if diarrhœa should be present. Animal food should not be used until convalescence is pretty well established.

VARIOLOID.

This is simply a modified form of small-pox. The treatment which has been given for that answers equally well for this disease.

VACCINATION.

As a preventive against small-pox, vaccination is favorably known, and practised by all civilized nations. Many persons object to vaccination, for fear that by this means some other disease may be introduced into the system. To avoid this, seek the aid of a physician whose integrity and ability are above suspicion. Vaccination, and revaccination from time to time, are considered by every physician as an imperative duty, and the only safeguard against the encroachment of one of the most loathsome and fearful of all diseases.

INTERTRIGO, EXCORIATIONS.

By " Intertrigo " is understood those superficial sores, excoriations, or gallings which sometimes appear behind the
ears, between the thighs, in the folds of the neck, or other
parts of the body where the skin folds back upon itself.
This troublesome disorder, as a general rule, is peculiar to
fat children. It is said to be caused by the mother or nurse
indulging in highly-seasoned or acrid food, particularly pork.
Fat children are particularly predisposed to the disease,
but, without doubt, anything which irritates the skin will
act as an exciting cause : a want of cleanliness, or the contrary, too frequent washing, especially with coarse soap ;
acrid perspiration, especially when combined with some of
the various " baby-powders " sold by druggists, materially
aid in the development of and even cause the disease, from
the fact that lycopodium, and other vegetable productions
having chemical properties, form the chief ingredient of
these powders. (Remedy No. 91.)

PIMPLES ON THE FACE, ACNE, PUNCTATA, COMEDONES.

We not unfrequently find upon the faces of children and
young persons, small, black-headed pimples, from which,
by pressing upon their sides, we can squeeze out a small,
vermiform or worm-like cylinder, about one-tenth of an
inch in length. The disease received the name of *comedones* from the fact that they were for a long time believed
to be small insects ; investigation has proved, however, that
the white cylinder which we squeeze out is nothing more
nor less than an accumulation of fatty matter in the follicles
of the skin, and the black head is caused by the dust which
adheres to it. The causes of *comedones* are anything which
obstructs the excretory ducts of the cutaneous follicles, or,

indeed, the secretion of itself may be of a morbid character, which is frequently the case in persons with a torpid skin; the contents of the oil tubes become too thick and dry to escape in the usual manner. The obstructed and distended tube sometimes inflames, even suppurates, and the pimples become very sore. (Remedies Nos. 92, 289.)

ABSCESSES.

By the term "abscess" is understood what, in popular language, is called a "gathering." A collection of pus, or matter, in any part of the body, resulting from inflammation, which may be either acute or chronic. Abscesses are of various kinds; we shall confine our present consideration to those lymphatic tumors and superficial gatherings, such as we so often meet with in children, especially about the head and neck.

An abscess is not an original disease, but always the result or termination of inflammatory action. Inflammation and suppuration of the cervical glands of the neck are frequently concomitants of other diseases. Scald-head, scarlet fever, measles, and many other diseases are frequently followed by glandular inflammation, which terminates in the formation of pus — true abscesses. There is about some children an hereditary dyscrasia or constitutional taint, — scrofula, or some kindred diseases for instance, — which predisposes to the disorder.

Symptoms. — *Acute abscesses* are preceded and accompanied by sensible and inflammatory action in the affected part; it is hot, tumefied, throbbing, and painful. The commencement of the suppurative process, that is, when the formation of matter takes place, is to be known, or at least suspected, by the change in the character of the pain which takes place at this time, and by the appearance of the skin. The pain, which has previously been acute, loses its intensity, becomes dull and throbbing, the skin changes from a red to a livid color. The tumor presents a somewhat conical shape; and the skin over its apex becomes thin and of a dark livid color. At this point, if left alone, the abscess will burst, and allow its contents to escape. In abscesses of any magnitude, during the suppurative process, we have usually more or less definitely marked rigors and chills, succeeded in turn by increase of fever. After an abscess is fully formed, provided it is not too deeply seated, fluctuation in the tumor is always perceptible.

Chronic Abscesses. — Although all abscesses are the result of inflammation, the inflammatory action in chronic abscesses is sometimes of so low a grade as to be almost imperceptible; indeed, during the first stage of the disease, it is entirely so, and were it not for the swelling, which always becomes apparent before it reaches any great magnitude, we would scarcely know that anything ailed the child. The entire absence of all local and constitutional symptoms renders the disease obscure, until it begins to approach the surface and form an external swelling. An acute abscess readily heals, as soon as the pus is freely evacuated. Not so with a chronic abscess: the latter, instead of contracting and filling up with healthy granulations, that is, portions of new flesh, remains open and discharges copiously of thin, acrid matter; and this state, if continued any great length of time, results in the production of hectic fever; or, in other words, the patient goes into a decline.

TREATMENT.

As abscesses do not always end in suppuration, but sometimes in resolution, — that is, the inflammation and swelling subside without the formation of pus, the tumor not gathering, — it is not *always* advisable to apply poultices, as this may cause it to gather, when it otherwise would not. Should a swelling appear anywhere upon the surface of the body, which we apprehend may terminate in an abscess, our first endeavor should be to cut short the inflammation before it reaches the point of suppuration. This can best be done by the external application of cold-water bandages and the internal administration of Remedy 93. This treatment is especially recommended when there is considerable constitutional disturbance, with intense pain and extensive inflammation of the parts. Should this treatment fail to arrest the disease, the next best thing to be done is to hasten suppuration, or bring the abscess " to a head " by the external application of hot fomentations and ground flaxseed poultices. As soon as the abscess points or comes to a head, the skin becoming livid and thin, and there is distinct fluctuation, it is advisable to make a free incision into the tumor, and evacuate the matter. The sooner it is discharged the sooner will the abscess heal. I see nothing gained by waiting; it is but prolonging the patient's suffering and retarding the cure. After the abscess has been opened, and the matter freely discharged, the poultices should be discontinued and simple dressing substituted. (Remedy No. 93.)

Diet. — In acute abscesses, where there is considerable fever, the diet should be about the same as in fevers. During the long and tedious course of some exhausting chronic abscesses, it will be found necessary to select such a diet as will nourish and strengthen the patient. The food should be nutritious and of easy digestion. Broiled steak, mutton-chop, meat broths, rice and barley gruel, etc., may be allowed.

BOILS.

A boil consists of a round, cone-shaped, inflammatory, and very painful swelling immediately under the skin. It varies in size from a pin's head to a pigeon's egg. It always has a central "core," as it is called, and is mostly found in strong and vigorous children. A boil always suppurates, and sooner or later discharges its contents, the matter being at first mixed with blood, and afterwards composed of pus. A boil never discharges freely, and never heals until the core comes away. The causes of these annoying excrescences are certainly impurity, fermentation, and impoverishment of the blood; the treatment is similar to that prescribed for abscesses. Apply a poultice early, and bring the tumor to a head as soon as possible. After the matter has been discharged, wash the parts clean, and dress with lint and simple salve. The lint should be placed *next* to the sore, and the salve *over* the lint. To eradicate the predisposition to boils, a dose of sulphur, twice a week, will be found very efficacious. (Remedy No. 94.)

SCALD-HEAD, TINEA CAPITIS, FAVUS.

Tinea capitis is a contagious eruptive disease of the scalp. It is characterized at first by small yellow pustules, situated on an inflamed ground. The pustules are of a peculiar shape, depressed in the centre, and scarcely raised above the level of the skin. Each pustule, as a general thing, surrounds a hair. Probably, the whole disease consists in an inflammation of the hair follicles. The disease is comparatively rare in America, much more so than ringworm of the scalp or milk-crust. Among the upper or middle classes of society it is seldom, if ever, met with. I have not seen more than two cases in this city (Boston), except in dispensary practice. There is very little doubt but that this disease is contagious, and may be propagated by direct contact of a diseased with a healthy skin, or by means of combs, brushes, towels, etc. Although chiefly found in children, it is by no means exclusively confined to them. Children living in low, damp, and ill-ventilated dwellings, and those subjected to an unwholesome or an insufficient diet, are most prone to it. The feature which distinguishes this disease from other eruptions of the scalp is the peculiar shape of the scabs or crusts. Commencing as a small yellow pustule, scarcely raised above the level of the skin, it gradually increases to perhaps an inch in circumference. As it spreads, the watery portions of the pustule dry up, leaving a large yellow crust with inverted edges and a depressed centre. This cup-formed yellow crust, *pierced by a hair*, is peculiar to this disease, and distinguishes it from all other eruptions of the scalp. At first, when the pustules are small, they are usually isolated; but as they increase in diameter, their edges come in contact, and thus a number of pustules, blended together, form irregular patches of larger or smaller size. When the crusts have been removed, the surface beneath is seen to be red and moist, having the appearance of ulceration. By no other eruptive disease of the scalp with which I am acquainted is there a permanent loss of hair. In this disease the hair falls out, and the scalp is left shining and uneven. The hair seldom, if ever, reappears; if it does, it is short, woolly, and unhealthy.

TREATMENT.

Until you can secure the services of a good physician, follow the treatment given for " milk-crust." The first essential step is to remove the hair. This may be sufficiently well done with a sharp pair of scissors; shaving the head is scarcely practicable. No attempt whatever should be made to remove the crusts. Strict attention should be paid to cleanliness. A good and soothing wash for the head is bran-water. (Remedies Nos. 95, 186, 247, 248, 251.)

CRUSTA LACTEA, MILK-CRUST, IMPETIGO.

This is almost exclusively a disease of infancy. It is characterized by an eruption of small, round, yellow, flattened pustules, which are crowded together upon a red surface. The pustules end by the drying up of their contents into thick, rough, and yellow scabs. The eruption may appear upon the forehead, cheeks, or scalp, the latter place being the more frequent seat of the disease. Like most other varieties of infantile eruptive diseases, the real cause is very imperfectly understood. Many suppose the cutting of the teeth to be the cause. Others ascribe it to unhealthy hygienic conditions, as, for instance, improper or unwholesome food, want of cleanliness, damp or ill-ventilated apartments. Not a few think it arises from some constitutional taint existing within the child, as scrofula, or some kindred disease, and that it more frequently manifests itself in fair, fat children. My own opinion is, that children so affected possess a constitutional tendency to the disease, and that the exciting cause, in nine cases out of ten, is some gastric derangement.

Symptoms. — In some cases the eruption is confined entirely to the face; in others, entirely to the scalp; or, again, it may implicate both, extending up the side of the face, affecting the ear, neck, and portions of the scalp. The disease may be either acute or chronic in its nature. When acute, it is not unfrequently attended with severe inflammation of the skin. It appears in all grades of severity; in some cases it is very light, extending over a small surface, remaining stationary, or quickly drying up and disappearing ; or, when severe, the whole scalp may become completely scabbed over, presenting an offensive and disgusting appearance. As a general thing it attacks but a small spot at first, and then gradually spreads to the surrounding parts. When they first appear the pustules are numerous, small in size, of a light yellowish or straw color, and not unfrequently attended with severe burning or itching. These soon break or get broken, and discharge a sticky fluid, which glues the hair together, and forms into thick uneven crusts. The successive discharges from the surface beneath, constantly add to the thickness of the crust, and as the fluid escapes from under the crusts, it irritates or inoculates the parts with which it comes into contact, and thus extends the disease until, in some cases, the whole scalp is covered with a thick, rough, brownish-yellow crust. In warm weather, or from the warmth of the head and exposure to the air, these crusts sometimes undergo partial decomposition, and exhale a sickening and most offensive odor. When the crusts are removed, the surface beneath is found inflamed and wet, the secretion which oozes from them plainly visi ble ; little excoriated points soon form new crusts similar to the one that has been removed. The disease, as it appears about the face, passes through about the same course as when appearing upon the scalp, except that the large crusts are seldom allowed to form. The severe itching attending the disease causes the child constantly to scratch or rub the part, sometimes to such an extent as not only to prevent the scabs from forming, but to cause the surface to bleed quite freely. In most cases the general health of the patient remains good ; sometimes, when the inflammation and itching are severe, it makes the child cross and peevish, disturbs its sleep, and makes it feverish. The glands situated upon the neck, and especially behind the ears, not unfrequently inflame, become hard and painful, and finally gather and break. The duration of the disease depends upon the severity of the case, and the treatment which is instituted for its removal. Some cases yield in a few weeks; others, more stubborn, may continue for months, and, if improperly treated, even for years. The whole course of the disease may not be of the same severity; it not unfrequently subsides to such an extent that the mother is already congratulating herself upon the speedy return of her child's health, when a fit of indigestion, the cutting of a new tooth, or even some change in the weather, may bring it back with renewed violence.

TREATMENT.

I frequently meet with children who have had the disease for months, their parents refusing to do anything for its removal, under the impression that the attempt to cure it would be attended with serious risk to the health, and even the life of the patient. Now, the idea that the disease is useful, and beneficial to the future health of the child, is preposterous. Perhaps it originated from the fact that a sudden suppression of the disease, by active, external means, has been followed by dangerous and even fatal symptoms. But then it should be remembered that suppressing is not *curing* a disease. I believe it to be unsafe to procure, by the employment of external means, the suppression of any eruptive diseases. We are all aware that alarming and dangerous symptoms frequently follow the " striking in," as it is called, of measles and scarlet-fever. Every physician can call to mind cases of acute disease of the brain, resulting from the sudden drying-up of this very disease by the application of some one of the numerous specific ointments. This is a literal " striking-in " of the disease, or a translation from the scalp to the brain. The idea that the disease is in any way beneficial to the health of the child, cannot be entertained by any one who has had much acquaintance with the suffering it produces. The dreadful itching induces restlessness, crying, and sleeplessness ; in fact, it keeps the infant in a constant state of actual suffering, which cannot continue for months or years without seriously injuring the constitution of the child. I consider it, therefore, an entirely mistaken act of kindness which permits the disease to continue a single day without an endeavor to arrest it, under the impression that the child is thereby being permanently benefited. Active treatment should be instituted as soon as the first symptom of the disease is observed. Apply nothing externally but a little glycerine. Keep the head clean by washing with weak soapsuds of Castile soap. (Remedies Nos. 96, 185, 186.)

CHAPTER XII.

DISEASES OF THE BRAIN AND NERVOUS SYSTEM.

INFLAMMATION OF THE BRAIN.

INFLAMMATION of the brain itself is called *encephalitis;* inflammation of the membranes which invest the brain is called *meningitis.* Inflammation of the brain and of its investing membranes has no fixed uniform train of symptoms by which it declares itself; perhaps the most common and striking phenomenon is a sudden and long-continued attack of general convulsions. Still, convulsions, especially in children, frequently arise from various other causes; for instance, from teething, from overloading the stomach, or from worms. The attack may come on with but slight pain in the head, with vomiting and impatience to light. More commonly, however, there are severe pains over the entire head, throbbing of the arteries of the neck and temples, fits of shivering, vertigo, sleeplessness or restless sleep, disturbed dreams, unsteady gait, quick pulse. In those cases which are occasioned by blows or falls upon the head, the patient may recover entirely from the shock and external wound, if there be one, and remain for a certain period, to all appearances, perfectly well. But, after some days, or even weeks, he begins to complain; may come in from his play with headache and chilliness; the skin soon becomes hot and dry; he is restless; cannot sleep; his countenance becomes flushed; his eyes red and fiery; the pulse is hard and frequent; nausea and vomiting supervenes;

the substance thrown up is generally a greenish or yellow-ish fluid; and, as the case draws to a close, delirium, con-vulsions, or profound stupor takes place.   Inflammation of the substance of the brain, either when it invades the whole organ at once, or begins in one part of either or of all the membranes and extends rapidly to all the rest, is always attended with high excitement, much fever, and great delirium.   The face is red and bloated, the eyes are blood-shot and brilliant, the pupils contracted, great sensitiveness to light and noise.   The deeper the interior of the brain is affected, the more the senses are stupefied, until the patient becomes entirely unconscious ; he can neither see nor hear ; the pulse is small, frequent, and tremulous.   Owing to the fact that the organization of a child's brain is much more delicate, and, therefore, more sensitive, than that of adults, its disease must necessarily be more frequent and dan-gerous.   It is well, therefore, promptly to heed, and notice critically, every symptom, no matter how trivial, which points a finger of suspicion toward the brain.

Generally speaking, these diseases are uncommon as a *primary* disorder; but they are frequently met with as a *secondary* affection, and in the following manner : A child suffering from a discharge, either acute or chronic, from the ear, takes cold; the discharge stops, or from any other cause is suddenly suppressed ; the inflammation, so to speak, travels inward, or, in other words, is transmitted to the membranes covering the brain.   The patient becomes dull and drowsy ; sometimes, when there is high fever, he is delirious ; he puts his hands up to his head, or bores it into the pillow, and by degrees sinks into a complete stupor, from which he may never recover.

There is no doubt that catarrhal difficulties of the head and throat are frequently transmitted to the brain.   Or the inflammation creeps along from one membrane to another, until it finally reaches the brain, and produces fatal results. This is especially the case when astringent lotions and injections are made use of for the cure of such complaints. In eruptive fevers, especially when the eruption does not come out well upon the surface, or, after being well out, it suddenly strikes in; also, during difficult dentition, or even in some forms of severe colds, the child will complain of chilliness, with alternate flushes of heat. There will be pain in the head, manifested by the child putting his hands up to it; he moans, becomes drowsy, stupid, or restless; rolls his head and screams. Any general irritation may bring on the disease. It sometimes supervenes upon the drying up of eruptions, such as scald-head or sores behind the ears, especially in scrofulous children. Diseases of the brain during infancy are much more frequent among those born of parents who are either suffering from some tubercular disease themselves, or in whose families such complaints have existed to a greater or less extent.

Perhaps the most frequent form of brain disease in childhood is that known as " tubercular meningitis," or acute hydrocephalus, or more commonly, water on the brain. This is very insidious in its attack; it steals upon the patient before he is aware of its approach. He loses his appetite; becomes capricious; sometimes he appears to dislike his food, and sometimes devours it voraciously; his tongue is foul, breath offensive, his abdomen enlarges, and is sometimes tender; his bowels are torpid, and the evacuations from them unnatural; the stools are pale, and contain but little bile, or are dark, fetid, sour-smelling, lumpy; and the child loses his former healthy aspect. There are obscure indications of cerebral derangement, and he shows unsteadiness and tottering in his gait. When these symptoms are observed in a child who has any hereditary tendency to scrofula, or in a child who is precocious or particularly clever, there will be much reason to apprehend that mischief is brewing in his head. The pain is usually located just over the brows, but may extend all over the head ; at the beginning of an attack, there is often pain and stiffness at the back of the neck, extreme tenderness of the scalp, and the child cries and shrieks when taken up. Vomiting is nearly a constant symptom, and is often excited by raising the child to an erect position ; the headache and vomiting are both aggravated by motion ; there is a total loss of appetite, the tongue is coated white, the breath is offensive. Constipation is almost always a prominent symptom. Diarrhœa is rare ; the constipation is generally obstinate for the first week or ten days of the disease, and then, toward the termination of the case, gives way to a diarrhœa with involuntary stools. The head is usually hot, the pulse variable, the senses of sight and hearing become painfully acute, the patient is excessively sensitive to light, and the slightest noise or jar, even a person walking across the room, irritates and distresses him. The next stage evidences a marked change in his aspect and symptoms : noises no longer disturb him, he is in a half comatose condition, convulsions frequently occur, sometimes paralysis. The urine and stools are passed unconsciously ; he is incessantly picking at his lips, nostrils, ears, and bedclothes. This stage may last a week or two, with intermissions of intelligence, apparently restored health and vivacity, and regaining the use of his senses, but only to sink into a deeper stupor than before. He now enters upon the last stage. He rolls his head perpetually from side to side ; palsy frequently occurs ; the circulation is unequal ; intermittent fever supervenes ; he is alternately raving and insensible ; the pulse gets weaker and weaker ; and death, in many instances, takes place in the midst of a convulsion.

No treatment of a temporary nature can be prescribed ; the most skilful and patient watchfulness of an experienced physician can be alone relied upon. The diet and regimen should be the same as in fevers. (Remedies Nos. 97, 132, 133, 134, 135.)

CHRONIC HYDROCEPHALUS.

Chronic hydrocephalus is an actual dropsy of the brain. The disease is generally congenital, the child being born with a head out of all proportion to the rest of its body. From some cause not well understood, a watery fluid collects within the brain, and the skull, being but imperfectly developed, yields to the inward pressure, and the head is augmented in some cases to an enormous size. When this accumulation of water takes place, as it frequently does, while the child is yet in the womb, it is sometimes impossible for it to pass through the natural outlets into the world. In such cases the mother's life must be saved and the infant's sacrificed. In a large number of cases, however, the child is brought into the world entire and unhurt, and lives for a longer or shorter period. Sometimes, however, the accumulation of water does not take place until after birth; but when it does take place, which may be in a few days, weeks, or even months, it will be perceived that the head enlarges with great rapidity, is quite disproportionate to other parts of the body, and of course gives the child a very strange aspect. The greater part of those afflicted with this form of dropsy either recover or die during infancy; they seldom grow up in this condition; there are, however, isolated and rare instances where they have reached adult life and possessed a fair amount of intellect. (Remedies Nos. 129, 130, 131, 230.)

CONVULSIONS, SPASMS, OR FITS.

These terms are used indifferently and indiscriminately, to indicate a violent and involuntary agitation of a part or of the whole body. These agitations consist in alternate contractions and relaxations of the muscles of the part affected. Convulsions may be either general or partial. When general, the muscles of the face and body, as well as those of the extremities, are involved. When partial, the spasmodic action is confined to one particular part. All convulsive diseases consist in affections of the true spinal system of nerves.

Causes. — Among the predisposing causes of the disease may be mentioned a highly susceptible, irritable, or nervous temperament. It has been stated that convulsions are more common in girls than in boys. Whether there is any truth in this or not, I am unable to say. It is also said that delicate children are more subject to them than robust ones. This may be so. We frequently meet with families in which all the children, during infancy, are afflicted more or less with spasms. This may be owing to a similarity of nervous temperament inherited from the parents. Convulsions occur most frequently in children under seven years of age, and particularly during first dentition. The most common causes are irritation of the bowels, difficult teething, and worms. A dangerous form of convulsions is often produced by overloading the stomach, or by eating heavy or indigestible substances. The most alarming variety, however, — because of frequently terminating unfavorably, — is that occasioned by heavy blows or falls upon the head.

Spasms in children are frequently occasioned by the inordinate use of drugs. When we come to consider how delicate the organization of an infant's nervous system must be, it is not to be wondered at that the enormous quantities of patent and domestic remedies which children are compelled to take frequently derange the equilibrium. The only wonder is, that they ever recover from convulsions thus produced. Excessive joy, sorrow, anger, fear, or any other passion, undue exposure to cold or heat, severe pain, and repelled eruptions, frequently cause convulsions. When convulsions usher in eruptive fevers, they seldom have a fatal termination, but when they appear during or at the termination of those diseases, the result is usually fatal.

Symptoms. — Convulsions are not, as a rule, preceded by any premonitory symptoms. The attack usually commences with the eyes, which are at first fixed and staring; as the case advances, they become agitated, and are turned up beneath the upper eyelid, leaving only the whites visible; the eyelids are sometimes open, and sometimes shut; the eyes are frequently crossed, the pupils being either excessively contracted or dilated. The muscles of the face next become affected, and the contractions produce at times most horrid contortions. There are sometimes only slight twitchings of the muscles of the face, with alternate contractions and relaxations. The mouth is distorted into various shapes: the corners are drawn down and fixed in this position, or the muscles of one side may contract while the others relax, and so keep the parts in a constant state of agitation. The tongue, when it can be seen, will be observed to be in constant motion. It not unfrequently gets between the teeth, and is severely bitten. Sometimes the jaws are firmly set, and at other times in violent motion. In rare cases there is foaming at the mouth; there is always a blue shade around the eyes and mouth, and often the whole surface of the head becomes violet-colored. In severe cases, the movements and distortions are much more violent. In *all cases*, mild or severe, though consciousness is entirely destroyed, the child is still sensible to external impressions, that is, will open his mouth when a spoon is put against his lips, and swallow anything given to him. The duration of the fit is extremely uncer-

tain ; it may be of only a few moments' duration, or it may
continue for hours. The average range of the paroxysm is
from two minutes to half an hour. When the spasm is
protracted, it is usually broken by brief intermissions. The
duration and recurrence of an attack depends entirely upon
the *cause* of the disorder. So long as the disease continues
in action, we can expect no permanent improvement. Our
first effort is naturally to remove the cause, but it is often
so difficult of detection, that the cure is necessarily much
retarded.

TREATMENT.

In this work, we must of course confine ourselves to *general treatment*, the special and particular treatment for the
disease which occasions it, and the disordered condition of
the nervous system, being solely the work of the physician.
The first thing to be done in a case of convulsions, is to put
the child into a warm bath, about 96° ; it is almost certain to
allay the spasm. In slight cases, a foot-bath, with a little
mustard in it, will be found sufficient. The patient should
be kept in the water from ten to twenty minutes, or until
the convulsion ceases. When he is taken out, you should
not stop to wipe or dress him, but just wrap him in a warm
flannel or woollen blanket. *Cold water* applied to the head
is also an excellent auxiliary. During the application of the
foot-bath, cold applications can be made, and should be
continued until the head feels quite cool. The best way to
apply it is to pour cold water from the nozzle of a small
watering-pot, held two or three feet above the child's head.
This process must be repeated as often as the head begins
to get warm again ; it should not be allowed to get hot. If
possible, the child should be placed in a large, well-ventilated room, where the air is pure. When this cannot be
done, the next best thing is to expose the patient to fresh

air at an open window. Above all, do not torture the child by the application of mustard plasters, onion-draughts, and the legion of disgusting nostrums advised by well-meaning but indiscreet friends. (*See Remedies Nos.* 98, 184.)

Indigestion. — If the convulsions be caused by overloading the child's stomach, or the presence of some indigestible substance, evacuate the stomach at once by an emetic. Lukewarm water is as good as anything, or tickling the throat with a feather. The feet and legs should be placed up to the knees in hot water, as hot as can be borne. Should this fail, immerse the entire body in hot water, at the same time pouring cold water on the head. If the bowels are constipated, or if you cannot excite vomiting, give an injection of sweet-oil and warm milk.

Teething. — When the spasms arise from this cause, as they very frequently do, take a sharp penknife and make an incision in the gum wherever the seat of the irritation appears to be. A warm bath is, of course, essential.

Mechanical Injuries. — When convulsions arise from blows or falls upon the head, a solution of arnica should be at once applied, and a physician summoned.

Fright, Suppressed Eruption, or Catarrh, or Unknown Causes. — In convulsions having such origin, the warm bath or cold douché are indispensable.

Diet. — After a convulsive seizure it is best that the child should be kept upon a low or rather non-stimulating diet for a few days. Especial care should be taken in this respect if the spasm has been caused by indigestion.

CHOREA — ST. VITUS'S DANCE.

Chorea is essentially a disease of the nervous system, of a spasmodic nature. It is characterized by tremulous, irregular, and, in some cases, most ludicrous motions of all or any of the voluntary muscles. These contractions are partially involuntary, are more marked upon the left side than the right, affect females more than males, occur chiefly in persons between six and fifteen years of age, and are not accompanied by pain.

Causes. — It is said to occur most frequently in children of a nervous, delicate, excitable temperament, and is frequently hereditary. No doubt but that a disordered condition of the digestive system, as well as uterine affections, predispose to the disease. Among the exciting causes may be mentioned anything which makes a forcible impression upon the nervous system : strong mental emotions, of which fright is the most common ; injuries of the head and back ; the improper employment of lead, mercury, and other metallic poisons ; suppressed eruptions or discharges of any kind ; the extension of rheumatism to the membrane of the spinal cord ; cutting of the permanent teeth ; anxiety ; suppressed emotion of any kind ; excitement of the passions, and retained or difficult menstruation.

Symptoms. — Its approach may be either sudden or gradual, and is indicated by a variable condition of health for two or three weeks, imperfect digestion, constipation of the bowels, loss of appetite, and general derangement of the digestive and menstrual functions. This disease, unlike many other convulsive diseases, does not render the subject unconscious, neither does it affect volition. The patient knows perfectly well what he is about, and what he wants to do ; but he cannot always do exactly as he wishes. The ordinary movements of the body, to a certain extent, can be performed ; but there is some other power besides the will at work, and this power is constantly interfering with all the movements, misdirecting the hand that is put out to seize something, or jerking it back and giving it a new direction, rendering unsteady and imperfect every act, bringing into play muscles that should be quiet, and arresting those which the will has set at work. The muscles of the face are jerked about with an agility that is truly surprising, drawing the face into all sorts of shapes and grimaces. The hands and arms are twisted and jerked into every conceivable position. The inferior extremities are affected in the same way. In fact, it seems sometimes as though the whole muscular system had gone crazy. If you ask the patient to put out her tongue, she will have to make many attempts before she can accomplish it. She cannot keep her limbs in one position for half a minute. Walking is always more or less difficult. There is no saying in what direction the hand or foot may be moved when it is once lifted. The patient progresses in a zigzag direction, totters from side to side, going by fits and starts, even standing being frequently impossible. Articulation, too, is not uncommonly arrested, and mastication so materially interrupted by the irregular contraction of the muscles which move the jaw, as to render feeding by the ordinary method impracticable.

TREATMENT. — Remedies Nos. 99, 165, 166, 167.

Diet and Regimen. — The diet should be perfectly plain and nutritious; all articles of pastry, and rich or highly seasoned dishes should be avoided. Coffee and tea are unmistakably injurious. Out-door exercise — plenty of it — is decidedly beneficial.

HEADACHE.

By this general term is understood a pain of any description in the head. It is usually accompanied by an intolerance of noise and light, and always with incapability of mental exertion. As headache appears in many forms, it has, consequently, as many causes, and an accurate knowledge of the *cause* of the pain is essential to the successful choice of the remedy. It may arise from nervous irritability or excitement, uterine or abdominal derangement, over-exertion, fasting, decayed teeth, excessive menstruation, nursing, impure air, sudden suppression of discharges or eruptions, mental excitement of any kind, severe cold, etc. The symptoms of the various descriptions of pain: neuralgic, rheumatic, nervous, sick, bilious, congestive, catarrhal, etc., are too well known to need enumeration, and the treatment of the various phases of this affection will be found in detail under its appropriate head in Remedies Nos. 100, 216, 217, 218.

The diet during an attack should be light and low, and all articles of a stimulating nature should be avoided, and especially tea, coffee, or alcoholic drinks.

RHEUMATISM

is an inflammation of a peculiar character, being caused by acid or poisonous matter in the blood, and having for its seat the *fibrous tissue*, or that thready texture which enters largely into the composition of the cords and muscles of the human body. The synovial, or lining membrane of joints, is also peculiarly subject to rheumatic inflammation. Hence the terms *fibrous rheumatism* and *synovial rheumatism*. There are also acute and chronic rheumatism. *Acute rheumatism* is a very painful affection. It is most frequently brought on by exposure to wet and cold after fatiguing exercise to the muscles. Women are usually very reckless in this respect, dabbling about in water, both warm and cold, and rushing about from kitchen or scullery to dining-room, parlor, or drying-ground, half clad, and encountering a constant alteration or change of temperature. The only wonder is that they manage to retain their health at all under such opposite conditions of the atmosphere.

Symptoms. — Its principal characteristics are high fever, full, bounding pulse, furred tongue, profuse sweat, with a sour smell, the bodily weakness increasing without relieving the pain; scanty and high-colored urine, with brick-dust sediment, and swelling of the joints, with slight redness, great tenderness, and severe pain, which is particularly agonizing when the patient attempts to move. This affection often changes suddenly from one part of the body to another, or from one set of joints to another. This sudden shifting, termed metastasis, is peculiarly dangerous, for sometimes the inflammation, seeming to regard the constantly moving heart as a large central point, suddenly seizes upon its lining membrane, and occasionally results in sudden death.

Chronic Rheumatism sometimes follows as the sequence or result of a severe attack of acute rheumatism, but it is much more often an independent disease. It differs from the acute form in being seldom attended by fever. It frequently lasts for many years, and causes excruciating suffering. The symptoms are varied, according to the temperament and constitution of the patient, but usually comprise pain, lameness, stiffness, etc., in the joints and other parts. The joints are frequently swollen, but not nearly so much as in the acute disease. It is peculiar to this form of the complaint that when the patient remains at rest for a time he will have pain and stiffness in the affected part on beginning to move, but as he grows warm both will disappear.

Treatment. — Chronic rheumatism is often palliated, and sometimes cured, by passing a current of electro-magnetism through the affected parts, both internally and externally. Remedies Nos. 297, 298, 299, will be found efficacious. It is well to wear a piece of oiled silk over the affected part. It keeps up a gentle perspiration from the rheumatic surface, and materially hastens a cure. To bathe the affected joint at bedtime with hot sweet-oil, and then envelope it in cotton batting, to be kept on through the night, will often give much relief. The bowels must be kept regular, and all exposure to wet feet or clothes, and to currents of cool air when sweating, must be carefully shunned.

NEURALGIA.

The meaning of this term is ·pain in the nerve. When occurring in particular parts of the system it is frequently confounded with rheumatism, though it differs materially from that disease. Rheumatism is a specific kind of inflammation, affecting particular tissues of the body; while neuralgia is quite independent of inflammation, and is simply a pain in the nerve, unaccompanied by fever or any noticeable change of structure in the affected part. The pain of neuralgia is often severe, sometimes excruciating; it occurs in paroxysms of irregular duration, and, after either regular or irregular intervals, affects various parts of the body, and attacks males as well as females.

Causes. — In not a few cases the causes of neuralgia are very obscure. One very great difficulty in making out the causes and origin of these pains is, that they are so frequently occasioned by some source of irritation situated in a part distant from where the pain is felt. For instance, you strike your elbow in a certain way, and you produce a tingling sensation, not in the part struck, but at a distance, — in your little finger. The same thing is constantly happening in disease. Something taken into the stomach, which arrests digestion, may cause pain in a remote part;

some affection of the brain or spinal cord may cause it. Damp and cold, in any form, dwellings, clothing, exposure to inclement weather, etc., most frequently cause neuralgia. Facial neuralgia is often occasioned by decayed teeth, while the tooth itself may be perfectly free from pain.

The symptoms are so familiar to us all that they need not be enumerated. The attacks are very variable in duration, as they may last a minute, an hour, a day, or a week. Tic douloureux, angina pectoris, and sciatica are purely neuralgic affections.

Treatment. — As an external application, perhaps nothing is better than cold water, or, when that cannot be borne, warm or tepid water. Frequent bathing and plenty of out-door exercise are very beneficial. Internally, take prescriptions Nos. 101, 280, 281, 282, 283.

Angina Pectoris is a neuralgic affection of the heart. It is sometimes called *breast-pang*, and is a painful, suffocative sensation in the breast, which comes on suddenly during walking, but ceases as soon as one stands still. During the paroxysm the patient should remain perfectly quiet, in an erect position, with all the clothing loosened. For remedies, *see* Nos. 280, 281, 282, 283.

Tic-Douloureux, or neuralgia of the face, is an excruciating affection, too well known to need comment. The most effective remedies are Nos. 280, 281, 282, 283.

Sciatica. — This is neuralgia of the great sciatic nerve. The pain starts in the region of the hip-joint and extends to the knee or even to the foot, accurately following the course of the great sciatic nerve. The pain is sometimes so severe as not only to impede the motion of the foot but to deprive the patient of all rest. It frequently produces stiffness and contraction of the limb. As diseases of the hip and knee joint not unfrequently result in serious deformity, it is always best, when these parts are threatened, to place the child under the care of an experienced physician. For remedies, *see* Nos. 101, 280, 281.

Diet and Regimen. — The diet should be plain and nutritious. Coffee and green tea should be specially avoided. As gastric disturbance is frequently the exciting cause, care should be taken to avoid all indigestible and other descriptions of food likely to disagree with the constitution.

DISEASES OF THE EYES, EARS, AND NOSE IN CHILDREN.

Diseases of the Eyes. — The delicate structure of the eye renders it extremely liable to accidents of various kinds and diseases of various forms; and, what is indeed fortunate for all these diseases and accidents, every one whom the patient meets has a *certain* cure, — one never known to fail. It is not asserting too much to say that, without a doubt, more permanent injury has been done to the eyes by the use of local applications than has ever been done by natural disease. Slight ailments, which would have been but trifles under rational treatment, have been aggravated into serious diseases by irritant washes and lotions. I would advise every one to abjure all eye-waters, lotions, salves, ointments, and the like, and adhere to pure cold, or, in cases of peculiar sensitiveness, warm water.

When **Erysipelas** affects the eyes (which will be known by the surrounding redness), nothing wet should be applied, but, instead, dry and warm applications should be used.

SORE EYES OF YOUNG INFANTS.

This affection is very common among young infants, setting in frequently when the child is but a few days old. Generally the eyelids are first affected, but the eye proper soon becomes implicated if the disease continues long or is neglected. It is occasioned, either by some irritating substance getting into the eye — soap, for instance — when the child is being washed, or by cold. The child's eyes should be washed with lukewarm water, giving the child internally Nos. 103, 242.

STY ON THE EYELID.

A sty is simply a small boil on the margin of the eyelid. They are quite painful, suppurate slowly, and show no tendency to burst. Give remedies Nos. 104, 242. When the remedy does not check the advance of the disease, and suppuration is about to take place, or when there is considerable redness, with throbbing pain, a warm poultice should be applied to facilitate its breaking.

SQUINTING-STRABISMUS.

This is an affection of the eyes, in which they are drawn out of their natural position. It may be spasmodic, caused by some affection of the brain, or it may be occasioned by a permanent shortening of one of the lateral straight muscles of the eyeball. In the latter case an operation is necessary. Drugs seldom have any beneficial effect on this affection, though sometimes Nos. 105, 242, render material aid in conjunction with the mechanical means.

DISEASES OF THE EAR — INFLAMMATION OF THE EAR.

This is a very painful disease. The inflammation affects the passage or tube of the ear, sometimes causing it to swell to such an extent as to close it entirely, and at times occasioning such severe pain as to scarcely allow the affected member to be touched. The symptoms indicative of this disease are : violent burning, itching, beating pains, deep in the ear ; and, finally, swelling and redness, both internally and externally, great sensibility to noise, and more or less fever. As a general rule, this disease can be controlled by one of the following remedies : Nos. 106, 242.

It is neither judicious nor safe to be constantly introducing oil, laudanum, and the like, into the ear. Relief is often afforded by covering the ear with cotton to protect it from the air and noise ; and warm water, applied with a soft linen rag, will often ease the pain without doing injury.

EARACHE.

This is a very frequent affection of young children, and, although resembling inflammation of the ear, is quite a different disease, the one being accompanied by fever, and the other not. The pain of earache is of a neuralgic or rheumatic character, and generally arises from taking cold. The attacks come on suddenly, and are of short duration.

Treatment. — *See* remedies Nos. 107, 242.

In regard to external applications, which are so frequently used, and sometimes with advantage, I have no serious objection to offer. For my own part I would rather trust to the remedies I have mentioned, because the parts may be so injured that restoration will be almost impossible. The safest way is to use nothing, unless it be a little olive-oil, or tepid water. Sometimes a sponge, or soft muslin, dipped in water and applied to the ear, will migitate the pain.

RUNNING OF THE EARS.

This troublesome and sometimes exceedingly offensive disorder arises from various causes. It frequently remains after inflammation, gatherings in the head, etc. Perhaps the worst form, and that which is most difficult to cure, is that resulting from scarlet-fever. The attempts which are frequently made to arrest the discharge by some local applications are greatly to be reprehended, as the most insignificant discharge may, when suddenly suppressed, produce most dangerous consequences. It is best to bear patiently with the affliction until a cure can be effected with the proper remedies. Never tamper with the eye or ear. The principal remedies are Nos. 108, 242.

BLEEDING FROM THE NOSE. EPISTAXIS.

This is quite a common occurrence among young people, and seldom amounts to more than a temporary inconvenience, rarely needing any remedial assistance. A slight blow, a fit of sneezing, or the summer heat, is sufficient, with many, to bring it on. This is owing to an undue fulness of the bloodvessels of the head. In young girls it sometimes comes on periodically, with or at the time the menses should appear, and frequently in fevers and other diseases. It often relieves or cures headache or vertigo. In young children it is almost always salutary, and may be left to work its own cure.

TREATMENT.

The nursery remedy is to slip a key or piece of cold metal down the back, or to sprinkle the face with cold water, which sometimes restrains the hemorrhage by producing a contraction of the bloodvessels. Very often a severe hemorrhage can be stopped by causing the patient to hold his hands high above his head. Remedies Nos. 109, 220, 221, will be found very efficacious.

DISEASES OF THE URINARY ORGANS. WETTING THE BED.

It is a general, but erroneous, opinion that this affection is simply a " bad habit," when it is in reality a disease.

The child is whipped for what is not its fault, — for a thing which is totally beyond its control. In the majority of cases it arises from a weakness of the parts, and, as the bladder fills up, the urine escapes involuntarily. Now, is the child to blame here?

TREATMENT.

All articles which have a tendency to increase the secretion of urine should be avoided. Tea, coffee, and all salt or sour articles of food are objectionable. The child should take a moderate supper, and as little drink as possible (cold water or milk being preferable) and should not be sent immediately to bed. Plenty of out-door exercise is always advisable. The child should be taken out every day, and permitted to run, hop, skip, jump, etc. When in bed, it should not be allowed to lie on its back. If the difficulty presents itself in young boys of ten or twelve years of age, there is reason to suppose that they are practising a *secret vice*, which should at once be arrested. The proper remedies are indicated at No. 110.

RETENTION OF URINE IN INFANTS.

As a general thing, new-born infants discharge the contents of the rectum and bladder shortly after birth; occasionally, however, it happens that the urine is retained for a longer period, not unfrequently producing symptoms which demand our immediate attention. It is uneasy, nervous, and cries, especially when pressure is made on the region of the bladder; there is more or less fever; it twists its body and draws up its legs. If relief is not soon afforded, convulsions and other dangerous symptoms follow.

Treatment. — A warm bath, or rubbing with the warm hand over the region of the bladder, is of great service. The

following remedies will generally afford a satisfactory result:
No. 111.

INFLAMMATION OF THE PRIVATES.

A great source of annoyance sometimes befalls young
girls (and occasionally boys, too) in the shape of an inflammatory swelling of the private parts. The lips of the
vagina become swollen, hard, red, and very sensitive to the
touch. In boys the prepuce or foreskin becomes swollen
or puffed up; there is always more or less fever, accompanied by burning or shooting pains. It may arise from
cold, excoriations or chafings, or mechanical injuries. In
women it is at times caused by rupture of the hymen, or
difficult labor.

CHAPTER XIII.

THE DOMESTIC MANAGEMENT OF THE SICK-ROOM.

THE most important element in the recovery of the sick is the hygienic and sanitary surroundings of the patient. If these be neglected, the efforts of the physician, however skilful or attentive he may be, will avail but little. The sleeping apartment of the sick should be large, airy, lofty, and well ventilated; if possible, it should have a northern aspect, so as to avoid the mid-day heat or the afternoon sun; and the windows should always be let down about an inch from the top, so as to allow the escape of the foul air and the admission of fresh, pure air. No room, however large, should be used as a sick-room, where it is possible, unless it has an open chimney. There should be no unnecessary articles of furniture, — two tables, a chair or two, a chest of drawers, and the bedstead and its necessary appurtenances, such as a wash-stand, basins, etc., being all that is requisite. A sofa or reclining-chair is, of course, essential to the patient's comfort. There should be no kettle or other culinary implement in the sick-room, — the odor, or even the noise and bustle, of culinary preparation being specially injurious and annoying to the patient. The room should be carpeted, in order that the movements of the attendants may not disturb the sick person; and on no account should there be annoyance or excitement by the sight or conversation of visitors.

Beds and Bedding. — French, or iron bedsteads, without curtains, are the most suitable. In every case of disease, especially when attended by fever, the patient should be kept cool, and the most perfect freedom for respiration afforded. The mattress (not one of *straw*, however) should be placed over the feather-bed, and the pillows firm and elastic. The bed-spreads usually placed upon the bed during the day, and often retained during the night, are much too heavy, and calculated rather to increase than to subdue fever. When the patient is suffering from fever, it would be much better if she could have *two* beds at her disposal, either in the same or adjoining room, so that she could be moved from one to another every twelve hours. It would promote sleep, and go far to ensure her personal cleanliness. But when there is only one bed, the linen should be changed every morning and evening, or at least once in twenty-four hours.

In infectious diseases especially, and, relatively so, in diseases of all kinds, thorough ventilation is an absolute necessity. It must be remembered that infection rarely extends above a few feet from the patient — even in the most malignant diseases not more than a few yards — if the room be well ventilated. If ventilation be neglected, the power of infection becomes greatly augmented, settling upon the clothes of the attendants, and even impregnating the upholstery and furniture of the room. Smooth and polished surfaces do not readily retain or receive the infectious matter; consequently, the nurses and attendants, in all infectious cases, should wear glazed gowns and aprons of oiled silk. It must also be borne in mind that infectious matter, even in the most virulent diseases, is not poisonous to every one; there must be a predisposition — a sort of receptive condition — in the person placed within the sphere of its influence. Hence, a *thoroughly* healthy person, taking the necessary precautions, will be wholly unaffected by the malarious or miasmatic influence of the disease.

In every case of infectious disease the attendants, even in the best-ventilated rooms, should stand on the windward, or that side of the sick-bed from which the current of air comes ; if they keep on the other side the infectious exhalations from the patient are blown upon them in a direct stream. They should never lean over the sick, or inhale their breath.

Next to ventilation the *temperature* of the room is to be considered. The extremes of depression or elevation should be studiously avoided ; but much depends on the nature of the disease as to the exact degree of temperature required. It is especially important in fevers, as it often does more good than any other remedial measures. The best average temperature is 60° Fahrenheit. In convalescence the air of the chamber should be frequently renewed ; the temperature in spring and autumn should be maintained, as near as possible, at 55° or 60° Fahrenheit, gradually lowering it as the patient acquires strength, so that she may be able to bear with impunity the atmospheric variations of the open air.

The absolute *cleanliness* of the sick-room itself, and everything in or about it, is of the first importance. It should be cleansed and arranged the *first thing* in the morning, with as little noise or bustle as possible. The moment any vessel or implement is used by the invalid, it should be removed from the apartment, thoroughly cleansed, and returned as soon as it is cleaned. Nothing in the form of a slop-basin or a slop-pail is admissible ; they only administer to the laziness of nurses. A glass or cup should never be used for medicine a second time without cleansing. It is a great mistake to suppose that the sick-room should be *darkened ;* a moderate amount of light should always be admitted.

The nurses or attendants should always be healthy, cheerful, good-tempered, neat, active, orderly, sober, and possessed of a practical knowledge of diseases and remedies, and good general education sufficient to understand and intelligently carry out the instructions of the physician.

One of the most important auxiliaries to the remedial efforts of the physician is the

COOKERY FOR THE SICK-ROOM.

BARLEY WATER. — Pearl barley, two ounces; boiling water, two quarts. Boil to one-half, and strain. A little lemon-juice and sugar may be added, if desirable. To be taken freely in inflammatory diseases.

RICE WATER. — Rice, two ounces; water, two quarts. Boil one hour and a half, and add sugar and nutmeg. Rice, when boiled for a considerable time, becomes a kind of jelly, and, mixed with milk, is an excellent diet for children. It has, to a certain extent, a constipating property, which is increased by boiling the milk.

DECOCTION OF BRAN. — New wheat bran, one pint; water, three quarts. Boil down one-third; strain off the liquor, and add sugar, honey, or molasses, according to the patient's taste. *Bran tea* may be made by using boiling water, and allowing the mixture to stand in a covered vessel for three or four hours.

SAGE TEA. — Dried sage leaves, half an ounce; boiling water, one quart. Infuse for half an hour, and strain; add sugar and lemon-juice to suit the taste. *Balm* and other teas are made in the same manner.

The above infusions form agreeable and useful drinks in fevers, and their diaphoretic powers may be increased by adding a little sweet spirits of nitre.

BARLEY COFFEE. — Roast one pint of common barley in the same way in which coffee is roasted. Add two large spoonfuls to a quart of boiling water; boil five minutes. Add a little sugar.

LEMON WATER. — Put two slices of lemon, thinly pared, into a tea-pot; a little bit of peel, and a bit of sugar. Pour in a pint of boiling water, and cover it close two hours.

A REFRESHING DRINK IN FEVERS. — A little sage, two sprigs of balm, and a little sorrel in a stone jug, — the herbs having first been washed and dried. Peel thin a small lemon, slice it, and put in a small piece of the peel; then pour in three pints of boiling water; sweeten, and cover close.

A VERY PLEASANT DRINK. — Pour a teacupful of cranberries into a cup of water, and wash them. In the meantime boil two quarts of water with one large spoonful of corn or oat meal, and a bit of lemon-peel, then add the cranberries. Add as much fine sugar as shall leave a smart flavor of the fruit, and a wineglassful of good sherry. Boil the whole gently for fifteen minutes, and strain.

FLAXSEED TEA. — Take of flaxseed, one ounce; boiling water, one pint. Pour the boiling water on the unbruised seed, cover the vessel, and let it stand near the fire for an hour or two. The seeds must not be crushed or boiled, lest the oil in the interior be extracted. Having strained the infusion, add to it a little lemon-juice, if no contra-indicating circumstances exist. Flaxseed is a mild demulcent, and is much used in diseases of the throat, chest and urinary passages.

OATMEAL GRUEL. — Of oatmeal (coarse is the best) two tablespoonsfuls; water, one quart. Boil for ten or fifteen minutes, and strain. Add a little salt, and sweeten to taste. If no reason to the contrary exists, the flavor of the gruel is much improved by adding some nutmeg, with a little wine or brandy.

TOAST WATER. — Cut a slice half an inch thick, from a loaf of stale bread; remove the crust, and carefully toast the slice on both sides. Place the toast and a small piece of orange or lemon-peel in a suitable vessel, add a pint of boiling water, cover the vessel, and, when cold, strain off the water. This forms an agreeable drink in fevers.

GUM–ARABIC WATER. — As an article of diet, the proper proportions are one ounce of gum-arabic to a pint of boiling water. The solution is allowed to cool before it is used.

Gum-arabic is very nutritive, and life can be sustained on it alone for some time.

LEMONADE. — Take of fresh lemon-juice four ounces; fresh and very thin lemon peel, one-half ounce; white sugar four ounces; boiling water, three pints. Let them stand until cold, then strain off for use. When employed in fevers, a little sweet spirit of nitre is sometimes added.

When fresh lemon-juice cannot be procured, an excellent lemonade can be prepared from lemon syrup, made as follows: dissolve ten drachms of tartaric or citric acid, and eight pounds of loaf sugar, in a gallon of water. Then rub from half a drachm to a drachm of fresh oil of lemon with a portion of the syrup, and afterward carefully mix it with the remainder.

Lemonade, when freely taken, sometimes produces pain in the bowels. It must, therefore, be used with some reserve as a daily drink.

ARROW–ROOT. — Arrow-root, one tablespoonful; sweet milk, half a pint; boiling water, half a pint; sweeten with loaf sugar. Excellent aliment for children when the bowels are irritable.

VEGETABLE SOUP. — Take one turnip, one carrot, two potatoes, and one onion. Let them be sliced, and boiled in one quart of water for an hour. Add as much salt as is agreeable, flavor with a small portion of pot-herbs, and pour the whole upon a piece of dry toast. This is an agreeable preparation, and may be given when animal food is inadmissible.

MUTTON BROTH. — Select two tender mutton chops, put

them into a saucepan, add one quart of cold water, and a little salt, cover the pan, and cook slowly for two hours ; then skim off the fat, and add a tablespoonful of rice, one white potato, one turnip, and a little parsley, chopped fine. Simmer for three quarters of an hour, then pour into a bowl, and remove the chops, with all of the remaining fat. This broth is nutritious and palatable.

BEEF TEA. — Take of lean beef, cut into shreds, one pound, cold water, one quart. Heat slowly to the boiling point, and then boil twenty minutes, taking off the scum as it rises. Strain the liquor, and add salt according to the taste. This preparation is more nourishing than ordinary broths, and very palatable.

LIEBIG'S BEEF TEA. — One pound of lean beef, free of fat, and separated from the bones, in the finely chopped state in which it is used for mince-meat, is uniformly mixed with its own weight of cold water, slowly heated to boiling, and the liquid, after boiling briskly for a minute or two, is strained through a towel, from the coagulated albumen and fibrine, now become hard and horny. Thus we obtain an equal weight of the most aromatic soup, of such strength as cannot be obtained, even by boiling for hours, from a piece of flesh. This is to be seasoned to taste.

ESSENCE OF BEEF. — Put a sufficient quantity of lean beef, sliced, into a porter-bottle to fill up its body ; cork it loosely, or not at all, and place it in a pot of cold water, attaching the neck to the handle of the vessel by means of a string. Boil for an hour and a half or two hours, then decant the liquid and skim it.

To this preparation may be added spices, salt, wine, brandy, etc., according to the taste of the patient and nature of the disease.

CALVES'-FEET JELLY. — Take a set of four feet, break them into small pieces, add to them one gallon of water, and reduce by boiling to one quart. Strain, and, when cold, skim the fat entirely off. Add to this the whites of six eggs, well beaten, a pint of wine, a pound of loaf-sugar, and the juice of four lemons, and let them be well mixed. Boil the whole for a few minutes, stirring constantly, and then pass the jelly through a flannel strainer. This forms a very nutritious article of diet for the sick and convalescent.

BLANC–MANGE. — Boil one ounce of shred gelatine in a quart of milk, for a few minutes, stirring constantly. Sweeten to the taste, flavor with peach-water or essence of vanilla, and strain into a mould.

CHICKEN WATER. — Take one half of a chicken, divested of all fat, and break the bones; add to this half a gallon of water, and boil for half an hour. Strain, and season with salt. A nutritious drink.

MUCILAGE OF STARCH. — Take of starch, one ounce, powdered cinnamon, one drachm, gum-arabic, one ounce, boiling water, three pints. Boil until reduced one-third, and strain. The above may be taken for a common drink in dysentery.

MULLED WINE. — Take two drachms of bruised cinnamon, half a nutmeg, grated, ten bruised cloves, and half a pint of boiling water. Infuse one hour, strain, and add of hot port or sherry wine (or good domestic wine), one pint, and white sugar, one ounce. Mix. This is a mild, stimulant drink, used in convalescence from low forms of disease.

WINE WHEY. — Heat half a pint of milk to the boiling point, and, when boiling, add a gill of sherry or Madeira wine. Let it boil again, and then remove from the fire, and let it stand a few minutes. Then remove the curd, pour the whey into a bowl, and sweeten it. A little sugar and nutmeg may be added if desired. This is a mild and very useful stimulant, and may be used in diseases where there is great debility; the dose to be regulated by the circumstances of the case; from a gill to a pint or more may be taken during the day.

FLAXSEED MEAL, *and other Poultices.* — Take of the flaxseed meal, or other material, sufficient to serve the intended purpose; pour boiling water over it, stirring briskly until it becomes a thick mass; spread it upon an old piece of linen or cotton cloth, and place a small piece of netting or lace over it, so that it does not soil the clothing. Apply to the seat of pain.

CHAPTER XIV.

CASUALTIES — EXTERNAL INJURIES — BURNS AND SCALDS.

BURNS AND SCALDS.

BURNS and scalds, unless superficial and of small extent, are always troublesome to manage. When covering a large surface, or deep, they are both dangerous and troublesome. In the treatment of burns there are two essential points: First, careful attention to the constitutional symptoms in severe cases ; and, second, the prevention of adhesions and contractions during the process of healing.   When the shock is severe, and there is great constitutional depression, it will be necessary to give stimulants ; these, however, should be given sparingly and with discrimination, for, when reaction takes place, it may even proceed to inflammation. The healing surfaces of fingers or other portions of the extremities should be widely separated by splints and bandages to prevent deformity.

The **Treatment** for superficial burns, of slight extent, is extremely simple.   The most convenient and effectual is first to evacuate the serum or contents of the blisters, cover the part with raw cotton, and apply a bandage firmly over it.   Other excellent remedies are detailed at Nos. 112, 143, 144.

CONCUSSION OF THE BRAIN.

This may arise from a fall or blow upon the head, or from some violent shock to the body.   The symptoms will usually depend upon the severity of the shock.   In cases where the violence has been comparatively slight the disturbance of the intellectual functions will be transient. There will usually be some vertigo, dimness of sight, trembling of the limbs, and sickness of the stomach. In severe cases loss of sensation may exist for many hours, and finally be followed by reaction, which, if not controlled by treatment, increases to inflammation.

TREATMENT. — Remedies Nos. 113, 152, 153.

SPRAINS.

These are caused by falls, lifting heavy weights, jerks, false steps, etc. Sprains of joints, when severe, often arise from momentary displacement of the bones, which strains or perhaps partially tears the ligaments surrounding the joint. They are often troublesome, and require rest and bandaging a long time after the occurrence of the accident.

The **Treatment** consists in bandaging the part with cloths wrung out in cold water, to which a little tincture of arnica has been added. Internally, Remedy No. 114.

WOUNDS.

Wounds are classified into incised, contused, lacerated, punctured, poisoned, and gunshot wounds. *Incised wounds* are clean cuts made in the soft parts with a sharp instrument. The troublesome feature of such wounds is hemorrhage or bleeding. If an artery has been cut, the blood spouts in jets, and is of a bright-red color; if a vein, the flow of blood is gradual, and of a purple color. The treatment consists in cleaning the wound by removing anything that may be left in it, arresting the hemorrhage, bringing the cut surfaces and edges in close contact, and retaining them in such position. Small superficial wounds seldom require anything more than a bandage, snugly applied after the edges have been nicely adjusted. Wounds of greater depth sometimes require a stitch or two ; but, as a general thing, small strips of adhesive or arnica plaster properly applied will answer every purpose. No more dressing should be applied than is actually necessary to keep the parts together. After a wound is dressed, the injured part should be kept in such a position that the wound will not gap. To accomplish this, it is sometimes necessary to apply a splint. Surgeons nowadays apply nothing to wounds for the purpose of healing them, except cold water. Ointments, salves, and a host of other things once used, have long since been abandoned. After a proper dressing has been applied to the wound, if bleeding still continues, apply cold water or pounded ice, nothing more. Should this not arrest it, and the blood be of a bright-red color, spouting out at intervals as the pulse beats, endeavor to compress the artery between the heart and the wound. Feel for the artery on the interior part of the limb ; you will know it by its beating when found. Place over it a large-sized cork, or a compress made by folding up a piece of cloth about two inches square, and as thick as your finger, and bind it down firmly with a roller. This will arrest the bleeding until you can procure professional assistance. (No. 115.)

Contused Wounds, Bruises, etc. — These are occasioned by blunt surfaces, falls, or forcibly coming in contact with some object. There is generally no break or division of the external surface, consequently the hemorrhage is comparatively slight and internal, forming what is called a " black and blue " spot. If the wound is severe there is little pain, the life of the part being destroyed ; if the wound be slight the pain is often intense. There is usually swelling, and discoloration of the skin, Treatment consists in the prompt and continued application of cold water. The water can best be applied by saturating linen cloths with it, and keeping them applied to the parts. In severe cases, where suppuration is about to take place, it should be hurried forward by poultices. When all the dead flesh separates and comes away, this application should be changed to simple dressing of water, perhaps it will also be necessary to support the parts by adhesive strips.

Lacerated Wounds are those where the soft parts are torn or rent asunder by violence, leaving a ragged, irregular wound. Hemorrhage is usually slight. Treatment consists in cleaning out the wound, and adjusting the parts as near to their natural position as possible, and securing them by as little dressing as practicable. As these wounds generally suppurate, spaces should be left between the adhesive strips to allow the matter to escape, and thus prevent abscesses from forming. Cold water, to which a little calendula has been added, should be constantly applied. Lacerated wounds are prone to inflammation.

Punctured Wounds. — These are made by some sharp, narrow instrument, as a needle, pin, thorn, splinter, piece of glass, etc. Slight wounds of this kind are seldom troublesome, provided the substance can be removed, unless it extend deep down among the tendons and nerves, where matter may form, causing great pain, and even deformity. When a person runs a nail or splinters of glass into the foot, they should be removed. When this cannot be done, all that need be applied is a little Canada balsam. This should be renewed every day. If there is much inflammation, apply cold water.

Poisoned Wounds. — Bites and stings of insects, — bees, spiders, bugs, mosquitos. The bites and stings of insects, though seldom dangerous, are often exceedingly troublesome. The treatment for wounds of this kind consists in removing the sting of the insect, when it remains in the part, applying a plaster of damp earth, and keep it wet afterward with a mixture of *arnica* and water. Should inflammation and fever ensue give internally *aconite*. For mosquito-bites apply *spirits of camphor* or lemon-juice.

DISLOCATION OF JOINTS.

To be skilful and successful in the reduction of dislocations, a perfect knowledge of the anatomy of the joints is indispensable. A dislocation may be recognized by the following symptoms: in addition to the pain there will be loss of motion, swelling, alteration in the shape, length, and direction of the limb. The treatment consists in reducing the luxation as speedily as possible. This, however, a layman can seldom do. Still you can make a trial. If you do not succeed apply a solution of arnica — a spoonful to half a pint of water to the injured part. Give arnica internally (five to ten drops in a wineglass of water), and await the arrival of a competent surgeon.

FRACTURES.

Fractures of bones may be recognized by the deformity, which, by comparing the sound limb with the injured one, will be readily recognized. The most certain sign, perhaps, is that of *crepitation*, which is a peculiar grating sound, distinctly heard, when the two broken surfaces of the bone are rubbed together. In all cases of suspected fracture place the limb in the most comfortable position, and keep it constantly bathed with a solution of arnica, after which send for a competent surgeon. Do not get so excited as to accept the first doctor you can get, without any knowledge of his abilities. The case will take no harm if a whole day should elapse before you obtain assistance. If the patient should be faint and weak give an occasional dose of camphorated spirit, or aconite (about ten drops in a wineglass of water).

FOREIGN SUBSTANCES IN THE EYE, EAR, NOSE, AND THROAT.

In the Eye. — No matter what has gotten into the eye, washing with cold water will always be beneficial. Rubbing the eye only increases the irritation, and should, therefore, always be avoided. When lime or ashes enter it a little cream or sour milk is the best remedy. If a hard subject or an insect has got into the eye draw the eyelids apart, and turn the upper one over the lower one a couple of times until it is felt that the substance is removed. If particles of iron have entered, and have become fixed, bathe with arnica lotion, ten drops in a teacupful of water, until you can have it extracted. Should there be much inflammation use a tincture of aconite (a teaspoonful in half a pint of water) as a fomentation.

In the Ear. — Insects sometimes find their way into children's ears; in such cases lean the head to one side, and fill the ear in which the insect is with sweet-oil. This floats it to the surface, when it can easily be removed. If a bean or any other substance which will swell by heat and moisture gets into the ear, the best way to remove it is to make a hook by bending a hair-pin into the right shape. This should be cautiously introduced behind the substance, and an effort made gradually to extract it. After the operation wash the ear out with a lotion of arnica.

In the Nose. — Foreign substances may be removed from the nose with a small pair of forceps, or the same instrument recommended for the ear. First, endeavor to eject it by sneezing, which may be excited either with snuff, or by tickling the nose with a feather. Sometimes the obstruction may be pushed back so as to fall into the mouth. If these means fail apply to a surgeon.

In the Throat or Windpipe. — If a foreign substance lodges in the throat, first examine closely, and, if within sight, endeavor to extract it with the fingers. If it is not visible excite vomiting immediately, by tickling the throat with a feather, or by putting mustard or snuff far back upon the tongue. Foreign substances have been removed from the windpipe by gently turning the patient upside down.

CHAPTER XV.

POISONS, AND THEIR ANTIDOTES.

SUBSTANCES which derange the vital functions and produce death by an action not mechanical are denominated poisons ; and the most eminent authorities on this subject have divided these substances into three classes, viz. : —

Irritant Poisons, or those which produce irritation and inflammation, such as mineral acids, — arsenic, copper, etc.

Narcotic Poisons. — Those producing stupor or delirium, and specially affecting the brain and nervous system, as opium, prussic acid, etc.

Narcotic-acrid Poisons. — Those producing irritation or narcotism, and sometimes both together, — all of which are derived from the vegetable kingdom, — as strychnia, nux vomica, etc.

As most everybody is compulsorily so situated that, through mistake or accident, poison of some kind is deposited in various accessible places about their dwelling, and may, therefore, at any time, be taken into the system, it is desirable that every housekeeper or mother of a family should be thoroughly versed in the use of antidotes and the treatment of the supervening irritation or inflammation.

If *Sulphuric, Nitric, or Hydrochloric Acid* is taken into the stomach in poisonous doses, give chalk or magnesia ; or, if this is not at hand, take the wall-plastering, and make it into a thin paste with water, and take it plentifully. Soap-suds is another antidote. Drink freely of milk, or any other mild fluid, both before and after the administration of the antidote. For the supervening inflammation use the same remedies as for gastric disturbances generally.

In the event of *Oxalic Acid* being taken by mistake, as it easily may be, from its close resemblance to Epsom salts, you must at once excite vomiting by tickling the throat with a feather, or by an emetic; afterwards administering the remedies we have given for nitric and sulphuric acids. Take freely of mucilaginous drinks.

For poisoning with Arsenic the first thing necessary is to thoroughly evacuate the stomach, and for this purpose give an emetic, and then follow with iron-rust, or the sesqui-oxide of iron largely diffused in water; and let the patient drink freely of mucilaginous, farinaceous, or albuminous drinks, and milk.

For poisoning with Mercury, and Corrosive Sublimate. — This is the usual form of mercury employed for committing suicide; and as it is used for bedbug poison, it is liable to be taken through carelessness or mistake. It is therefore requisite to point out some antidote the nearest at hand. Give whites of eggs in abundance, or gluten, if eggs cannot be had; or else milk. The most usual form of the supervening irritation is salivation or mercury fever. For these give nitric acid, and obtain for the patient fresh air and a nourishing diet as soon as possible.

For poisoning with Copper. — The blue vitriol, or sulphate of copper, and verdigris, are the forms in which this poison is most common; and when taken into the system, the most ready antidote is the whites of eggs. Great care should be taken to exclude vinegar, for this acid would add virulence to the poison. To allay or cure the remaining irritation, give sulphur twice a day.

Poisoning with Antimony, or Tartar emetic, is liable to occur, as this substance is sometimes used in medicine as an emetic. When given to excess it produces vomiting, attended with burning pain at the pit of the stomach, followed by purging and colic, stricture of the throat, and cramps. As an antidote give large draughts of warm water, and tickle the throat with a feather to induce vomiting ; also the decoction of any bark containing tannin, as oak, hemlock, or cherry-tree bark.

For poisoning with Zinc, or white vitriol, which is denoted by violent vomiting, sunken eyes, and pale face, cold extremities, and fluttering pulse, give the infusion of any of the substances containing tannin, and feed the patient with cream, butter, and chalk quite freely.

For poisoning with Lead. — Red lead and white lead, as well as sugar of lead, are liable to be taken by mistake ; and their poisonous effects are denoted by obstinate colic, spasms of the muscles, and sometimes apoplexy. As an antidote give water of ammonia, or hartshorn, or pearlash-water, or any of the earthy sulphates, as lime, etc. For lead paralysis give sulphur.

For poisoning with Cantharides, which is denoted by intense burning in the pit of the stomach, and pain in the lower abdomen, feeble voice, laborious breathing, strangury, and tenesmus of the bladder, headache, and delirium. To remove the cantharides from the stomach excite vomiting in the most speedy way, — by tickling the throat with a feather, or drinking a strong mustard-tea, or by taking snuff upon the tongue. Spirits of camphor is the best antidote.

For poisoning with *Morphine*, which is denoted by giddiness and stupor, and insensibility to external impressions, the first thing necessary is to remove the poison from the stomach by vomiting. Then keep the patient from sleeping, and in vigorous action ; administer strong coffee.

For poisoning with *Prussic Acid*, when not fatal, resort to cold shower-baths, and inhalation of diluted water of ammonia vapor; also a solution of carbonate of potash (of course as a vapor).

For poisoning with *Charcoal-gas,* or with any of the poisonous gases, cold affusion should be immediately resorted to.

Poisoning by *Strychnia* or *Nux Vomica* is denoted by strong convulsions, with great agitation and anxiety during the fits, rigidity of the entire body and limbs, lividity of the face and hands, etc. If taken in poisonous doses it generally proves fatal, in spite of treatment. If emetics are given, and the stomach is made to disgorge its contents sufficiently early, and the patient is not attacked with convulsions in two hours, he will generally be safe.

Alcohol. — Large quantities of alcoholic liquors, taken by persons unaccustomed to their use, often produce fatal effects.

The treatment in such cases is to evacuate the stomach as soon as possible; but from the rapidity of absorption this may not always be effectual. The patient must be kept roused, and cold affusion plentifully used. Large draughts of tea or coffee ought to be taken until the stomach is thoroughly evacuated.

Vinegar is an antidote for alkaline poisons, and obviates the ill effects of aconite, opium, poisonous mushrooms, belladonna, etc.

Coffee. — Strong black coffee, made of the berry lightly roasted, and drank hot, is an antidote to opium, nux vomica, belladonna, narcotics, mushrooms, poisonous sumach, bitter almonds, and all those substances containing prussic acid. It must be borne in mind, however, that the cause must be removed, if possible, first.

Camphor antidotes the ill effects of poisonous insects, and especially cantharides, whether administered internally or externally. It also antidotes the toxical effects of phosphorus, spigelia, and santonin. It is very useful for the after-effects of acids, salts, metals, etc., after the poisonous substance itself has been removed from the stomach by means of vomiting, etc.

Milk is frequently alluded to as an antidote for poisons; but it has no real merit. Mucilaginous substances are better, and much to be preferred.

Olive-oil ranks with milk, and is much less useful than is believed. It is useless in metallic poisons, and even hurtful in poisoning with arsenic. It is of some service, however, in case of poisoning either with *nitric, sulphuric,* or *phosphoric acid.* Olive-oil and vinegar, administered in alternation, have proved serviceable in cases of poisoning with alkalies.

Soap. — Castile soap, dissolved in four times its bulk of hot water, and drank, will antidote many cases of poisoning with *corrosive sublimate,* and also with arsenic, or with any of the numerous forms of lead. Soapsuds is likewise a valuable antidote for poisoning with *sulphuric* and *nitric acid.* Soap is hurtful in cases of poisoning with alkalies.

Sugar, or sugar-water, is quite as good as any of the antidotes, and much to be preferred in cases of poisoning with paint, verdigris, copper, sulphate of copper, alum, etc. In cases of corrosive sublimate, in solution, being taken into the stomach by mistake, as it has been, sugar-water may be given before the white of an egg. Sugar is also an excellent antidote in cases of poisoning with *arsenic.* The other antidotes are : —

Ammoniacal gas, or the volatile odor of spirits of hartshorn, for poisoning with alcohol, bitter almonds, or prussic acid.

Iron-rust, or the sesqui-oxide of iron, for poisoning with arsenic.

Epsom salt, for the various alkaline poisons.

Charcoal, for poisoning with putrid fish, meat, or mush-rooms, or mussels.

Kitchen salt, for poisoning with nitrate of silver, and oisonous wounds.

Magnesia, for poisoning with any of the mineral acids.

Potash and sweet almond oil are also good antidotes for acids.

Starch, in solution, is the best antidote for poisoning with iodine.

Strong tea is a good antidote for poisoning with honey; and so is wine for noxious vapors and poisonous mush-rooms.

A general antidote for poisoning, in which the nature of the poison is unknown, consists of equal parts of calcined magnesia, pulverized charcoal, and hydrated peroxide of iron, which are to be diffused in water, and given freely. Although these articles are simple and innocent in their operation on the system, they will probably prove efficient, as one or another of them is an antidote to most of the mineral poisons.

The albumen of eggs and tannic acid are also general antidotes of great value. But, if we should know the poison to be mineral or metallic, the first resort may be to white of egg, sugar-water, soap-water, or soapsuds, and for the remaining effects give sulphur, which has been ascertained to be a real antidote to the effects of metallic poisons. If it should be known that acids or corrosive substances have been taken give Castile soap mixed with four times its bulk in warm water, or magnesia dissolved in water, or powdered chalk stirred up in water, or a solution of saleratus, pearlash, or super-carbonate of soda, in spoonful doses, after each paroxysm of vomiting, as long as it continues; and after-wards give mucilaginous drinks, and, alternately, coffee and opium. In case it should be known that alkaline substances have been swallowed in poisonous quantities vinegar and water may be liberally given, and also lemon-juice or cranberry-sauce, or sour milk, without sugar ; and for the secondary effects of poisoning with potash give coffee or powdered vegetable charcoal. If the poisoning is known to be the effects of spirits of hartshorn give *sulphur ;* if it should be known the patient had been inhaling *noxious vapors* bathe him freely with vinegar and water, and let him inhale the vapor of a solution of hydrochloric acid ; and, after a return to full consciousness, give a strong decoction of partially-charred coffee.

The *vapors of coal,* when having a poisonous effect, may be antidoted by copious draughts of vinegar and water ; and for poisoning by any substance not mentioned first follow the rule of removing the poison from the stomach by vomiting ; if it is known that any of the *animal poisons* have been taken into the stomach give a teaspoonful of powdered charcoal in half a tumbler of water at a dose, and repeat after each vomiting until the stomach becomes quiet ; and if a poison of this kind has come in contact with the eyes give aconite ; and, finally to guard against the infection of poisonous wounds, when touched with the fingers or hand, let them be held in strong heat, as strong as can be borne, for ten or fifteen minutes, and afterwards wash them with soap.

CHAPTER XVI.

UTERINE DISPLACEMENTS, FUNGOID GROWTHS, TUMORS, ETC.

WE had thought, at the first inception of this work, that, in deference to the inherent delicacy of our fair readers, and their natural reluctance to enter upon the discussion of affections connected with the female reproductive organs, we would omit any special reference to them in this volume ; but, in view of the lamentable fact that, from want of proper acquaintance with the peculiarities of their own system, from negligence or from constitutional weakness, fully five-sixths of the female sex throughout the world are suffering, to a greater or less extent, *from various affections of the womb;* and that to this cause alone is to be attributed the unhealthy, enervated condition and susceptibility to disease evinced in both parents and children, — for this reason, we, after consultation with several of our most experienced medical brethren, determined that, as a matter of common humanity and sympathy, it was our duty to present, in the most decorous, unobjectionable, and easily understood method that language would permit, a brief explanation of the nature, symptoms, and hygienic treatment of these affections, so that they might, to a great extent, prevent, or, at any rate, lessen, the inconvenience and suffering so generally and so uncomplainingly sustained by women. We are indebted to the generous courtesy of our eminent and talented brother, Dr. Ephraim Cutter, of New York, for the use of the admirable diagrams with which this chapter is embellished (with the exception of the first figure). Feeling convinced that our lady readers will accept the few following remarks in the respectful, sympathetic, and trustful spirit in which they are written, and credit us with an earnest desire to promote their best interests, while avoiding every term or allusion which should offend their sense of propriety, we will now venture on the consideration of

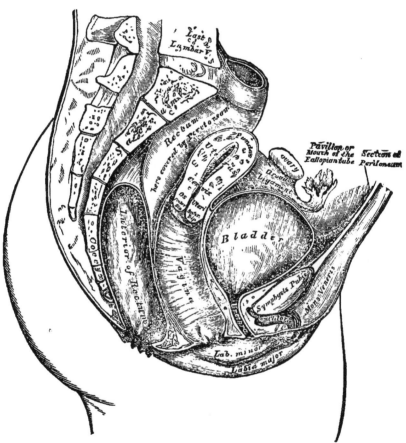

FIG. 1.—Pelvic Organs in position. Bladder distended. Womb virgin.

DISPLACEMENTS OF THE UTERUS OR WOMB.

The accompanying illustration (Fig. 1) represents the womb in its natural position in the uterine organism, with its adjacent organs and their relative action. But, from congenital or constitutional weakness, accidental or other causes, it is subject to various displacements, such as

Descent of the womb, or Prolapsus Uteri.

Retroversion of the uterus, or falling backwards of the womb.

Retroflexion of the uterus, or bending backward of the womb.

Retroversion and flexion, or bending and falling backward.

Anteversion of the uterus, or falling forward of the womb.

Anteflexion of the uterus, or bending forward of the womb.

Anteversion and anteflexion, or bending and falling forward.

Latero-version and latero-flexion or displacement sideways.

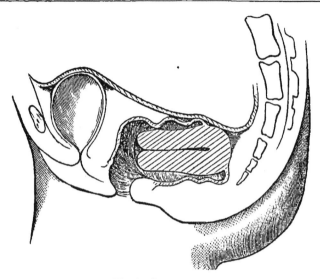

Fig. 2. RETROVERSION.

RETROVERSION OF THE WOMB.

This form of displacement is (excepting prolapse, or falling of the womb) the most frequent in occurrence. It rarely occurs as a disease, but is usually an accompaniment or symptom of the presence of inflammatory action in the uterus, or of an increase in its size and weight; in other words, it is usually caused by parturition, general muscular debility, or habits of indolence and inactivity. It frequently occurs as a consequence of pregnancy, congestion, the presence of fibroid tumors, a fall or other injury, or the pressure of the viscera down upon the fundus by tight clothing or muscular efforts. The absurd practice of tight bandaging after parturition, and making the patient lie upon her back continually, instead of on her side, is almost certain to cause this displacement. Retroversion may either be *partial* or *complete.*

Symptoms. — The most strongly marked symptom indicative of retroversion of the uterus is found in the difficulty of emptying the bladder, accompanied with pain and tenesmus, always more or less present. Next to this is the pressure of the rectum, and consequent frequent calls to stool, with difficulty or impossibility of evacuation. These two symptoms, especially the retention of urine, render the displacement dangerous and painfully distressing; while the accumulation of fecal matter renders the restoration to its natural position a matter of extreme difficulty. A gnawing or other pain in the back, backache, difficulty in walking, inability to stand for any length of time, nausea and vomiting, even to a violent degree, may set in, and unless the patient is promptly relieved she may sink under the accumulation of her sufferings.

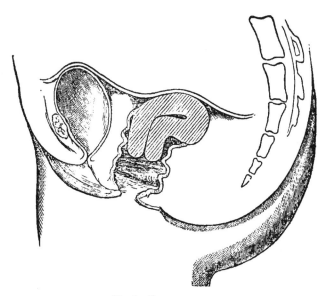

Fig. 3.   RETROFLEXION.

RETROFLEXION OF THE WOMB.

This form of displacement consists in the bending backward of the fundus and body of the womb toward the hollow of the spinal column, the womb being bent in such a way that the cervix or neck is not removed from its normal position, or is but slightly deviated from it. It is generally the result of a weakness of the uterine tissues, in consequence of which the body of the womb, either through its inherent weight, whether natural or preternatural, or in consequence of some force or pressure applied to it, is bent at the junction of the neck or cervix. Retroflexion occurs most frequently in women who have borne children, and seldom in the virgin uterus.

Symptoms. — Irritability of the rectum is one of the chief symptoms; and if the pressure upon the intestines be great, retention of stool will be a natural consequence. Neuralgia of the womb may occur as a result of the congestion and nervous compression, and so-called uterine colic may occur from a retention of the secretion of the intra-uterine mucous membrane. If the retroflexion exist in a marked degree, so as to close up the uterine canal, dysmenorrhœa and sterility will be the consequence.

TREATMENT OF RETROVERSION AND RETROFLEXION.

There are certain cases of Retroflexion which are unmanageable, and entirely beyond the control of the physician or the patient — for instance, those which have become chronic, and are surrounded with aggravated circumstances, causing incurable insanity, a permanent and agonizing derangement of the nervous system.

In ordinary or curable cases of Retroversion or Retroflexion, it is essential to cure that the uterus and vagina should be restored to their normal position and maintained in that position by means mechanically adapted to the parts allowing of the normal contraction of the vaginal fibres, and permitting a natural degree of mobility. Special care should be taken that the instruments employed should be manageable by the patient, and that general health should be maintained by ample nutriment. The catheter should be used regularly twice per day, until the uterus rises above the pelvis. The catheter should be small, flat, and curved considerably more than common, in consequence of the distorted course of the urethra. The bowels must be kept open, and absolute rest, in a recumbent posture, be enjoined. Should it be impracticable to draw off the urine, attempts must be made to replace the uterus by the medical practitioner who has the case in charge.

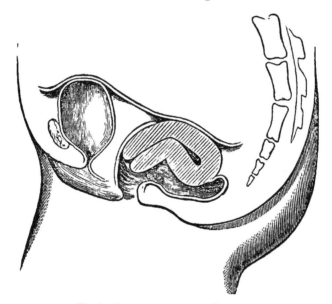

Fig. 4. Retroversion and Flexion.

RETROVERSION AND RETROFLEXION COMBINED.

Here the vagina is so much relaxed that the retroflexed uterus is thrown down as much as the sacrum and rectum will allow. This combination is very frequently found. It interferes with defecation to a greater extent than the two varieties previously described. It is more difficult to treat. In all these cases the patient should recline on a bed raised at the foot. The treatment should be similar to that already given.

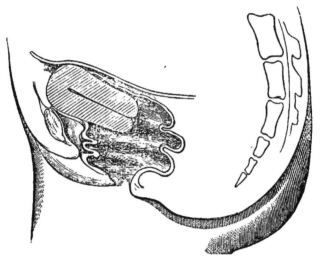

Fig. 5. ANTEVERSION.

ANTEVERSION OF THE WOMB.

This displacement, although it does sometimes occur, is comparatively rare in single women or those who have not borne children. The normal position of the womb is slightly forward, so that the natural tendency, in the event of pregnancy or undue exertion, is to fall forward rather than backward. But it is sometimes so morbidly increased as to give rise to some very unpleasant and painful symptoms. It may be considered in a condition of unnatural anteversion when it lies across the pelvis, with its neck in the hollow of the sacrum, and the fundus encroaching upon the bladder. Anteversion may be combined with flexion at the junction of the cervix with the body, in which case the fundus is thrown still more forward and downward.

Symptoms.—This form of displacement may come on gradually, and can be distinguished readily from prolapse by the vaginal touch, the body of the womb lying in a horizontal line across the pelvis. It may occur suddenly; and in this case the symptoms closely resemble those of prolapsus, or falling of the womb. It gives rise to very unpleasant urinary symptoms, and occasions more or less irritation of the rectal passage, constipation, and hemorrhoids. In some cases walking is rendered extremely difficult and even impossible. But the urinary disturbance is the most annoying, and prompts the woman to call in the aid of her physician.

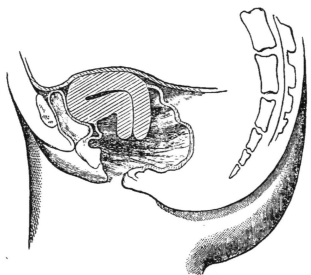

Fig. 6. ANTEFLEXION.

ANTEFLEXION OF THE WOMB.

This derangement is of common occurrence, and differs from anteversion, in that while the fundus and body of the uterus are directed downward and forward, the neck or cervix retains its proper position, or nearly so, although flexion of the uterus forward may comprise a bending in that direction of both body and neck at the point of junction of the two, the bending being in some cases nearly at an acute angle. Or, in other cases, while the body of the uterus retains its proper position, or nearly so, the cervix or neck becomes bent and extended forward toward the pubes.

Symptoms. — The symptoms produced by anteflexion are similar to those of anteversion, though generally rectal and vesical irritation are not so great. In consequence, however, of the bending of the womb, very serious symptoms may arise, which are the results of various diseases which this abnormal position may originate. Dysmenorrhœa is almost a constant attendant ; congestion of the womb may occur, and occasion a variety of diseased conditions, corporeal and cervical endometritis may be set up, and from pressure or other cause even peritonitis may result.

The same causes which occasion anteversion may operate to produce anteflexion, previously existing weakness of the uterine tissue at the junction of the cervix and body of the uterus being premised.

Treatment of Anteversion and Anteflexion. — In the treatment of these forms of displacement the patient must remain quiet, in a reclining position as much as possible, and, in severe cases, keep her bed and lie upon her back for a few days. In those cases only in which the fundus is actually thrown beneath the arch of the pubes will there be any necessity for manual interference. A patient suffering from this affection should not urinate too often, as by moderately distending the bladder she will greatly assist nature in the replacement of the womb. The health being once restored, the womb will naturally resume its normal position.

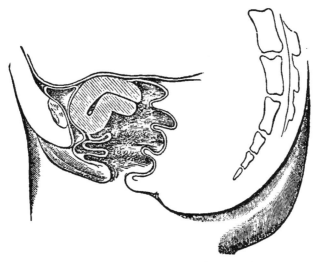

Fig. 7. ANTEVERSION AND FLEXION.

ANTEVERSION AND ANTEFLEXION COMBINED.

These constitute a formidable combination against successful treatment. The uterus forms a portion of the vaginal dome, reaching from the pubes in front, upwards, to beyond the middle of the highest point of the vagina. The uterus being so excessively flabby and yielding, every part is so much out of line as to render its normal replacement a matter of great difficulty. If the uterus is hypertrophied remedies should first be applied to reduce the size of the organ. Depletion or blood-letting (by applying two or three leeches) is one of the most prompt and successful agents in restoring contractility. The most frequent sequences to this combination are the irritable intra-uterine ulcer, and the doubling of the uterus over the pessary (if that instrument be used), which we disapprove.

It frequently occurs that the uterus shifts its position so readily that the patient may suffer alternately from *all* the displacements we have mentioned, the ligaments being so much relaxed that the womb *cannot* remain in one position for any length of time. In such cases it is best to convert it into a retroversion at once, allowing the organ to be slightly turned back, so as to be less liable to be thrown forwards during the body movements.

LATERO-VERSION.

There are two varieties of this displacement, but they are rare. They may be found associated with prolapsus (or falling), or with the backward and forward displacements, and result either in consequence of inflammatory conditions, unnatural weight, or direct pressure. The patient should be directed to lie upon the side opposite to that of the obliquity, and to keep at rest as much as possible.

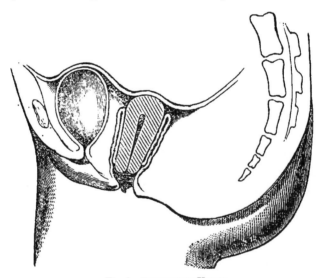

Fig. 8. PROLAPSUS UTERI.

PROLAPSUS UTERI, OR FALLING OF THE WOMB.

This is one of the most common forms of uterine displacement. It occurs in three different degrees, which we will briefly describe. Thus, *relaxation* or *simple descent of the womb* is understood to indicate the first and least displacement downward, and to consist only in a simple bearing-down of the womb upon the upper portion of the vagina. In *prolapsus uteri* the organ comes still lower down, and may even present itself at the lower orifice of the vagina. In *procidentia uteri* there is actual protrusion of the organ, even the entire body of the womb being in some cases extended from the vulva. These are but different degrees of the descent of the uterus in the line of the vagina. Upon examination of the same displaced uterus at different times of the day it will be found more or less prolapsed, according to the condition of active exercise or quiet in which the parts may have been for some hours previous.

Symptoms. — The principal indications are, dragging and aching pains in the small of the back; pulling and bearing-down pains in the lower part of the abdomen; a sensation of expulsion from the vagina; sufferings much greater after walking or other exercise; frequent calls to urinate, and incontinence or retention of the urine.

The primary cause of prolapse or descent is, of course, the excessive relaxation of the peritoneum and broad ligaments, by which the womb is retained in its proper position. (See Fig. 4, p. 25.) The system generally being much debilitated, the slightest strain or extra exertion increases the descent more and more, and renders recovery more doubtful. Persons of a scrofulitic temperament are especially liable to the affection; and, far more than all other causes combined, carelessness or improper treatment at and after confinement, are direct agents in producing this displacement. The want of prompt attention to the primary symptoms, and omitting to seek medical advice, would speedily transform the affection into a chronic displacement.

INVERSION OF THE UTERUS

is the most formidable and dangerous, as it is fortunately the rarest, form of displacement. It occurs principally as an accident in connection with delivery. When the inversion is complete the womb is turned entirely inside out, and protrudes beyond the orifice of the vulva, the mucous lining membrane consequently becoming the external covering. This naturally occasions a very serious complication of all the uterine organs, dragging the Fallopian tubes, ovaries, and bladder out of their natural position (see Fig. 4, p. 25), causing excruciating distress and derangement in each of the organs involved, and profound depression of the nervous system and vital strength. The presence of polypi, tumors, or malignant growths considerably increase the danger, not only from the unhealthy nature of the cause itself, and the fact that it may only be a development of still more serious forms of disease, but from the hemorrhage it may occasion.

TREATMENT.

The inverted womb should be carefully reduced, and the whole organ replaced in the pelvis, with as little and as gentle manipulation as possible. The patient should remain in bed in the position most comfortable to her, receiving such medicines as are best calculated to contribute to the functional and structural rehabilitation.

[In all affections of the womb the physician should be summoned without delay, the points of treatment we have mentioned being only of a temporary or provisional character. In an organism so delicate, sensitive, and complicated in structure, the slightest act of ignorance, recklessness, or nervous anxiety on the part of the nurse or friend in attendance (other than the physician) may be attended with the most serious or fatal consequences.]

THE IMMEDIATE EFFECTS OF THESE DISPLACEMENTS

are by no means confined to the uterus. They drag or force out of position every other organ within the trunk. A general derangement and disorganization ensue, and liver, spleen, heart, intestines, and kidneys are all involved, sympathetically and mechanically, in the disturbance. The machinery is entirely put out of order by this one neglected displacement, and, if not instantly attended to, will, in all probability, result in the most serious if not fatal, organic difficulty, and ultimate prostration or death.

CANCERS, TUMORS, FUNGOID GROWTHS, ABSCESSES, ETC., IN THE ORGANS OF GENERATION.

The reproductive organs are so constructed as to render them liable to a variety of complaints, which are apt to excite alarm in the mind of the female, however free from actual danger some of them may in reality be. Women, especially those who are married, when laboring under affections of these parts, are often kept in a state of great anxiety, until they are made acquainted with the exact nature of the difficulty, and that it will not be attended with danger. On this account, if for no other reason, every woman should be acquainted with the diseases to which they are incident, and thus, in a measure, be enabled to preserve their peace of mind, avoid the causes which tend to them, and be enabled to obtain a timely application of the necessary remedies.

The attention of our readers will naturally first be drawn to

AFFECTIONS OF THE EXTERNAL GENERATIVE ORGANS.

The **Labia Majora,** or **Outer Lips,** are, on account of their looseness of texture, subject to considerable enlargement or swelling, from comparatively slight irritation. It frequently happens with persons who neglect the daily washing of the parts during and after the menstrual flow, producing acrid secretions and intolerable itching; and, if scratching or rubbing is used for the purpose of allaying the irritation, the parts become much swollen and inflamed. The same remarks will apply to the Nymphæ, or inner lips.

ABSCESSES OF THE LABIA

sometimes occur, which are extremely painful, and are usually caused by blows, falls, forcible intercourse, or casual injuries of any kind, and sometimes from an inflammatory condition of the body, without any other cause. The symptoms are heat, swelling, redness, and throbbing pain in the part, extending to the groin and down the thigh; there is a circumscribed hardness, and the part is exquisitely tender. The treatment comprises rest and quiet, the application of leeches or poultices, according to the special symptoms, and brisk purgatives. Another affection to which they are subject is the formation of

ENCYSTED AND WARTY TUMORS,

both of which vary in size from that of a pea to a turkey's egg. The warty tumors are distinguished by having a pedicle or stem, and are apt to spread internally; but they are neither painful nor tender, and are only inconvenient from their size. In many cases they are of venereal origin, and frequently degenerate into unhealthy sores or ulcers. The treatment for their removal is necessarily surgical.

THE CLITORIS

is subject to certain diseased or abnormal conditions, and local inflammations. It is sometimes the seat of

CANCEROUS GROWTHS AND GLANDULAR ENLARGEMENT.

If cancers exist there is no alternative but excision by the knife; if there be malformation or enlargement of the part it is frequently found necessary to resort to amputation of the organ; but simple and uncomplicated inflammatory conditions will soon succumb to cooling and astringent lotions.

TUMORS, ABSCESSES, AND THICKENING OF THE MEMBRANE

also invade the orifice of the urethra, the pelvis, exterior to the vaginal canal, and the space between the vagina and rectum.

THE INTERNAL ORGANS OF GENERATION

are subject to a much greater and more appalling variety of affections and complications; and, from their being so intimately connected with every other section of the human system, so entirely disorganize and derange the functions of the body, that they may truly be said to originate three-fourths of the diseases of which the female frame is susceptible.

Married women frequently suffer from

ELEVATION, IMMOBILITY, AND HERNIA OF THE UTERUS AND OVARIES.

Elevation of the uterus is caused by abscess or insufficiency of the ligaments, dropsy of the ovaries, displacements of the uterus, and the presence of hydatids and other growths. *Immobility* is caused by peritoneal adhesions, or inflammation of the neighboring organs. Scrofulous women, girls of a lymphatic temperament, those who practise masturbation, and those who suffer from severe constipation, are especially liable to it. *Hernia* of the womb is fortunately a very rare affection, the tumor being hard, of a roundish form, and scarcely capable of reduction. There are three descriptions of this hernial trouble, called the *inguinal*, when it makes its appearance in the groin; the *crural*, when it appears on the thigh; and the *ventral*, when it takes place in the gravid uterus, through an accidental separation of the abdominal muscles. It originates from relaxation or weakness of the ligaments, a violent blow, or contusion, etc. The ovaries are subject to six different kinds of hernia, viz. : the *inguinal*, *crural*, and *ventral;* the *ischiatic*, when it appears near the loins; the *umbilical*, when the protrusion takes place at the navel, and the *vaginal* or *rectal*, when it takes place in those organs. These tumors are caused by tight-lacing, undue compression, wounds and abscesses, and violent emotion or crying. The treatment is purely surgical.

DISPLACEMENT OF THE BLADDER,

or rather its protrusion into the vagina and external labia, is a very frequent and annoying affection. Fig. 1, in our chapter on uterine displacements, will plainly show the connection between the vagina and bladder, and the means by which the accident may arise. The predisposing causes are : an excessively large pelvis, numerous pregnancies, violent exertion of the muscles, leucorrhœa, retention of urine, an abuse of sexual indulgence, or warm bathing, tight-lacing and the use of busks, general debility, and peculiar conformation. The principal exciting causes are : parturition, violent exertion of any kind, coughing, vomiting, excessive dancing, etc. It is more common in women who have borne children.

The tumor presents itself usually within the vaginal canal, or just outside the vulva ; it is bluish, with a polished surface, when the bladder is distended ; and at other times is wrinkled and soft. It is frequently complicated with various displacements and inflammations of the womb. When it occurs in a pregnant woman it may become so large as to prevent the delivery of the infant.

DISPLACEMENT OF THE INTESTINES INTO THE VAGINA,

arises from the same causes as prolapse of the bladder ; but is much more amenable to treatment, and is not attended with such disagreeable or disastrous consequences.

PROLAPSUS OF THE VAGINA

means a turning inside out of the lining membrane of that organ. It may be either *complete* or *incomplete*. In the former the tumor projects more or less beyond the vulva ; in the latter it remains within the vaginal orifice. The predisposing causes are lymphatic temperament, chronic leucorrhœa, frequent childbearing or abortion, abuse of hot bathing and of warm, relaxing drinks, bad nutrition, and general debility. It may be immediately brought on by abuse of marital privileges, injury during labor, external violence, excessive efforts of any kind, etc. It is a very troublesome affection, and may become dangerous during labor.

MOLES, HYDATIDS, AND FALSE CONCEPTIONS

are growths within the cavity of the uterus, originating in the destruction of the fœtus a short time after conception. The causes of this destruction may be found in general debility, mental and bodily shocks, and irregularities of the nervous and circulatory systems. The fœtus is seldom retained in the womb more than two or three months, but if not then expelled, it becomes transformed into the fleshy mole. These moles and hydatids are found, sometimes singly, varying in size from a pea to a grape; and sometimes in clusters, attached by a stem, to the outside of the ovum and placenta. They are necessarily consequent upon sexual intercourse and impregnation, and are simply blighted or imperfect conceptions. The symptoms exactly resemble those of pregnancy, and can only be distinguished therefrom by the absence of fœtal movement and pulsation.

FIBROUS TUMORS OF UTERUS AND OVARIES.

This appellation is applied to a species of fleshy tumor attached to the body of the uterus, and consisting of a mass of irregular fibres, bound together by cellular tissue, which do not ulcerate or become malignant. They vary in size from that of a pea to a man's head, and have been known to weigh as much as forty pounds. They are usually found in persons of a lymphatic or scrofulous temperament, and may be caused by chronic leucorrhœa, celibacy, barrenness, abortion, and from too close application to sedentary employments. They are mostly met with between the ages of thirty and fifty. When these are present, conception is possible, but abortion at the third or fourth month will most probably follow ; if pregnancy proceeds to the full term there is great danger of flooding.

POLYPUS OF THE UTERUS.

This description of tumor is similar in structure to the fibrous, but differs in form, situation, and symptoms.   They may be attached to *any part* of the uterus, vary in size, shape, and color, and are accompanied by a leucorrhœal or hemorrhagic discharge, according to the stage of growth. The appetite becomes impaired, bowels relaxed, and dropsy of the extremities is frequently induced.   They are oval or pyramidal in shape, and appear both singly and in clusters.

CAULIFLOWER EXCRESCENCES OF THE UTERUS

are usually found at its mouth, and are met with in women of all ages, temperaments and conditions of life, married and unmarried, without regard to habits or residence.   They are mostly hereditary or congenital in origin, very vascular, of a bright, fleshy color, with a smooth or slightly granulated surface, upon which are numerous small projections.   They vary in size from a strawberry to a bulk sufficient to fill the entire vaginal orifice, and are always attended by more or less hemorrhage and watery discharge.   They bleed freely upon being touched, and if extirpated, grow again with great rapidity.   The stomach and bowels become much disordered, and the patient is liable at any time to die from dropsy, or effusion into some of the great cavities of the body.

Cancers and corroding ulcers of the uterus and ovaries are diseases of the most serious and excruciating character, which, unfortunately for the sufferers, are seldom detected and treated in their earlier stages ; consequently they have been classed among the incurable diseases. They are malignant and contagious in their character, poisoning the glands, tissues, and fluids of the body and adjacent organs as they come in contact with them. They are generally accompanied by hemorrhage and leucorrhœa, and, from the severity and critical character of the affection, demand the most skilful treatment that medical science can afford. They may be found in almost any part of the uterine, ovarian, and abdominal regions, and are so varied in their appearance, form, and character, that even to attempt a description of them would require space far beyond the limits of this volume.

CHAPTER XVII.

THE CHANGE OF LIFE.

By this phrase, or the " critical period," as it is sometimes termed, is understood the final cessation of menstruation, and, consequently, the capacity for childbearing. Popular opinion fixes the time for this change at about the forty-fifth year, and all women anticipate its occurrence about that time; but many are disappointed, for some women have been known to cease to flow at the thirtieth or thirty-second year, while others have continued to menstruate to the fiftieth year, or even later; but, as a general thing, the average period of its occurrence may fairly be fixed at the age of forty-five.

Women of delicate constitutions, and those who have been in the habit of living well, enjoying the good things of life, and whose occupations have been sedentary, who have been confined to the house, and especially to warm rooms, experience the change earlier than those of a more robust organization, or those who have led a temperate, active life, avoiding all dissipation.

This period of life, which is rightly considered a critical one for every female, may pass without a single untoward symptom, the monthly evacuation gradually ceasing, without being attended by any unpleasant consequence, and leaving the patient enjoying better health than she ever experienced before. On the contrary, it may be fraught with peril, through which she can be safely conducted only by a skilful and experienced physician. It is, therefore, highly important that all the unpleasant sensations which may be experienced during this time should receive a careful consideration, and not be hushed up with the unsatisfactory reply that such complaints are owing to the " change of life," and likely to vanish whenever that change shall become complete.

If proper attention is not paid to the various affections which may and frequently do manifest themselves during this period, the seeds of endless miseries and even early death will be allowed to germinate and cut short a life that, by proper foresight and care, might have been conducted to a ripe old age. As the change approaches, the menses gradually become irregular, both in regard to the time of their recurrence, and the quantity discharged. They may return too soon, or be delayed beyond the usual time. The quantity discharged is at times much less than common. Sometimes the discharge returns every two weeks, then ceases for several weeks, or even months, and afterward recurs for a few periods as regularly as ever, and then ceases altogether.

Perhaps, in the majority of women, while this change, which usually extends over a year or a year and a half, is in progress, there is more or less disturbance of the general health. It is often difficult, and even impossible, to say exactly what is the matter with the patient, except that she is generally out of health. A host of symptoms present themselves ; the patient complains of headache, vertigo, biliousness, indigestion, flatulency, acidity of the stomach, diarrhœa, costiveness, irregularity in the urinary discharge, piles, pruritus, violent itching of the privates, cramps and colics in the abdomen, palpitation of the heart, nervousness, pains in the back and loins, swelling of the abdomen and the extremities, paleness and general debility. To unravel all these, and to decide what is the best plan of treatment to pursue, requires the skill of an intelligent physician. Remedies Nos. 117 and 157 will be found effective. It is very important in these cases to pay strict attention to dress, diet, and exercise. The diet should be light and easily digested. Everything of a stimulating nature, unless ordered by the attending physician, should be studiously avoided. Daily exercise in the open air, either by walking or riding, will be found highly beneficial. The clothing should be warm and comfortable, and changed to suit the weather.

OLD AGE AND ITS DISEASES.

Life, like the natural day, has its morning, its noon, and its evening; and its decline, unlike the period of youth and manhood, is marked by a calm, quiet stillness and peaceful repose. The decline of old age reverses the natural order of growth, the waste of tissue and nerve force gradually but surely outstripping the addition resulting from the food taken into the body daily. Declining age may be said to extend from fifty to sixty. Incipient old age from sixty to seventy. Ripe old age from seventy to eighty, and decrepitude or second infancy from eighty to the close of life. During all these periods, particularly the latter, important structural and other changes are occurring in the human system, and piles, apoplexy, paralysis, diseases of the liver, kidneys, and bladder, with organic changes of the heart, dropsy, chronic affections of the respiratory organs, gout, etc., frequently present themselves.

There is an *hygiene* for old age as well as for youth,— a means of preserving health, which may lengthen out their days to the utmost limit without the intervention of any of these affections which we have named. The first essential is *regular habits.* Any *sudden* cessation of a practice long indulged in, such as the use of tobacco or any narcotic, ardent spirits, or malt liquors, will often prove fatal to the aged. Even the hours of taking meals should not be changed; nor should old people transfer their residence to new climates, or form new social relations, their decreased vitality being insufficient to sustain the altered physical and mental conditions.

Diet and Regimen. —This is by far the most essential item in maintaining the healthy condition of those who have passed life's meridian. We need scarcely say that special attention in the selection of food easy of digestion is a primary requisite. Plain boiled or roasted meats are frequently unsuitable, —they should always be finely chopped, and thoroughly cooked. Soups and milk are excellent articles of diet; and, in *moderation*, ripe fruits, and plain puddings and pies are not objectionable. Vegetables should be sparingly partaken of. Wine, good and pure, if judiciously administered, will be found beneficial in cases of unduly diminished vitality. But, above all things, the aged should be cautious never to eat to excess or repletion; they should eat slowly, and chew their food very thoroughly.

Aged people suffer very much from cold hands and feet, and indeed from languid circulation and low temperature generally. The heart and muscular system usually becomes feeble and relaxed. The clothing of elderly people should therefore be much thicker and warmer than that worn in youth and manhood. We must prevent the escape of what little animal heat there is by flannel worn next the skin, and by woollen clothes generally, they being bad conductors of heat. It is during winter nights that the old are apt to suffer most from cold. If the bed clothing is not sufficient to create the required amount of warmth, artificial heat, in the shape of a heated stone, or bottle of warm water, should be applied.

A far greater number of aged people die in winter than in summer or fall. For this reason they should be very careful how they expose themselves to frequent and sudden changes of temperature, or inclement weather.

Important as the care of the skin is to the health of persons at all periods of life, it is especially so in old age. In the decline of life, the scarf-skin exhibits a tendency to become dry and peel off; this may be prevented by frequent and regular ablution in tepid water, followed by brisk rubbing, or if the skin be too tender for the application of water, friction alone can be employed, either by the naked hand, a piece of soft flannel, or a flesh-brush. In rubbing the abdomen, the rubbing should be across, from right to left, in order to remove or obviate constipation or flatulence. Exercise of any kind — walking, riding or working — should be used in the greatest moderation, as great fatigue would always prove injurious. Aged people should get as much sleep as they feel to want, from eight to ten hours out of the twenty-four being by no means too much. They should always retire early. Many aged people suffer considerably from inability to sleep ; but, for all that, the use of narcotics should be avoided as much as possible. A great deal may be often effected by taking early and light suppers. Early rising and exercise in the open air will generally produce refreshing sleep at night. With many persons electricity, moderately indulged in, of course under the advice of a physician, will be found to have an extremely beneficial effect.

APPENDICES.

APPENDICES.

Appendices A and B are intended as an intelligent and infallible guide to the interpretation of symptoms and the temporary remedial agencies for the numerous affections — chronic, acute, or incidental — to which women and children are especially liable at all ages, and under all circumstances. For instance, supposing a child to present the premonitory symptoms of measles, scarlet-fever, or colic, the mother or nurse, from inexperience, ignorance of the symptoms, or the natural anxiety of the parents, is uncertain as to the nature of the affection by which the child is threatened. By reference to Appendix A, she will be able to form a tolerably correct judgment as to the probable cause of the physical disturbance, and by referring to the disease mentioned in the table of remedies in Appendix B, will at once have it in her power to intelligently and safely administer the appropriate temporary remedy, which may mitigate the patient's suffering, and arrest the progress of the disease until skilful professional aid can be obtained.

APPENDIX A.

DISEASES AND THEIR SYMPTOMS.

The following table is designed to enable the mother, wife, nurse, or attendant, to form some idea of the disease or affection under which the patient may be suffering by the symptoms which may present themselves, and thereby to intelligently administer local or temporary treatment, and mitigate the suffering : —

DISEASE.	SYMPTOMS.
Dropsy — Inflammation or Obstruction of Bowels..........	*Abdomen.* — Increased size. Can only lie upon the back, pinched countenance, tongue furred and dry, increased thirst, diminished perspiration, diminished secretion of urine, small pulse.
Dyspepsia	Cold hands and feet, red or yellow deposits in urine, voracious appetite (occasionally), black flecks floating before the eyes, tongue white or furry; when *acute*, coating of tongue peels off, fetid breath, clammy mouth, bitter taste, eructations, weight at pit of stomach.
Disease of the Heart or Lungs, interfering with breathing	Maintaining the sitting posture only, or lying upon the unaffected side only, forcible and rapid dilation of the nostrils, debility, cough harsh and concentrated, diminished appetite.
Pleurisy	Pain in parts moved in breathing, harsh and concentrated cough, breathing diminished in rapidity, ability to lie upon one side only, depression or retraction of one side of chest; sharp, tearing pain below the nipple, enlargement of one side of chest.

DISEASES AND THEIR SYMPTOMS.— *Continued.*

DISEASE.	SYMPTOMS.
Convulsions	Head bent on one side, temporary spasm.
Consumption	Hands and feet hot, night-sweats, fat in stools, constant pain between shoulders, pain darting from front part of chest to between shoulder-blades, mucopurulent expectoration, or expectoration of pus, harsh and concentrated cough, depression or retraction of one side of chest, breathing slower, paleness of face, faint and sweetish smell of expectoration, hollow, barking cough.
Erysipelas or Small-pox........	Swollen scalp, frequent pulse, diminished secretion of urine, external local heat.
Nervous diseases generally	Cold hands and feet, white sediment in urine (in severe cases), small, weak pulse, hollow and barking cough, trembling, exaltation of vision.
Paralysis of one half of the body..	Distorted features, altered position and impaired motion of limbs, head bent to one side, eyelids remain open, limbs immovable and diminished in size, retention of urine in the bladder.
Fevers	General heat of surface, chills, fetid smelling sweats, red or yellow sand deposits in urine, tongue trembling, dry, and diminished in size (in low fevers), morbidly increased sensation, tongue white, afterwards clean, red, and dry, with much thirst; paleness of face (in cold stage), restlessness and tossings.
Typhus Fever	Bluish tint of head, face, and neck, tongue white and loaded, or dry, parched, and black, difficult of protrusion and trembling, dull hearing, debility, loss of moral sensibility, acrid heat, burning the hand when applied.

Diseases and their Symptoms. — *Continued.*

. Disease.	Symptoms.
Typhoid Fever	Perspiration, smelling like ammonia, retention of urine in the bladder, tongue trembling, dry and diminished in size, dull expression of face.
Various kinds of Colics	Lying upon the face; hard, sharp, contracted pulse, vomiting, pain relieved by pressure, hard and lumpy stools, cramps, abdomen diminished in size.
Cerebral, or Brain Diseases	Hot scalp, absence of thirst, full pulse, temporary spasm, rigidity of extremities, exaltation of vision, black flecks floating before eyes, morbidly increased sensation, contracted limbs, violent pulsation of carotid arteries, perpetual motion of eyelids, head increased in size, full, red face, blood-vessels of eyes injected, restlessness and tossing, ability to lie only on the back, painfully acute hearing, vomiting.
Insanity, Mania, Delirium	Sweat having odor of mice, voracious appetite, great and unnatural boldness, and many of cerebral symptoms previously noted; increase of strength and loss of moral sensibility.
St. Vitus's Dance..............	Irregular and perpetual motion.
Catalepsy	Entire and absolute loss of power of motion.
Apoplexy	Ability to lie only upon back; full or slow pulse.
Acute Diseases generally, and progress of Chronic Complaints.	Great and unusual languor, contusive pains, debility, diminished appetite, abdomen diminished in size.
Diabetes....................	Diminished perspiration, sugar in urine, increase in amount of urine, fat with stools.

Diseases and their Symptoms. — *Continued.*

DISEASE.	SYMPTOMS.
Rheumatism, Gout, etc.........	Profuse perspiration, sour-smelling sweats, sand deposits in urine, boring pains, swelling of joints, ability to lie only on the back.
Neuralgia....................	Shooting, tearing pains in part affected.
Measles, Scarlet-Fever, etc.	Sweat with mouldy odor, forcible closure of eyelids, tongue loaded with white, through which numerous elongated red papillæ protrude their points.
Hysteria	Increase in amount of urine, retention of urine in bladder, small pulse, voracious appetite, cough in paroxysms, cramps, painfully acute hearing, morbidly increased sensation, enlargement in epigastrium, increased rise in abdomen.
Bilious disorders..............	Dark-green, yellow or dark-brown stools, surface of tongue covered with white, soft, mucous substance, clammy mouth.
Disorders of the Bowels — Diarrhœa, (1); Dysentery, (2); Constipation, (3); Cholera, (4); etc.	Shreds of false membranes in stools (1 and 2) ; stools red and bloody (2) ; hard and lumpy stools (3) ; urgent desire for stool (1 and 2) ; watery stool (1 and 4) ; increased thirst, small pulse, pain in left shoulder, rice-water stools (4) ; bluish tint of face, etc. (4).
Congestion or Inflammation of Liver.....................	Strong pulse, pain in right shoulder, tongue covered with yellow fur.
Bronchitis and Bronchial affections	Dull, heavy, aching pain at base of chest, soreness of breastbone and between shoulders, faint, sweetish smell of expectoration, (yellow in color, and mucous) ; hollow, barking cough.

Disease.	Symptoms.
Pregnancy	Vomiting, voracious appetite, cramp.
Cancerous disease	Citron tint of countenance, enlargement in epigastrium, shooting, tearing pains.
Asthma	Wheezing cough, spasm of muscles of chest, jerking respiration, breathing increased in rapidity, clammy mouth, bitter taste, fetid breath.
Inflammations generally; Hemorrhages	Vomiting, external local heat, diminished secretion of urine, frequent pulse, breathing diminished in rapidity, paleness of face, tongue furred and dry.
Chronic affections	Hectic flush, at first, afterwards changing to paleness, tongue white and furry, hoarse, hollow, or barking cough, weak pulse, constant pain between shoulders, abdomen diminished in size.
Diseases of Bladder	Mucus in urine, small pulse, pain in region of bladder.
Bright's Disease	Albumen in urine, paleness of face, great debility.
Worms	Shreds of false membrane in stool, itching of nostrils.

APPENDIX B.

REMEDIES FOR THE DISEASES.

I. HOMŒOPATHIC REMEDIES.

These remedies are given in the order of symptoms mentioned in the description of the disease.

Prolapsus Uteri (Falling of the Womb).

1. *Aurum,* or *Belladonna,* in cramping pains through abdomen, pelvis, and spinal column; great sensibility and irritability, and when accompanied by leucorrhœa and menorrhagia.

2. *Nux Vomica,* for congestion of the womb, with pressure downwards; great heat and weight in vagina and womb; dragging pains, abdominal spasms, tendency to miscarriage, profuse or irregular menstruation, and fetid leucorrhœal discharges.

3. *Sepia,* in suppressed or irregular menstruation, contractile and expulsive pains in back and abdomen, frequent urination, and itching, excoriating leucorrhœa.

4. *Calcarea-carb.,* for weakness or laxity of muscular system, scrofulous habit of body, and exhaustive or profuse menstruation.

5. *Secale cornutum,* in prolonged bearing-down, forcing pains, profuse menstruation, depression, and deficient contraction after miscarriage.

Of either of these remedies, one dose of *five pellets* should be taken every four hours, and be continued for one week; during the next week *no medicine* should be taken, and so on, in alternation, until a cure is effected.

Leucorrhœa, or Whites.

6. *Pulsatilla,* when the discharge is thin and acrid ; *Sepia,* when the patient is sensitive and delicate ; *Alumina,* when it appears just before and after the menses, is profuse and transparent during the day, and is of a corrosive character ; *Calcarea-carb.,* for itching, burning leucorrhœa, and too profuse or too frequent menstruation, especially for persons of light complexion ; *Kreosotum,* when smarting, itching, of a whitish hue, accompanied by great pain and weakness, falling of womb, etc. ; *Nitric Acid,* for fetid, brownish, greenish, or flesh-colored discharges ; *Mercurius,* when the discharge is purulent ; *Cocculus,* if watery and bloody, during pregnancy, — for scanty menstruation, with leucorrhœa between the periods, or leucorrhœa instead of the menses ; *Conium,* for excruciating leucorrhœa, with pinching colic, lameness in small of back, and excessive itching ; *Sulphur,* in stubborn cases ; and *Silicea,* when milky, acrid, and accompanied by itching.

Of these remedies, six pills every morning and evening, until five doses are taken; then suspend for four days, and repeat treatment.

Chlorosis, or Green Sickness.

7. *Pulsatilla,* for females of mild, easy, melancholy disposition ; *Bryonia,* in alternation, when congestion, constipation, and fever are present ; *Ferrum,* in great debility and dropsical tendency ; *Sulphur,* for obstinate cases, pain in back of head, emaciation, constipation, drowsiness, and sense of pressure, especially for irritable persons ; *Calcarea-carb.,* in alternation, for difficulty of breathing, and excessive emaciation, palpitation, etc. ; *Belladonna,* for pressing or bearing-down pains, scanty and painful menses, preceded by colic ; *China,* when it occurs after a severe fit of sickness or severe hemorrhages.

Five or six globules once in six hours until improvement takes place; afterwards gradually lengthen the intervals to twelve hours and two or three days.

DISEASES OF PREGNANCY AND PARTURITION.

Continued Menstruation.

8. *Cocculus*, for severe spasmodic pain low down in the abdomen ; *Crocus*, when the discharge is dark and copious ; *Phosphorus*, *Platina*, and *Sulphur* are also serviceable.

Headache and Vertigo during Pregnancy.

9. *Aconitum*, *Belladonna*, *Opium*, *Nux Vomica*, *Coffea*, *Ignatia*, or *Pulsatilla*, for the respective symptoms of headache, vertigo, sparks before the eyes, sleeplessness, and sleepiness.

Morning-Sickness.

10. *Ipecacuanha*, for bilious vomiting and relaxed bowels ; *Arsenicum* for excessive vomiting, fainting, and emaciation ; *Nux Vomica*, for constipation, vomiting, and nausea in the morning or after eating, irritability of stomach, etc. ; *Pulsatilla*, for depraved appetite and obstinate sickness.

Dissolve twelve globules of either of these remedies in eight teaspoonfuls of water, and take one teaspoonful every three hours.

Constipation during Pregnancy.

11. *Nux Vomica*, *Opium*, *Lycopodium*, or *Sulphur*, one dose of five pellets every night and morning, for three or four days. On no account whatever should cathartics of any kind be given.

Diarrhœa during Pregnancy.

12.   *Chamomilla, Pulsatilla,* or *Dulcamara,* for violent colic, watery, or greenish stools, or severe cold.   Six pills at a dose, every two or three hours.

Hysteria, or Fainting Fits.

13.   *Chamomilla, Belladonna,* or *Aconitum,* when arising from anger, or in cases of congestion in the head.   Dose as in No. 12.

Palpitation of the Heart.

14.   *Chamomilla,* when caused by anger ; *Veratrum,* when by fear ; *Coffea,* when by joy ; *Opium,* when by sudden fright. *Ignatia, Coffea,* or *Chamomilla,* for nervous persons ; *Aconitum* and *Belladonna,* for plethoric persons.

Dissolve twelve globules in twelve teaspoonfuls of water, and take one every hour, or half-hour if the attack be severe.

Toothache.

15.   *Pulsatilla,* when pain flies about from one tooth to another ; *Antimonium, Mercurius,* or *Sulphur,* for carious teeth ; for violent and sudden paroxysms, *Coffea* or *Belladonna ;* for nervous toothache, *Ignatia, Hyoscyamus,* or *Sepia.* Six globules at intervals of one to six hours, according to severity of pain.

Neuralgia.

16.   *Belladonna, Aconite, Coffea,* or *Bryonia,* six globules at intervals of one to four hours, according to severity of pain.

Pains in Back and Side.

17. *Rhus, Belladonna,* or *Nux Vomica,* for pain in back ; *Aconite, Pulsatilla, Mercurius,* or *Sulphur,* for pain in side. Dose same as No. 16.

Cramps in Limbs, Back, or Abdomen.

18. *Veratrum,* or *Sulphur,* for the limbs ; *Nux Vomica, Belladonna,* or *Pulsatilla,* for the abdomen ; *Ignatia* or *Rhus,* for the back. Dose same as No. 16.

Varicose Veins, or Swelling of the Veins.

19. *Nux Vomica,* when attended with hemorrhoids, constipation, and bearing-down pains ; *Pulsatilla,* when much pain, inflammation, and swelling ; *Arsenicum,* when swelling is of livid hue, with burning pain ; *Lycopodium* for inveterate cases. Dose, twelve globules in twelve teaspoonfuls of water, a teaspoonful every four hours.

Hemorrhoids, or Piles.

20. *Pulsatilla, Nux Vomica,* and *Sulphur,* especially the two latter. Take ten globules, dry, upon the tongue, night and morning ; unless in severe cases, when the remedy should be repeated every hour.

Jaundice, or Icterus.

21. Commence with *Mercurius,* six globules every three hours for three days, followed by *Hepar Sulphur,* or *Lachesis.* Two doses of six globules daily, night and morning.

Incontinence of Urine.

22.   *Pulsatilla, Sepia, Belladonna,* or *Hyoscyamus.*   Six pills, dry, upon the tongue, once in three or four hours.

Dysury, or Strangury (Difficult or Scanty Urination).

23.   *Pulsatilla* or *Nux Vomica.*   Six globules every two hours.

Flooding during Pregnancy.

24.   *Tincture of Cinnamon.*   Three drops in half-tumbler of water, a teaspoonful every quarter or half hour, until physician can be summoned.

Miscarriage, or Abortion.

25.   *Arnica, Cinnamon, Secale Cornutum,* or *Belladonna,* as temporary remedies.   Twelve globules in twelve teaspoonfuls of water, one teaspoonful every fifteen, thirty, or sixty minutes, according to severity of case.

False Pains.

26.   *Bryonia, Nux Vomica, Pulsatilla,* or *Aconite.*   Twelve globules in twelve teaspoonfuls of water, one teaspoonful every half hour or hour.

Constipation after Confinement.

27.   *Bryonia, Nux Vomica,* or *Sulphur.*   One or two doses of six pills each.

Sore Nipples.

28.   *Chamomilla, Nux Vomica, Mercurius, Graphites,* or *Silicea.*   Same doses as No. 26, every six hours.

Gathered or Broken Breasts.

29. *Bryonia* or *Belladonna.* In same doses as No. 28, every hour.

Child-bed Fever, or Puerperal Peritonitis.

30. *Aconite, Belladonna, Bryonia,* or *Pulsatilla.* Ten or twelve globules of either two, in alternation, at intervals of one, two, three, or four hours, according to severity.

Milk-leg, or Crural Phlebitis.

31. *Aconite, Arnica, Belladonna, Bryonia,* or *Pulsatilla.* Six globules, dry, upon the tongue, once in two hours.

Nursing Sore Mouth.

32. *Mercurius, Borax, Nitric Acid,* or *Sulphur,* according to severity. Twelve globules·in twelve teaspoonfuls of water, one teaspoonful every four or six hours.

Perspiration after Delivery.

33. *Dulcamara, Bryonia, Belladonna,* or *Sulphuric Acid.* In same doses as No. 32, every three or four hours.

Excessive Perspiration.

34. *China, Sambucus,* or *Sulphuric Acid.* Six globules every three hours.

Asphyxia.

35. *Tartar Emetic* or *Opium.* One or two globules, dissolved or dry, upon the tongue, every ten or fifteen minutes.

DISEASES OF CHILDREN.

Coryza, Snuffles, or Cold in the Head.

36. *Arsenicum, Nux Vomica, Chamomilla, Belladonna, Mercurius, Pulsatilla,* or *Sulphur*, according to symptoms. In same doses as No. 32, every one, two, or three hours.

Cough or Tussis.

37. *Aconite, Belladonna, Bryonia, Chamomilla, Nux Vomica,* etc. Same doses as No. 32, at intervals of one to four hours.

Bronchitis.

38. *Aconite, Pulsatilla, Phosphorus, Tartar Emetic, Chamomilla,* given dry or in solution. If dry, three to six pills for a dose. If in solution, dissolve twelve globules in twelve teaspoonfuls of water, a teaspoonful for a dose. Doses to be repeated every two to four hours.

Pleurisy.

39. *Aconite* and *Bryonia* are the two principal remedies, and in most cases will be all that is necessary to complete a cure. *Mercurius, Arnica,* and *Arsenicum* are sometimes used in severe cases. Doses as in No. 38, every half-hour, hour, or two hours, according to the severity of symptoms.

Pneumonia.

40. In the first stages *Aconite* is the most prominent remedy ; *Belladonna* and *Bryonia* are also used in alternation. In severe cases, *Phosphorus, Tartar Emetic, Pulsa-*

tilla, *Arnica*, *Mercurius*, and *Arsenicum* are administered, according as the peculiar symptoms may indicate. Dose as in Nos. 38 and 39.

Hoarseness or Raucitus.

41. For excessive acrid discharges from nose, *Arsenicum*. If combined with influenza, catarrh, or chronic hoarseness, *Causticum*. For tickling or crawling at nose, violent cough, and smarting in throat, *Capsicum*. If fever, accumulation of mucus, and pain in throat, and great irritability, *Chamomilla*. *Carbo-veg.*, *Mercurius*, *Nux Vomica*, *Phosphorus*, *Pulsatilla*, and *Sulphur*, are also used in severe cases, especially when the affection is the sequel or result of other diseases. Dose as in No. 38.

Spasmodic Croup.

42. *Aconite, Hepar Sulph., Spongia*, or *Tartar Emetic*, are effective remedies in all ordinary cases of croup. Dose as in No. 38, repeated every half-hour, hour, or two hours, according to circumstances.

Membranous Croup.

43. *Aconite* is the first remedy, in alternation with *Spongia;* if this has no effect, institute *Hepar Sulph.* If the case be very violent, *Kali Bichrom., Bromine, Lachesis*, or *Phosphorus*, will prove efficacious. Dose as in No. 38.

Whooping-Cough, or Pertussis.

44. *Corallia*, one dose every four hours, as in No. 38; *Drosera*, a dose every six hours, and an occasional dose of *Causticum*, will generally have the desired effect. In convalescence, *Hepar Sulph.* should be administered.

Asthma of Miller.

45. The principal remedies are *Sambucus, Ipecac.,* and *Arsenicum,* the first two especially. A dose of five pellets every ten or fifteen minutes. If these fail, *Phosphorus* or *Belladonna* may be tried.

Laryngitis, or Inflammation of the Larynx.

46. *Aconite* is specially indicated. *Spongia, Belladonna, Hepar, Tartar Emetic, Phosphorus,* and *Lachesis,* are also very influential in severe cases, according to the prominent symptom in each case, in doses of five pellets from one to two hours apart.

Colds.

47. When they result in COUGH, *Aconite;* for cold in the head, *Belladonna* and *Nux Vomica;* when very severe, *Mercurius, Sepia, Arsenicum,* and *Pulsatilla;* for HEADACHE, *Belladonna;* for EARACHE, *Rhus, Dulcamara, Mercurius, Bryonia,* or *Sulphur;* for TOOTHACHE, *Aconite, Bryonia, Rhus, Nux Moschata,* or *Mercurius;* for SORE THROAT, *Belladonna* or *Mercurius;* for DIARRHŒA, *Arsenicum, Bryonia, Dulcamara,* or *Glonoine.* Repeat the dose as in No. 38 as frequently as the exigency of the case demands.

Thrush, or Aphthæ.

48. For ordinary cases, *Mercurius* and *Sulphur* internally, with a gargle of *Borax; Arsenicum* and *Nitric Acid,* in very severe cases.

Canker of the Mouth.

49. *Mercurius* may always be given, followed, if necessary, by *Hepar Sulphur* or *Nitric Acid.* In very severe

cases, *Natrum Muriaticum*, and *Nux Vomica*. *Sulphur* has been found extremely beneficial, and is now generally used. The dose of five pellets, dry or in solution, should be repeated every two, three, or four hours, according to urgency of the case.

Ptyalism, or Salivation.

50. If from use of Mercury, *Hepar*, *Lachesis*, *Belladonna*, or *Sulphur*, will be the appropriate remedies. When caused by cold, give *Mercurius*. Dose as in No. 49, both as to time and quantity.

Ranula, Swelling Under Tongue.

51. *Mercurius*, *Calcarea Carb.*, *Thuja*, and *Sulphur*, are the principal remedies, three or four globules being given, dry, upon the tongue, night and morning.

Gumboils, Abscess in the Gums.

52. *Aconite* and *Belladonna*, in alternation, every two hours, when caused by decayed teeth ; *Mercurius*, when they fail to afford relief. If there be swelling of the jaw with suppuration, *Silicea* is the appropriate remedy. During the inflammatory stage, the remedies may be given every hour, gradually coming to two or three hours interval. When *Silicea* or *Calcarea*, three pills night and morning.

Mumps, or Parotitis.

53. *Mercurius* is the principal remedy, and often the only one required ; two or three doses in most cases will effect a cure. A dose of four pills every night until four doses are taken. When it has an erysipelatous appearance,

or affects the brain, *Belladonna* or *Hyoscyamus*, three globules every hour. Should it suddenly disappear and affect the ovaries, *Pulsatilla*, same quantity, every two or three hours, will prove beneficial.

Inflammation and Swelling of the Tongue, Glossitis.

54. *Aconite* should be administered at the commencement, followed by *Mercurius* and *Belladonna* in alternation, at intervals of from one to four hours, according to circumstances. Dose, four globules, dry, upon the tongue.

Dentition, or Teething.

55. *Aconite, Belladonna,* or *Chamomilla,* for the ordinary nervous derangements; *Cina,* when attended with cough; *Coffea,* for fever; *Ignatia,* for threatened convulsions; *Sulphur, Magnesia Calc.,* or *Mercurius,* for diarrhœa; *Ipecacuanha,* when nausea and vomiting are combined with diarrhœa; *Nux Vomica* in alternation with *Bryonia,* for obstinate constipation; *Calcarea Carb.,* to hasten the process of dentition. Three pills, dry, to the dose, every one, two, three, or four hours, according to circumstances.

Toothache, or Odontalgia.

56. For feverishness, violent pain, congestion, and heat, *Aconite,* followed by *Belladonna* or *Chamomilla.* When the pain arises from mechanical injury, as extraction or plugging, *Arnica.* For hollow or decayed teeth, *Antimonium crude.* For drawing, jerking pain, *Belladonna, Bryonia,* or *Chamomilla.* When gums are swollen or congested, *Kreosotum* and *Mercurius.* For toothache in pregnancy, *Nux Moschata* or *Nux Vomica.* *Pulsatilla* is most suitable

for young girls or children, and *Sulphur* for tearing and pulsative pain in carious teeth. Twelve globules in twelve teaspoonfuls of water, a teaspoonful at a dose, at intervals of fifteen minutes, one hour, or two hours, according to severity of pain.

Sore Throat, or Quinsy.

57. *Aconite* and *Belladonna* in the first stages will generally effect a cure. In the sore throat of scarlet-fever, *Belladonna* or *Mercurius*, *Bryonia* or *Chamomilla*, when resulting from cold. In dryness and extensive swelling, *Lachesis*. For women and young persons, *Pulsatilla* and *Rhus*, in ordinary cases, and *Sulphur* and *Silicea* in prolonged and severe cases. The doses, either dry, or in solution (three globules at a time), may be given at intervals of one, two, or three hours, according to urgency of symptoms.. Lengthen the intervals as the severity subsides.

Malignant, or Putrid Sore Throat.

58. If there be inflammatory fever, a few doses of *Aconite* should be given, followed by *Belladonna* and *Mercurius* in alternation. If the ulcers increase in size, and become painful, *Nitric Acid*. For gangrenous sore throat, *Arsenicum*. Dose, same as in No. 56, every hour at commencement, gradually increasing the interval as improvement takes place.

Tonsillitis.—Inflammation of the Tonsils.

59. *Belladonna, Causticum, Graphites, Lachesis*, or *Sulphur*, in doses of three globules, every other night, for about six weeks, will usually eradicate the affection.

Falling of the Palate.

60. *Nux Vomica* (three globules every two hours) will usually prove efficacious. Should it fail, *Mercurius*, *Belladonna*, or *Sulphur* will doubtless result successfully.

Diphtheria.

61. *Aconite* at the commencement of an attack, in alternation with *Bryonia*, especially if there is considerable fever; *Belladonna* when the inflammation is of a bright scarlet, and extends uniformly over the entire mucous membrane; *Rhus tox.* when it is of a dark-red color; *Arsenicum* in aggravated conditions; and *Kali-chlor.* in extreme depression and septic tendencies. Dissolve twelve globules in six teaspoonfuls of water; one teaspoonful every six hours.

Pyrosis, Heartburn, Water-Brash, Sour Stomach.

62. For water-brash, *Nux Vomica*, *Pulsatilla*, *Silicea*, *Chamomilla*. For heartburn, *Arsenicum*, *China*, *Sepia*, *Sulphur*. For flatulence, *Graphites*, *Phosphorus*, *Pulsatilla*, *Carbo-veg.* For sour stomach, *Chamomilla*, *Pulsatilla*, *Phosphorus*, *Sulphur*. Twelve globules, dissolved in twelve teaspoonfuls of water; one teaspoonful every hour, for children. An adult may take six or eight globules every hour.

Nausea, Vomiting, and Regurgitation of Milk.

63. *Ipecacuanha* is generally the only remedy required. If much flatulence, *Pulsatilla* and *Antimonium crude*, in alternation. When attended with diarrhœa or convulsions, *Chamomilla*. In cases accompanied by vomiting and con-

stipation, *Nux Vomica* and *Bryonia*, in alternation. Chronic cases of long standing require *Calcarea* or *Sulphur*. For vomiting caused by worms, *Cina*, *Mercurius*, or *Ferrum*. Three globules at a dose, or a teaspoonful of solution (as in No. 62) every four hours, for an infant. In severe cases of vomiting repeat the dose every fifteen minutes or half-hour.

Biliousness.

64. When accompanied by chilliness, fever, headache, etc., use *Bryonia*. If caused by eating fat or greasy substances, and accompanied with offensive eructations, *Pulsatilla*. For ordinary cases, *Ipecacuanha* and *Mercurius* will effect a cure. Dose, six globules every one or two hours.

Offensive Breath.

65. If only in the morning, *Belladonna*, *Nux Vomica*, or *Sulphur*. If at morning and night, *Pulsatilla*. If after a meal, *Sulphur* or *Chamomilla*. If in young girls at the age of puberty, *Aurum*, *Pulsatilla*, *Belladonna*, *Sepia*. If caused by worms, *Cina* or *Sulphur*. If caused by salivation with calomel, *Carbo-veg.*, *Hepar Sulph.*, *Nitric Acid*. One dose may be given every night and morning, either dry, or dissolved (as in No. 62). Dose, if dry, six or eight pills.

Colic.

66. SPASMODIC COLIC, *Colocynth*, *Chamomilla*, *Belladonna*, or *Nux Vomica*, five or six pills every few minutes. BILIOUS COLIC, *Nux Vomica*, *Colocynth*, *Mercurius*, *Pulsatilla*, *Chamomilla*, and *Plumbum*, every five or ten minutes. The COLIC of INFANTS, *Chamomilla* will answer in ordinary cases. If there is nausea, vomiting, and diarrhœa, *Pulsa-*

tilla. If the evacuations are fermented, and have a putrid odor, *Ipecacuanha.* If constipated, *Nux Vomica.* If caused by excitement on the part of the mother, *Ignatia.* Colic caused by worms, *Cina, Sulphur,* or *Mercurius.* For Colic in Pregnant Women, *Chamomilla, Nux Vomica, Pulsatilla.* For Menstrual Colic, *Pulsatilla, Coffea, Belladonna, Cocculus.* Dose for infants, two or three globules, dry, upon the tongue; for adults and older children, five or six pills every fifteen or thirty minutes, hour, or two hours, according to the severity of the case.

Cholera Morbus.

67. If the attack be induced by excitement, and there are severe pains and cramps, *Chamomilla.* When vomiting predominates, with severe pains in the abdomen, *Ipecacuanha* and *Veratrum,* in alternation. For violent cramps and constrictions, or cutting pains, *Colocynth.* In cases of rapid prostration, *Arsenicum.* In most cases, even the most severe, *Veratrum* will prove an almost certain cure. *Cuprum,* when there are severe spasms of the limbs; and *Cinchona,* for the debility which invariably accompanies the disease. Dose, six pills every few minutes.

Cholera Infantum.

68. *Ipecacuanha* is usually the only remedy required. When mucus or sour vomiting occurs, with green and slimy evacuations and colicky pains, use *Chamomilla,* especially during dentition. When the diarrhœa is chronic, *Magnesia carb.* Should the evacuations be light-colored, offensive, and frothy, the child moan and toss in its sleep, and have cramp-like pains in the abdomen, use *Podophyllum.* If accompanied with colic and straining, the stools slimy and

mixed with blood, *Mercurius.* For diarrhœa immediately
after eating, and loss of appetite, give *Cinchona.* In extreme cases, with great prostration, nausea, and vomiting,
Arsenicum. Sulphur is a valuable remedy for protracted
cases. If head symptoms manifest themselves, give *Aconite, Bryonia,* or *Helleborus.* Doses, same as in No. 62,
repeated every fifteen or thirty minutes, until the severe
symptoms have subsided, after which lengthen the
intervals.

Dyspepsia, or Indigestion.

69. DYSPEPSIA OF ADULTS. — *Arnica,* when caused by a
fall, blow upon the stomach, or lifting heavy weights;
Aconite, when considerable fever, thirst, and nausea; *Antimonium crudum,* when from an overloading of the stomach;
Belladonna, in painful distention of the abdomen; *Arsenicum,* in curious chronic cases; *Bryonia, Cepa, Carbo veg.,
Calcarea carb., Chamomilla, China,* and *Hepar Sulph.,* are
given in other complications of the disorder. For DIS-
PEPSIA IN CHILDREN, *Ipecacuanha* is the chief remedy, as it
generally arises from imperfect mastication, or improper
food. Either for CHILDREN or ADULTS, *Lachesis, Mercurius,
Nux Vomica, Pulsatilla, Phosphorus, Sepia, Sulphur,* and
Veratrum, are singularly effectual in severe cases of dis-
peptic derangement, according to the predominating symptoms. For recent attacks, there should be intervals of half
an hour, until relief is attained, gradually lengthening until
one to three hours apart. In chronic cases, doses three
times per day; adults, ten globules at a dose; an infant,
two globules.

Constipation.

70. *Nux Vomica* or *Bryonia,* with *Opium* in alternation,
for irritability, distention, loss of appetite, and tendency to

vomit; *Platinum* and *Magnesia mur.*, with occasional dose of *Lycopodium*, when there is impaction of fæces, shuddering after evacuation, severe bearing-down, or inability to pass the fæces. *Sulphur* and *Plumbum*, for obstinate constipation, accompanied by piles. For constipation of pregnant women, *Nux Vomica*, *Opium*, and *Sepia;* for lying-in women, *Bryonia* and *Nux Vomica;* for nursing infants, *Bryonia*, *Nux Vomica*, *Opium*, and *Sulphur*. Dose, for adults, six pills once in four hours; for infants, two or three pills.

Diarrhœa.

71. When arising from cold, and without pain, *Dulcamara;* for frothy, fermented evacuations, offensive, with pain in rectum, *Ipecacuanha;* also for nursing infants, when caused by overloaded stomach, offensive, tinged with blood, accompanied with nausea and vomiting. For infants, where evacuations are slimy, green, or yellowish, the child drawing up its legs, fretting, worrying, and wanting to be carried all the time. For the diarrhœa of teething, *Nux Vomica*. Dose, three or four pills dry upon the tongue, every half-hour, hour, or two or three hours, according to the severity of the pain. For adults, six pills, at similar intervals.

Dysentery.

72. *Aconite*, for ordinary or inflammatory dysentery; *Arsenicum*, in severe cases; *Belladonna*, in severe pain and tenderness; *Chamomilla*, for thirst, headache, fever, and nausea; and *Colocynth*, for extreme pain, slimy, and bloody discharges; *Podophyllum*, for cramp-like pains, moaning, rolling of the eyes, etc. *Mercurius*, *Nux Vomica*, and *Sulphur* are exceedingly valuable and effective remedies.

Dose, in solution, twelve globules, in twelve teaspoonfuls of water, every half-hour or hour, until relief is obtained.

Prolapsus Ani, or Falling of the Body.

73. *Ignatia* and *Sulphur* are the principal remedies, the latter in alternation with *Nux Vomica; Calcarea*, and *Mercurius* are also very efficacious. Three globules every twelve hours.

Rupture, or Hernia.

74. *Nux Vomica* or *Sulphuric Acid.* Three globules every evening for about a week.

Jaundice, or Icterus.

75. *Mercurius* and *Cinchona* in alternation, six pills once in four hours. *Nux Vomica, Sulphur* or *Lachesis*, when accompanied by constipation or diarrhœa, or both. One dose of six pills every night and morning.

Jaundice of Infants.

76. *Chamomilla, China*, or *Nux Vomica.* One or two pills every four hours.

Worms.

77. *Aconite, Cina*, or *Nux Vomica*, are the chief remedies ; when complicated with colic, *Spigelia ;* with scrofulous eruptions, *Silicea ; Lycopodium, Sulphur*, and *Teucrium*, are each of them almost specific for the affection. Dose, three or four drops of the tincture in a tumbler half full of water, giving a large teaspoonful three times per day for two, three, or four days.

Epidemic Cholera.

78. PREVENTIVE TREATMENT: *Cuprum* and *Veratrum* alternately for every six or seven days. Dose, same as No. 72. CURATIVE TREATMENT: *Camphor.* Dose, same as No. 72, after each evacuation; for violent vomiting and purging, *Veratrum;* for intolerable burning in the bowels, *Arsenicum.* Same dose repeated every five minutes. *Cuprum* for rice-water discharges. *Carbo-veg.* for the state of collapse. *Secale Cornutum,* when the patient is aged.

Scarlet-Fever.

79. *Belladonna* and *Cuprum* are specifics for this disease in its various phases. *Aconite, Mercurius, Arsenicum,* or *Opium* may be found very useful, In malignant cases, *Crotalus, Phosphoric Acid, Lachesis,* or *Nitric Acid;* and in the after affections, *Sulphur, Rhus, Digitalis, Pulsatilla,* or *Lycopodium* will afford relief. Dose, two drops, or twelve globules, in twelve teaspoonfuls of water, a teaspoonful every half-hour, hour, or two hours, according to severity of the symptoms.

Scarlet-Rash.

80. *Aconite,* in alternation with *Coffea,* is the only remedy necessary in ordinary cases. Exceptionally, *Belladonna, Ipecacuanha, Pulsatilla,* or *Bryonia* are called for. Dose, as in No. 79, every two hours.

Measles.

81. *Aconite* is generally sufficient, but, in the subsequent complications, *Belladonna, Pulsatilla, Ipecacuanha, Bryonia, Euphrasia,* and *Rhus* will prove efficacious. Dose, as in No. 79.

Nettle-Rash, Hives, Urticaria.

82. *Aconite, Pulsatilla, Nux Vomica, Dulcamara,* and *Rhus,* but especially *Ledum Palustre,* which will cure all ordinary cases. Dose, five globules, dry, every three hours. In severe cases, dose same as No. 79.

Erysipelas, St. Anthony's Fire.

83. *Aconite* when there is high inflammatory fever; *Belladonna* is especially valuable for erysipelas of the face, with delirium, swollen eyes, great thirst, and dry skin; *Lachesis* when the entire face and glands are involved; *Arsenicum* when the eruption is of a dark color, and there is great prostration; *Pulsatilla* and *Graphites* when the eruption changes from one locality to another; *Mercurius* and *Hepar Sulph.* when it terminates in abscesses. Dose, as in No. 79, every two or three hours.

Itch, Psora, Scabies.

84. *Sulphur* ointment is the only specific remedy for this disease. An occasional dose of *Mercurius* or *Causticum,* three globules every four hours, will sometimes prove beneficial.

Itching of the Skin.

85. *Drosera, Sulphur,* or *Lycopodium* (same dose as No. 84) will effectually relieve this symptom, for it can scarcely be called a disease.

Herpes, or Tetter; Zoster, or Shingles; Circinatus, or Ringworm.

86. Ringworm will generally yield to *Sepia,* three globules every night for three nights, then omit for three days, and

repeat, and so on until cured. *Rhus, Calcarea, Graphites*, or *Sulphur* may sometimes be found necessary. For Shingles, give *Aconite* or *Tartar Emetic*.

Prickly-Heat.

87. *Aconite* and *Chamomilla* will usually afford relief, though *Rhus, Arsenicum*, and *Sulphur* are sometimes called for. Dose, same as No. 79, every two hours.

Strophulus, Red-Gum, White-Gum, Tooth-Rash.

88. *Coffea, Chamomilla, Aconite*, or *Belladonna* may be given, in similar doses to No. 79, when there is great restlessness.

Chicken-Pox.

89. *Aconite* and *Belladonna* are generally all that is required ; *Pulsatilla* will considerably shorten, if not entirely prevent, the disease. Dose, same as in No. 79.

Variola and Varioloid.

90. In the first stage, *Aconite, Belladonna, Bryonia, Rhus*, and *Tartar Emetic*. In the second, or eruptive stage, *Tartar Emetic, Thuja*, and *Stramonium*. The third, or suppurative stage, requires *Mercurius, Arsenicum, Muriatic Acid, Opium*, or *China*, according to symptoms. The fourth, or desquamative stage, requires only *Sulphur*. Dose, three globules every morning.

Intertrigo, Excoriations.

91. *Chamomilla, Ignatia, Pulsatilla*, or *Sulphur*, according to the circumstances, will generally prove effective.

Dose, three globules, night and morning. If fever exists give *Aconite*. In obstinate cases, *Sepia*, *Graphites*, *Sulphuric Acid*, or *Silicea*.

Pimples on Face, Acne Punctata, Comedones.

92. The remedies suitable for this disease are *Belladonna, Calcarea, Sulphur,* and *Nux Vomica*. Six globules in twelve teaspoonfuls of water ; one teaspoonful every six hours.

Abscesses.

93. *Mercurius, Hepar Sulph., Silicea, Calcarea, Lachesis, Phosphorus,* and *Sulphur* are suitable remedies, according to the cause and stage of the disease. Dose, same as No. 92, every four hours.

Boils.

94. *Arnica* to lessen pain ; *Belladonna* and *Mercurius* for fever and headache ; *Sulphur* as a preventive. Twelve globules in twelve teaspoonfuls of water ; a spoonful every three or four hours.

Scald Head, Tinea Capitis, Favus.

95. *Calcarea carb., Sulphur, Lycopodium, Sepia, Arsenicum,* and *Rhus*, are the prominent remedies for this disease. Dose as in No. 94.

Crustea Lactea, Milk-Crust, or Impetigo.

96. *Aconite*, either alone or in alternation with *Chamomilla*, once in two hours, is the best remedy, though circumstances sometimes necessitate the use of *Rhus, Viola tricolor,* and *Sulphur*. Nothing should be applied externally but a little *Glycerine*. Dose as in No. 94.

Inflammation of the Brain.

97. *Aconite* at the commencement of the attack. *Belladonna* is the most important and effective remedy, and in alternation with *Aconite*, *Stramonium*, or *Hyoscyamus*, is often attended with the happiest results. In very severe cases, where the head symptoms are the result of some other disease, *Helleborus*, *Bryonia*, *Zincum*, and *Opium* will be called for. Dose, twelve globules in twelve teaspoonfuls of water; one spoonful every one or two hours.

Convulsions.

98. If from indigestion, *Nux Vomica*, *Pulsatilla*, *Veratrum*, or *Aconite*. If from teething, *Belladonna*, *Coffea*, *Chamomilla*, or *Ignatia*. If caused by worms, *Cina*, *Mercurius*, or *Hyoscyamus*. If the result of repelled eruptions, *Tartar Emetic*, *Belladonna*, *Stramonium*, *Sulphur*, or *Bryonia*. If from fright, *Hyoscyamus*, *Coffea*, *Ignatia*, or *Stramonium*. If from mechanical injuries, *Arnica*. Dose, same as in No. 97, every ten, fifteen, or twenty minutes, until improvement takes place, when the intervals should be lengthened.

Chorea, or St. Vitus's Dance.

99. *Belladonna*, *Cocculus*, *Colchicum*, *Pulsatilla*, *Nux Vomica*, and *Sulphur*. Dose, same as No. 97, three times per day.

Headache.

100. Nervous or Neuralgic, *Aconite*, *Belladonna*, *Coffea*, *Nux Vomica*, *Ignatia*, and *Pulsatilla*, according to circumstances. If from suppression of menstruation, *Bryonia*,

Nux Moschata, or *Chamomilla*. If accompanied by irritability and hysteria, *Mercurius*, *Platina*, or *Hepar Sulph.* In stubborn, chronic cases, *China*, *Colocynth*, *Arsenicum*, *Veratrum*, *Silicea*, or *Sulphur*, according to accompanying symptoms ; *Sulphur*, especially, in cases of nausea. Sick or bilious headache, *Belladonna*; *Sanguinaria, Ipecac., Pulsatilla*, or *Spigelia*. Dose, same as No. 97, if in solution ; if dry, three or four globules, to be repeated every half-hour, one, two, three, or six hours, according to severity, until relief is obtained. Congestive headache, *Aconite*, *Belladonna, Bryonia, Rhus, Glonoine*, or *Pulsatilla*, in same doses. Rheumatic headache, *Chamomilla, Ignatia, Ipecac.*, or *Colocynth*, especially the latter. Catarrhal headache, *Hepar Sulph., Euphrasia, Arsenicum*, or *Aconite*. If from constipation, *Bryonia, Opium*, and *Lycopodium*.

Neuralgia.

101. *Aconite, Belladonna, Bryonia, China, Platinum, Spigelia, Staphysagria*, or *Kalmia*, according to locality and severity of pain. Dose, same as No. 97, repeated every quarter or half hour, hour or two hours, according to urgency of case, lengthening the intervals as the symptoms ameliorate.

Hysteria.

102. *Cocculus, Cuprum, Coffea, Ignatia, Lachesis, Conium, Natrum*, or *Veratrum*, according to origin and intensity of affection, in same dose as in Neuralgia.

Sore Eyes of Young Infants.

103. *Aconite, Belladonna, Chamomilla, Euphrasia, Rhus*, or *Sulphur*, according to symptoms. One globule every three hours.

Sty on the Eyelid.

104. *Staphysagria* or *Silicea*, three or four globules every four hours.

Squinting — Strabismus.

105. *Hyoscyamus* and *Belladonna*, in same dose as 97, every three hours, for two or three days, then discontinue for some length of time, and proceed as before.

Inflammation of Ear.

106. *Pulsatilla*, *Belladonna*, or *Aconite*, in same dose as 97, every half hour or hour, according to severity.

Earache.

107. *Chamomilla*, *Pulsatilla*, *Belladonna*, *Rhus*, or *Sulphur*, according to the cause of the affection. Dose, as in 97, every fifteen minutes, until better.

Running of the Ear.

108. *Pulsatilla*, *Lycopodium*, *Belladonna*, *Lachesis*, *Calcarea*, or *Sulphur*, in accordance with symptoms. Dose, as in 97, every four hours, for six days.

Bleeding from the Nose — Epistaxis.

109. When from a fall, *Arnica;* when from determination of blood to the head, *Aconite*, *Belladonna*, or *Bryonia;* when from over-exertion, *Rhus;* when from worms, *Cina* and *Mercurius*. Dose, same as 97, every ten or fifteen minutes, in profuse hemorrhage. When only periodical, every hour or two hours.

Wetting the Bed.

110. *Pulsatilla*, *Nux Vomica*, *Belladonna*, *Sulphur*, *Silicea*, or *Causticum*. Dose, three to six pills, every four hours, according to age of patient.

Retention of Urine in Infants.

111. *Aconite*, *Pulsatilla*, or *Ipecacuanha*, two globules every two hours. If in an older child, same dose as in 97.

Burns and Scalds.

112. *Linseed* and *Lime-water*, or a solution of *Cantharides*, ten drops to a half tumbler of water, applied by dipping linen cloths into the solution and placing them on the burnt surface. *Arnica* and *Tincture* of *Urtica Urens* are exceedingly efficacious remedies. To promote healing, use *Creosote-water;* if erysipelatous inflammation is threatened, *Belladonna* or *Rhus* internally. For the accompanying fever, *Aconite;* for pain and restlessness, *Carbo-veg.* and *Coffea.* If there is ulceration, *Causticum*, in water, externally, *Sulphur* or *Silicea*, internally, one dose every six hours.

Concussion of the Brain.

113. *Arnica* is the usual remedy. Where inflammation is imminent *Aconite* and *Belladonna* in alternation. Dose, as in 97, every hour.

Sprains.

114. Externally, *Tincture* of *Arnica*. Internally, *Rhus* or *Bryonia*, once in two hours. If there is sickness of the stomach, *Pulsatilla*.

Wounds.

115. SUPERFICIAL WOUNDS, *Staphysagria*, externally. CONTUSED WOUNDS, *Arnica* or *Hypericum per.* internally. *Hepar Sulph.* during the suppurative process. Bruises about the eye should be kept *constantly* wet with a solution of *Arnica*. LACERATED WOUNDS should be dressed with a solution of *Calendula* in cold water. If there is inflammation and fever, *Belladonna* or *Rhus*, and an occasional dose of *Aconite*. PUNCTURED WOUNDS only require an application of *Canada Balsam*. POISONED WOUNDS, — Externally, a solution of *Arnica;* internally, *Aconite* or *Apis*. For mosquito-bites, *Spirits* of *Camphor* or *Lemon-juice*.

II. ALLOPATHIC AND ECLECTIC REMEDIES.

116.   **For Acidity of Stomach.**

Liquor Potassa	20 minims.
Chalk Mixture	1 oz.
Tincture of Colombo	1 drachm.

Make a draught.

117.   **Alterative.**

Nitric Acid, diluted	$\frac{1}{2}$ drachm.
Hydrochloric Acid, diluted . . .	1 "
Spirit. Œther. Nitrici	$\frac{1}{2}$ oz.
Syrup of Sarsaparilla	1 "
Water, pure	$6\frac{1}{2}$ "

Two spoonfuls three times a day.

118. Amenorrhœa, or Suppression of Menses.

Powdered Aloes . . . . . 3 grains.
Powdered Tartrate of Antimony . . $\frac{3}{4}$ grain.
Cocoa Butter . . . . . . 2 drachms.

119.

Compound Aloes, pill, } of each . . 1 drachm.
Compound Iron, pill, }

Oil of Savin, } of each . . . . 3 drops.
Oil of Rue, }

Powdered Capsicum . . . . . 8 grains.
  Divided into 24 pills. One pill three times per day.

120.

Liquor of Ammonia . . . . . 3 scruples.
Cow's Milk . . . . . . 4 oz.
  An Injection — 1 oz. to be injected daily.

121. Anti-spasmodic.

Camphor, } of each . 3 grains.
Sesquicarbonate of Ammonia, }

Ipecacuanha Powder . . . . 1 grain.
Extract Hyoscyamus . . . . 4 grains.
Mucilage sufficient to form three pills. Dose, one or two.

122. Ascites, or Swelling of the Abdomen.

Gamboge . . . . 2 scruples.
Tartrate of Potash . . . . . 1 oz.
White Sugar . . . . . . 2 drachms.
Water . . . . 6 oz.
  A tablespoonful every two or three hours.

123.   **Asthma.**

Socotrine Aloes . . . . . 16 grains.
Mastick Root Powder . . . . 8 "
Ext. Gentian, ⎫
Compound Galbanum Pill, ⎬ of each . . 3 "
Oil of Anise . . . . . . sufficient.

Make into twelve pills. Take three every day on going to sleep.

124.

Decoction of Aloes . . . . . $6\frac{1}{2}$ oz.
Compound Tinct. of Senna . . . 1 "
Tinct. of Squills . . . . . 3 drachms.

Three table-spoonfuls to be taken occasionally.

125.   **Spasmodic Asthma.**

Sesquicarbonate of Ammonia . . . 1 drachm.
Rue Water . . . . . . 9 oz.
Syrup of Poppies . . . . . 1 "

A spoonful every ten minutes.

126.

Ext. Hyoscyamus . . . . . 4 grains.
Tinct. Squills . . . . . . 10 minims.
Nitric Acid . . . . . . 6 "
Water . . . . 10 drachms.

Make a draught, repeated every three hours.

127. Astringents.

Decoction of Oak Bark	1½ oz.
Powdered Nut-galls	10 grains.
Tinct. Catechu	½ drachm.
Tinct. Cardamoms (Compound) . .	1 "
Syrup of Orange	1 "

A draught to be taken twice a day.

128.

Outside Oak Bark, bruised . . .	1½ oz.
Aquafortis	1 pint.

Macerate for three hours and strain ; then add

Powdered Gall-nuts	2 drachms.
Tinct. Cardamoms (Compound) . .	2 oz.

A wineglass full for a dose.

129. Brain Diseases. — **Hydrocephalus.**

Tinct. Digitalis	1 oz.
Syrup of Squills	1 "

Ten drops for a child seven years old every four hours.

130.

Infusion of Digitalis	4 oz.
Acetate of Potash	2 drachms.
Sweet Spirits of Nitre . . .	2 "
Cinnamon Water	1½ oz.

A table-spoonful every four or five hours.

131.

Iodide of Potassium	1 drachm.
Water	½ oz.

Thirty drops to a child seven years old, every hour.

132.　　　　　　　　　　**Brain Fever.**

Pulv. Gamboge	12 grains.
Pulv. Scammony	12 "
Elaterium	2 "
Croton Oil	8 drops.
Ext. Stramonium	3 grains.

Make twelve pills. One pill repeated every hour until it operates.

133.

Pulv. Scammony	12 grains.
Pulv. Gamboge	12 "
Pulv. Colocynth	8 "
Castile Soap	4 "
Oil of Anise	5 drops.

Make twelve pills. One pill repeated every three hours till it operates.

134.

Powdered Antimony	4 grains.
Camphor Scrapings	4 "
Extract of Hyoscyamus	6 "
Syrup of Poppies	sufficient.

Three pills, to be taken just before bedtime.

135.　　　　　　　**Inflammation of Brain.**

Sulphuric Ether,	
Liquor of Acetate of Ammonia, } of each .	1½ oz.
Rectified Spirits of Wine,	
Rose Water	3½ "

An evaporating lotion.

136. **Hard and Inflamed Breasts.**

Liquor of Acetate of Ammonia   .   .   6 oz.
Rectified Spirit   .   .   .   .   .   2 "
   Make a lotion.

137. **Painful Affections of Breast.**

Chlorinated Ether   .   .   .   .   .   3 drachms.
Distilled Water   .   .   .   .   .   1 pint.
   Make a lotion.

138. **For Gathered Breasts.**

Fresh Tobacco Leaves, sliced   .   .   .   10 oz.
Diluted Acetic Acid .   .   .   .   .   4 pints.
   Boil the tobacco in the acid, strain, and evaporate the decoction to six ounces. Add this to thirteen ounces Basilicon Ointment, heated, and stir till cold.

139. **Bronchitis.**

Tartar Emetic .   .   .   .   .   .   1 grain.
Boiling Water .   .   .   .   .   .   10 drachms.
   One teaspoonful every two hours.

140.

Tincture Blood-root .   .   .   .   .   1 oz.
Sulphate of Morphia   .   .   .   . $1\frac{1}{2}$ grains.
Tinct. Digitalis   .   .   .   . $\frac{1}{2}$ .oz.
Wine of Antimony .   .   .   . $\frac{1}{2}$ "
Oil of Wintergreen .   .   .   .   10 drops.
   Dose, from twenty to forty drops twice or three times a day.

141.

Syrup of Tolu	1 oz.
Syrup of Squills	½ "
Wine of Ipecac	2 drachms.
Paregoric	3 "
Mucilage of Gum Arabic	1½ oz.

Take a teaspoonful occasionally.

142.

Carbonate of Soda	1 drachm.
Wine of Ipecacuanha	½ oz.
Tincture of Opium	1 drachm.
Syrup of Tolu	2 oz.
Water	1½ "

Half an ounce for a dose.

143. **Burns.**

Powdered Tragacanth	2 drachms.
Lime Water	3 oz.
Pure Glycerine	1 "
Rose Water	3 "

Make a liniment.

144. **Extensive Burns.**

Liquor Diacetate of Lead, } of each . .	1 oz.
Olive Oil, }	
Rose Water	4 "

Make a liniment.

145. **Canker of the Mouth, Ulcerated Throat, etc.**

Infusion of Cinchona	3 oz.
Chlorinate of Soda in solution . . .	1 "

    Make a mouth-wash.

146.

Sulphate of Copper	5 grains.
Oxymel	$\frac{1}{2}$ oz.

    Apply with a camel-hair brush.

147. **In Catarrhal Affections.**

Leaves of the Red Poppy Flower . .	2 oz.
Sulphuric Acid, diluted	15 drops.
Sugar	2 oz.
Decoction of Barley	1 pint.

    Infuse and strain. Drink freely, as often as you desire.

148.

Muriate of Ammonia	15 grains.
Gum Arabic	$\frac{1}{2}$ drachm.
Infusion of Chamomile	3 oz.
Antimony Wine	1 drachm.
Ext. Liquorice	2 drachms.

    Half a spoonful every two hours, to a child 5 or 6 years old.

149.

Muriate of Ammonia, } of each . .	1 drachm.
Extract Liquorice,	
Decoction of Marsh Mallow . . .	6 oz.
Oxymel of Squills	1 "

    Two table-spoonsfuls three times per day.

150.   **Common Catarrh.**

Acetate of Ammonia, Liquor, } of each   .   6 drachms.
Camphor Mixture,
Syrup of Poppies   .   .   .   .   .   1 drachm.
Antimony Wine (Tartrate of Potash).   .   20 minims.
   A draught to be taken before going to bed.

151.   **Cathartic.**

Magnesia,
Supertartrate of Potash,
Flour of Sulphur,   } of each   6 grains.
Powdered Rhubarb Root,
Powdered Chamomile Flowers,
Orange Syrup .   .   .   .   .   .   3 drachms.
Oil of Pimento .   .   .   .   .   .   2 minims.
   Make an electuary for a dose.

152.   **Cerebral Affections.**

Boracic Acid .   .   .   .   .   .   1 drachm.
Camphor Mixture .   .   .   .   .   4 oz.
Orange Syrup .   .   .   .   .   .   1 "
   Two teaspoonfuls every two or three hours.

153.

Camphor-powder .   .   .   .   .   3 grains.
James's powder .   .   .   .   .   4 "
Nitrate of Potash .   .   .   .   .   ½ scruple.
Extract of Hyoscyamus .   .   .   .   7 grains.
Conserve of Roses, sufficient to make a bolus.

154. **Laxative in Change of Life.**

Sulphuric Sublimate . . . . . 1 oz.
Bicarbonate of Soda . . . . . 1 drachm.
Powdered Ipecacuanha . . . . 5 grains.
    From 1 to 2 scruples, in milk, at bedtime.

155. **A Valuable Liniment for Chilblains.**

Sulphuric Acid . . . . . . 1 drachm.
Spirits Turpentine . . . . . 1 "
Olive Oil . . . . 3 "
    Mix the Oil and Turpentine first, gradually adding the
Sulphuric Acid. To be rubbed on two or three times a day.

156. **Chlorosis, Anæmia, etc.**

Decoction of Aloes . . . . . 2 oz.
Syrup of Crocus, } of each . . . ½ "
Syrup of Rhubarb, }
    Make a mixture, to be taken in two doses.

157.

Iron Filings . . . . . . 1 drachm.
Ext. Absinthe, sufficient to make into 4-grain pills. One
    to four, night and morning.

158.

Pulverized Iron, with Sesquioxide of Iron. ½ oz.
Nitrate of Bismuth . . . . . 5 drachms.
Extract of Opium, diluted . . . . 3 grains.
Syrup and Gum Acacia, sufficient to mix and divide into
    one hundred and twenty-five pills. From one to ten
    daily, during meals.

159.

Sesquioxide of Iron	½ scruple.
Valerian powder	½ "

Syrup of Ginger, sufficient to form a bolus.

160. **Cholera Infantum.**

Sub-muriate of Mercury	2 grains.
Acetate of Lead	1 grain.

Divide into four powders. One every three hours.

161.

Sulphate of Iron	2 grains.
Sulphuric Acid, diluted	10 drops.
White Sugar	1 drachm.
Water	1 oz.

One oz. three or four times per day.

162. **Cholera.**

Assafœtida,
Powdered Opium, } of each . . . 1½ grains.
Black Pepper,

Make a pill — to be bruised and taken in a glass of brandy and water every half or three-quarters of an hour.

163.

Potassio-tartrate of Antimony . . .	2 grains.
Sulphate of Magnesia	½ oz.
Water	10 "

For an adult, a table-spoonful; for a two-year old child, a teaspoonful every half hour.

164.

Pulv. Camphor	½ drachm.
Pulv. Opium	16 grains.
Pulv. Cayenne	½ drachm.

    Make sixteen pills. One every hour.

165.   **Chorea, Epilepsy, etc.**

Pulv. Senna	2 drachms.
Bicarbonate of Potassa	2 oz.
Pulv. Cayenne	10 grains.
Pulv. Jalap	1 drachm.

    Divide into twelve parts. One part every four hours until it operates.

166.

Leptandrin	1 drachm.
Podophyllin	1 scruple.
Scutillaria	2 drachms.
Pulv. Cayenne	1 scruple.
Pulv. Loaf Sugar	4 oz.

    Rub well together. Dose for an adult, one-sixteenth of the above.

167.

Ext. Skullcap	2 drachms.
Ext. Boneset	1 drachm.
Ext. Chamomile	2 drachms.
Quinine	1 drachm.
Pulv. Cayenne	1 scruple.
Oil of Valerian	½ drachm.

    Beat well together and make ninety pills. One pill every two or three hours.

168. **Cough.**

Nitrous Ether, Spirit, } of each . . 1 oz.
Syrup of Tolu, }
   A teaspoonful when troublesome.

169.   **Coughs of Children, without Inflammation.**

Liquor of Onions . . . . . 1 oz.
Sugar . . . . 1½ "
   Make into syrup — a teaspoonful occasionally.

170.

Alum . . . . 24 grains.
Sulphuric Acid, diluted . . . 12 minims.
Syrup of Red Poppy . . . . . 4 drachms.
Water . . . . 2½ oz.
   Take three drachms every six hours.

171.

Bicarbonate of Potassa . . . 15 grains.
Cochineal . . . . . . 8 "
Distilled Water . . . . . 6 oz.
   Rub up together, strain, and add
Hydrocyanic Acid, diluted . . . 10 minims.
   A teaspoonful to be taken when the cough is troublesome.

172. **Cough of Consumption.**

Mixture Acacia . . . . . 1 oz.
Distilled Water . . . . . 6½ "
Syrup Tolu . . . . . ½ oz.
Hydrocyanic Acid, diluted . . . 12 drops.
   A table-spoonful every three hours.

173. **Cough of Measles.**

Oil of Almonds . . . . . 2 oz.
Syrup of Poppies, } of each . . . 1 "
Syrup of Tolu, }
Powdered Sugar · . . . . 2 drachms.

Make a thick syrup, of which the patient may partake freely when the cough is troublesome.

174. **Croupy Cough.**

Hydrocyanic Acid diluted . . . . 5 minims.
Rose Water . . . . . . 5 oz.
Poppy Seed . . . . . . 3 drachms.

A teaspoonful every two or or three hours.

175, **Croup.**

Decoction of Seneka root . . . . 4 oz.
Antimony Wine . . . . 2 scruples.
Syrup of Marsh Mallow . . . . 1 oz.

A teaspoonful frequently.

176.

Decoction of Seneka root . . . . 5 oz.
Carbonate of Ammonia . . . . 8 grains.
Tinct. of Squills . . . . . 16 drops.
Syrup of Tolu . . . . . . 2 drachms.

Three drachms in milk, every 4th hour, for children three or four years old.

177.

Dover's Powders . . . . . 15 grains.
Calomel . . . . 5 "

Divide into ten powders. One every three hours for a child.

178. Condylomata or Fungous Growths.

Pulverized Savin	1 scruple.
Sulphate of Copper	1 "

To be sprinkled on the growths.

179. Constipation.

Leptandrin	1 drachm.
Podophyllin	1 scruple.
Apocynin	1 "
Ext. Nux Vomica	6 grains.
Castile Soap	1 drachm.

Make thirty pills. One pill every night.

180.

Compound Infusion of Senna . . .	4 oz.
Tartrate of Potassa	2 drachms.
Carraway Water	2 oz.
Manna	1 drachm.

For a child, a table-spoonful.

181. For Indigestion, with Costiveness.

Extract Aloes	1 scruple.
Powdered Ipecacuanha	8 grains.
Powdered Ginger	½ drachm.
Syrup	sufficient.

Make sixteen pills. One to be taken before dinner.

182.

Aloes and Myrrh Pill	1 drachm.
Compound Galbanum pill	2 drachms.

Divide into forty pills. Two pills to be taken three times a day.

183. **Obstinate Costiveness.**

Barbadoes Aloes . . . . . 24 grains.
Sulphuric Acid . . . . . . 6 drops.
    Divide into six pills. Two to be taken every four hours.

184. **Infantile Convulsions.**

Oil of Anise . . . . . . 4 drops.
White Sugar . . . . . . ½ scruple.
    Mix intimately, and add
Water . . . . 2 ounces.
Powdered Rhubarb . . . . . ½ scruple.
Carbonate of Magnesia . . . . 1 "
Tinct. Opium . . . . . . 4 drops.
Sulphuret of Ammonia . . . . 10 "
    A dessert-spoonful every third hour.

185. **Crusta Lactea — Milk Crust.**

Sulphuret of Zinc . . . . . 2 drachms.
Decoction of Marsh Mallow . . . 2 oz.
    Make a lotion.

186. **Cutaneous Eruptions of Infancy and Childhood.**

Green Iodide of Mercury . . . . 2 grains.
Mercury with Prepared Chalk . . . 12 "
Aromatic Powder . . . . . 9 "
    Divide into six powders. One every morning for a child
two years old.

187. **Diabetes.**

Phosphoric Acid diluted . . . . 1 drachm.
Decoction of Barley . . . . . 2 pints.
    To be used as an ordinary beverage.

188.

Sesquicarbonate of Ammonia . . .	½ drachm.
Sweet Spirit (Rum)	5 drachms.
Simple Syrup	5 "
Water	3 oz.

Half to be taken, morning and night.

189.

Liq. Arsenite of Potassium . . .	3 drachms.
Liq. Hydrosulphate of Ammonia . .	20 minims.
Tinct. Hyoscyamus	2 drachms.
Infusion of Buchu	8 oz.

Take a table-spoonful every fourth hour.

190.   **Diarrhœa.**

Pulverized Catechu	2 drachms.
Bruised Cinnamon	½ drachm.
Boiling Water	5 oz.

Steep in covered vessel for one hour and strain. A tea-spoonful every two, three, or four hours, according to age, etc.

191.

Tinct. Catechu	½ oz.
Laudanum	2 drachms.
Spirits of Camphor	2 "
Tinct. Myrrh	2 "
Tinct. Cayenne.	2 "

From half a teaspoonful to a teaspoonful when required.

192.

Syrup of Orange Peel	. . . .	1 oz.
Acetate of Morphia .	. . . .	2 grains.
Tinct. of Cinnamon .	. . . .	6 drachms.
Tinct. of Cardamoms	. . . .	2 "

A teaspoonful every one or two hours.

193. Diphtheria.

Hydrochloric Acid and Honey in equal parts.
Touch the fauces with the mixture.

194.

Powdered Alum	. . . . .	2½ drachms.
White Honey .	. . . . .	10 "

Half a spoonful every hour; powdered alum or sulphur to be blown into the throat every four hours.

195.

Iodide of Potassium .	. . . .	½ drachm.
Tincture of Orange .	. . . .	½ oz.
Syrup of Ginger	. . . .	½ "
Pure Water .	. . . . .	5 "

A tablespoonful in equal quantity of water three times per day.

196. Dropsy.

Tinct. Black Cohosh .	. . . .	1 oz.
Tinct. Myrrh .	. . . .	6 drachms.
Laudanum	. . . . .	1 drachm.
Tinct. Cayenne	. . . .	1 "

Thirty or forty drops four times per day.

197. **Dysmenorrhœa.**

Hydrochloric Acid, diluted, } each	.	.	30 minims.
Nitric Acid, diluted,		
Tinct. Camphor (Compound) .	.	.	4 drachms.
Tinct. Orange .	.	.	.	.	.	½ oz.
Syrup of Sarsaparilla	.	.	.	.	½ "
Rose Water	.	..	.	.	.	.	6 "

Take two teaspoonfuls twice a day.

198.

Hydrochloric Acid, diluted, } of each .   .   1 drachm.
Nitric Acid, diluted,

Extract Dandelion Root .   .   .   .   1 oz.

Compound Infusion of Gentian .   .   .   7 oz.

Two table-spoonfuls twice a day, before meals.

199

Hydrochloric Acid, diluted, } of each .   .   1 drachm.
Nitric Acid, diluted,

Infusion of Dandelion Root   .   .   .   1 oz.

Infusion of Cinchona   .   .   .   .   7 oz.

Two teaspoonfuls twice a day, before meals.

200. **Dysentery.**

Infusion Cascarilla .   .   .   .   .   6 drachms.

Cinnamon Water   3   "

Compound Powder of Kino   .   .   .   ½ scruple.

Syrup of Poppies   .   .   .   .   1 drachm.

A draught to be taken twice a day.

201.

Aromatic Confection   .   .   .   .   15 grains.

Lime-Water .   .   .   .   .   11 drachms.

Carbonate of Magnesia   .   .   .   6 grains.

Tinct. of Hops .   .   .   .   .   1 drachm.

A draught, three times per day.

202. Dyspepsia.

Sweet Tincture of Rhubarb . . . 4 oz.

Bicarbonate of Soda . . . . . 2 drachms.

   From a teaspoonful to a table-spoonful, as occasion may require.

203.

Pulv. Charcoal (Willow bark) . . . 1½ drachms.

Pulv. Rhubarb . . . . . . 2 scruples.

Pulv. Ipecac . . . . . . 6 grains.

Ext. Hyoscyamus . . . . . 12 "

   Divide into twelve portions. Give one every three or four hours.

204.

Pulverized Rhubarb . . . . . 2 oz.

Bicarbonate of Potassa . . . . 1 "

   Mix. Take sufficient to produce one movement of bowels per day.

205. Epilepsy.

Ox-gall,
Assafœtida, } of each . . . . 1 drachm.

Powdered Rhubarb . . . . . 1 scruple.

   Syrup, sufficient to make forty pills. Two, twice per day.

206. Erysipelas.

Wine of Colchicum . . . . . 30 minims.

Sulphate of Magnesia . . . . 1½ drachms.

Carbonate of Magnesia . . . . 1 scruple.

Peppermint Water . . . . . 10 drachms.

   Make a draught.

207.

Senna	3 drachms.
Salts	½ drachm.
Manna	½ "
Fennel Seed	1 "
Boiling Water	½ pint.

Macerate and strain. Teacupful every four hours unti
it operates.

208.

Calcined Magnesia	1 scruple.
Pulv. Rhubarb	1 "
Pulv. Ipecac	1 grain.

One-fourth of this daily.

209.

Nitrate of Silver	2 scruples.
Nitric Acid	12 drops.
Soft Water	1 oz.

Apply with a piece of lint tied to the end of a stick.

210.   Fevers — Typhus, Typhoid, Malignant, etc.

For Children.

Hydrochloric Acid	1 drachm.
Distilled Water	6 oz.
Syrup of Mulberry	1 "

A table-spoonful every four or five hours.

211.

For Adults.

Muriatic or Hydrochloric Acid . . .	1 drachm.
Decoction of Barley	1 pint.
White Sugar	½ oz.

Take from two to four ounces, twice or three times a day.

212.

Nitric Acid, diluted	1½ drachms.
Water	24 oz.
Sugar	1½ "

Take three ounces three times per day, through a glass tube.

213.

Mixt. Camphor	10 drachms.
Sulphuric Ether	1 drachm.
Aromatic Confection	1 scruple.
Compound Tinct. of Lavender . . .	½ oz.

In Sinking of Fevers. A draught, repeated every four hours, or oftener, if symptoms be urgent.

214. **Flatulence with Nausea.**

Angostura Bark, powdered, } of each . . 5 grains.
Rhubarb in powder,

To be taken an hour before dinner.

215. **Gout and Inveterate Rheumatism.**

Spirits of Hartshorn	4 drachms.
Succinic Acid, sufficient to saturate	
Sulphuric Ether	4 "

Twenty to forty drops in a glass of sugar and water two or three times a day.

216.   **Headache.**

Sulphate of Iron   .   .   .   .   .   2 grains.
Epsom Salts   .   .·   .   .   .   .   2 scruples.
Diluted Sulphuric Acid   .   .   .   .   10 drops.
Compound Tincture .   .   .   .   .   1 drachm.
Syrup of Poppies   .   .   .   .   .   1½ drachms.
Pimento Water   9   "
    To be taken at a draught twice a day.

217.

Comp. Infusion Senna   .   .   .   .   5 drachms.
Infusion of Rhubarb   5   "
Comp. Tinct. Cardamoms .   .   .   .   ½ drachm.
Syrup   .   .   .   .   1½ drachms.
    To be taken at a draught.   Excellent for the *headache of Dyspepsia.*

218.

Carbonate of Soda .   .   .   .   .   10 grains.
Aromatic Spirits of Ammonia   .   .   .   ½ drachm.
Tincture of Orange Peel   1   "
Syrup of Orange Peel   1   "
Compound Infusion of Gentian .   .   .   10 drachms.
    To be taken at a draught twice a day.

219.   **Hemorrhages.**

Acetate of Lead   .   .   .   .   .   3 grains.
Opium   .   .   .   .   1 grain.
Extract of Hemlock .   .   .   .   .   4 grains.
    Divide into two pills.   To be given twice a day.   Drink some acidulated draught afterwards, until it ceases.

220.   **Hemorrhages, Spasms, etc.**

Sulphuric Acid, diluted	15 minims.
Compound Infusion of Roses . · . .	1½ oz.
Syrup, simple	1 drachm.

Make a draught.

221.

Sulphuric Acid, } in equal parts.
Nitric Acid,

From five to ten drops in an ounce of water.

222.   **Hemorrhoids.**

Sulphate of Magnesia	3 drachms.
Carbonate of Magnesia	2 scruples.
Wine of Colchicum	1½ drachms.
Syrup of Red Poppy	½ oz.
Peppermint Water	4 "
Distilled Water	1½ "

Two tablespoonfuls twice a day.

223.

Morphine	2 grains.
Olive Oil	2 drachms.

Rub up together and add

Zinc Ointment	1 oz.
Powdered Nutgalls	1 drachm.

224.

Confection of Black Pepper . . .	1 drachm.
Assafœtida	5 grains.

Syrup of Ginger, sufficient to make a uniform confection.
Twice per day.

225.

Alum	1 drachm.
Fresh Butter	1 oz.

Make an Ointment.

226.

Confection of Senna	1½ oz.
Powdered Jalap	½ drachm.
Sulphur	½ "

Syrup of Senna, sufficient to make an electuary; taken three times a day, until bowels well opened.

227.

Slaked Lime	2 drachms.
Ointment of Colocynth	2 oz.
Wine of Opium	2 drachms.

Make an ointment.

228. Hemorrhoids, with Constipation.

Bitartrate of Potash,		
Extract of Horehound,	of each . .	2 drachms.
Honey Water		3 oz.

Half to be taken morning and evening.

229. Hydrocele of Children.

Muriate of Ammonia	1 drachm.
Liquor Acetate Ammonia	2 oz.
Water	4 "

Lotion: To be kept constantly applied.

230.   **Hydrocephalus.**

Chloride of Mercury .   .   .   .   .   1 grain.
Powdered Digitalis .   .   .   .   .   ½ "
Powdered Tragacanth (compound) .   .   6 grains.

 Make a powder One every sixth hour, for child of two or three years.

231.   **Hysteria.**

Tinct. Hyoscyamus .   .   .   .   .   ½ drachm.
Aromatic Spirit of Ammonia   .   .   .   ½ "
Syrup of Orange Peel   .   .   .   .   ½ "
Peppermint Water .   .   .   .   .   10 drachms.

232.

Aromatic Spirits of Ammonia   .   .   .   1½ drachm.
Sulphuric Ether   .   .   .   .   .   1 "
Syrup of Ginger   .   .   .   .   .   3 drachms.
Anise Water .   .   .   .   .   .   3½ oz.

 A third part frequently, as occasion requires.

233.

Ætheris Acetici   .   .   .   .   .   30 minims.
Camphor Mixture, with Magnesia   .   .   6 oz.

 To be taken immediately, and repeated every hour if necessary.

234.

Infusion of Quassia .   .   .   .   .   ½ oz.
Tinct. of Ammonio-Chloride of Iron .   .   ½ drachm.
Sesquicarbonate of Ammonia   .   .   .   6 grains.
Orange Syrup .   .   .   .   .   .   1 drachm.
Distilled Water   .   .   .   .   .   7 drachms.

 A draught to be taken two or three times a day.

235.

Strychnia	1 grain.
Compound Rhubarb Pill	1 drachm.
Calomel Pill	6 grains.
Oil of Peppermint	4 drops.

Divide into fifteen pills. One twice per day.

236. Icterus, Nettle Rash, and Scarlet Fever.

Pulverized Ipecac	1 scruple.
Pulverized Cayenne	10 grains.
Water	2 oz.

To be taken at a draught.

237.

Tartar Emetic	1 grain.
Pulverized Ipecac	1 scruple.

Mix. To be taken in a wineglassful of sweetened water.

238.

Pulverized Lobelia	1 oz.
Pulv. Bloodroot	½ "
Pulv. Seneka	1 scruple.
Pulv. Ipecac	6 drachms.
Pulv. Cayenne	4 scruples.

Half-teaspoonful in warm water, repeat every fifteen minutes, for one hour.

239. Indigestion.

Tincture of Musk Seed	1 oz.
Tincture of Hops	3 drachms.
Liquor of Potash	2 "
Infusion of Buchu	6 oz.

Three table-spoonfuls three times a day.

240.

Hydochloric Acid	2 drachms.
Infusion of Columbo Root. . . .	5½ oz.
Tinc. of Hops	½ "

One-sixth part for a dose.

241.

Hydrocyanic Acid	32 drops.
Muriate of Morphia, Tinct. . . .	3 drachms.
Mixt. Tinct. Sweet Almonds . . .	8 oz.

A table-spoonful three times a day.

242.   Inflammation of Eye or Ear.

Tartrate of Antimony	¾ grain.
Decoction of Barley	2 pints.

Dissolve and add

Syrup	3 oz.

To be taken by glassfuls in the course of the day.

243.   Inflammation of the Mucous Membranes.

Decoction of Barley (compound) . .	10 oz.
Syrup of Indian Sarsaparilla . . .	2 "

Two table-spoonfuls occasionally.

244. Inflammatory Complaints of Children — as an Emetic.

Potassia Tartrate of Antimony . . .	1 grain.
Distilled Water	1½ oz.
Simple Syrup	½ "

One, two, or three teaspoonfuls every quarter of an hour
until vomiting is produced.

245.   **Influenza.**

Citrate of Iron	1 drachm.
Sulphate of Quinine	1 scruple.
Ext. Nux Vomica	8 grains.

    Make thirty-two pills.   One pill three times per day.

246.   **Irritable and Acid Stomach.**

Hydrocyanic Acid, diluted . . .	4 minims.
Bicarbonate Potassa	10 grains.
Syrup of Ginger	½ drachm.
Anise Water	1½ oz.

    A draught to be taken twice per day.

247.   **Itching Eruptions.**

Hydrocyanic Acid, diluted . . .	2 drachms.
Bichloride of Mercury	2 grains.
Emulsion of Sweet Almonds . . .	6 oz.

    Mix.   Make a lotion.

248.

Cyanide of Potassium	12 grains.
Mixture of Sweet Almonds . . .	6 oz.

    Make a lotion.

249.

Hydrocyanic Acid, diluted . . .	4 drachms.
Decoction of Mallow Leaves . . .	1 lb.

    Make a lotion.

250.

Hydrocyanic Acid, diluted . . .	1½ drachms.
Distilled Water	7 oz.
Acetate of Lead	16 grains.
Rectified Spirit	2 drachms.

    Make a lotion.

251.   **In Tinea Capitis.**

Muriatic Acid, ⎫ each . .	½ oz.
Marsh Mallow Ointment, ⎭	
Ointment of Juniper	2 oz.

    To be applied twice a day.

252.   **Pruritus Vulvæ — Itching of the Privates.**

Biborate of Soda	½ oz.
Rose Water	6 "
Sulphate of Morphia	6 grains.

253.   **Jaundice.**

Calomel Pill	10 grains.
Powdered Ipecacuanha (compound) . .	5 "

    Divide into 3 pills ; one every 4 hours.

254.

Indian Hemp Seed	4 oz.
Light Beer	2 pints.

    Boil, strain, and add sugar, sufficient.
    Take half pint every morning.

255.　　　　　　　　　**Leucorrhœa, or Whites.**

Nitrate of Silver Crystals . . .	10 grains.
Corrosive Sublimate	5　"
Sugar of Lead	1½ drachms.
White Vitriol	1½　"
Soft Water	6 oz.

　　An injection.

256.

Powdered Cantharides	12 grains.
Ext. Hyoscyamus	1 drachm.
Nitrate of Silver	10 grains.
Sulphate of Quinia	2 scruples.

　　Make 40 pills ; one every night and morning.

257.

Cresote	20 minims.
Liquor Potassa	2 drachms.
White Sugar	2　"

　　Rub up together, and add

Distilled Water	8 oz.

　　Make an injection.

258.

Powdered Cubebs	1 oz.
Powdered Ergot	2 drachms.
Compound Powder of Cinnamon . .	2 scruples.
White Sugar	1 drachm.

　　Divide into 8 powders ; one three or four times per day.

259. **In Labor.**

Powder of Ergot	½ drachm.
Syrup	½ oz.
Peppermint	1 "

A third part every 20 minutes.

260. **Laxatives.**

Infusion of Cascarilla	1½ oz.
Sulphate of Magnesia	1 drachm.
Sulphuric Acid, diluted	15 minims.

Draught to be taken twice or thrice a day. Also a *Tonic*.

261.

Compound Decoction of Aloes . .	3 oz.
Bi-carbonate of Potass.	2 drachms.
Ammonia-tartrate of Iron . . .	½ "
Aromatic Spirits of Ammonia . . .	3 "
Water	6½ oz.

A sixth part twice per day. Also a *Tonic*.

262. **Liver Derangement.**

Compound Mixture of Gentian . .	10 drachms.
Sulphate of Magnesia	3 "
Tinct. Jalap	1 "
Aromatic Spirits of Ammonia . . .	½ "

A draught to be taken in the morning.

263.

Chloride of Mercury	4 grains.
Compound Ext. Colocynth . . .	8 "

Two pills to be taken at bedtime.

264.

Powdered Jalap	12 grains.
Submuriate of Mercury	3 "
Sulphate of Potash	7 "

Make into a mass.   Dose, 20 to 30 grains.

265.

Sulphate of Magnesia	$\frac{1}{2}$ oz.
Tinct. Jalap	1 drachm.
Nitric Acid	2 minims.
Green Peppermint Water	2 oz.

Make a draught.

266.   Mania, with Torpid Bowels.

Tinct. Black Hellebore	1$\frac{1}{2}$ drachms.
Compound Infusion of Senna . . .	1 oz.
Syrup of Ginger	2 drachms.

A draught to be taken early in the morning.

267.   Measles.

Fluid Extract Senna	1 drachm.
Compound Fluid Extract Gentian . .	$\frac{1}{2}$ "
Fluid Extract of Ginger	$\frac{1}{2}$ "
Aromatic Spirits of Ammonia . . .	$\frac{1}{2}$ "

To be taken in wineglass of sweetened water.

268.

Tinct. Lobelia	$\frac{1}{2}$ oz.
Syrup of Squills	$\frac{1}{2}$ "

Twenty drops four or five times a day for a child two years old.

269.

Tinct. of American Hellebore . . . 1 drachm.
Tinct. of Black Cohosh . . . . 2 oz.
    One tea-spoonful three to six times per day.

270. **As a Stimulant.**

Muriate of Ammonia . . . . 1 oz.
Soft water . . . . . . 9 "
    One table-spoonful three or four times per day.

271. **Painful Menstruation.**

Ext. Belladonna . . . . . $\frac{1}{2}$ drachm.
Lard . . . . $\frac{1}{2}$ oz.
    To be rubbed on the neck of the womb.

272. **Profuse Menstruation.**

Wine of Spurred Rye . . . . 2 oz.
    One teaspoonful three times a day.

273. **Nausea and Vomiting.**

Lime Water,
New Milk, } each 4 oz.
    A table-spoonful every half-hour, hour, or two hours.

274. **Nervous Affections.**

Assafœtida,
Powdered Valerian, } in equal parts.
    Syrup and Tinct. Valerian in sufficient quantity to make a mass, and divide into 5-grain pills. Take two twice per day.

275. **In Nervous Palpitations.**

Tinct. Digitalis	12 minims.
Mixture of Camphor	1 oz.
Orange Syrup	2 drachms.
Hydrocyanic Acid, diluted . . .	1 minim.

Make a draught. To be taken two or three times per day.

276.

Solution of Magnesia (with Carbonic Acid)	1½ oz.
Tincture Muriate of Iron	25 minims.

A draught to be taken three times a day, and immediately followed by a wineglass of cold or tepid water.

277. **Nervous Debility.**

Disulphate of Quinia	12 grains.
Powdered Sugar	2 drachms.

Divide into six powders. One night and morning.

278.

Disulphate of Quinia	15 grains.
Cinnamon Powder	½ drachm.
Extract of Cinchona	20 grains.

To make thirty pills. Four every second, third, or fourth hour.

279.

Sulphate of Quinine	32 grains.
Simple Syrup	8 oz.

Make a Quinine Syrup. Two teaspoonfuls two or three times per day.

280. **Neuralgia of the Bladder, etc.**

Arsenious Acid	2 grains.
Powdered Opium	5 "
Strychnia	1 "
Extract Aconite	8 "

Divide into sixteen pills. One every six hours; if nausea ensues, take only half a pill.

281. **Neuralgia.**

Distilled Water	2 oz.
Valerianic Acid	1 "

Subcarbonate of Ammonia enough to neutralize the acid; then add Alcoholic Extract of Valerian, 2 scruples. A teaspoonful three times per day.

282.

Ext. Hyoscyamus	½ drachm.
Sulphate of Morphia	3 grains.
Strychnine	2 "
Pulv. Cayenne	½ drachm.
Sulphate of Zinc	15 grains.

Make thirty pills. Take one four times per day.

283. **For Neuralgia of the Face.**

Ext. of Hyoscyamus	½ drachm.
Valerianate of Zinc	1 scruple.

Make thirty pills. Take one two or three times per day.

284. **Sore Nipples.**

Powdered Gum Arabic	½ drachm.
Powdered Alum	5 grains.

Make a powder. Apply as often as necessary.

285.   **Nurse's Sore Mouth.**

Nitrate of Silver . . . . .   8 grains.

Distilled water . . . . . .   4 oz.

    Make a gargle.

286.   **Offensive Breath.**

Charcoal of White Wood, powdered . .   1 oz.

Orange Syrup . . . . . .   3 "

    Make an electuary. One or two teaspoonfuls every two hours.

287.   **Paralysis, Rheumatism, etc.**

Flowers of Arnica . . . . .   1 oz.

Water . . . . . .   2½ pints.

    Boil to a pint and a half, and add

Syrup of Ginger . . . . .   2 oz.

    Take from two to three ounces every second hour.

288.   **Paralysis.**

Dried Juice of Spurge . . . .   1 drachm.

Olive-oil . . . .   10 drachms.

    Digest for ten days and strain. Used in rubbing.

289.   **Pimply Eruptions.**

Nitric Acid, diluted . . . . .   ½ drachm.

Decoction of Barley . . . .   1 pint.

    A wine-glassful to be taken three times per day.

290.   **Pleurisy, Colic.**

Compound Extract of Colocynth . .   ½ drachm.

Extract of Jalap . . . . .   15 grains.

    Mix. Make twelve pills. Two or three pills will produce active operation of the bowels.

291.

Fluid Ext. Sarsaparilla	4 oz.
Fluid Ext. Pipsissewa	1 "
Water	1 quart.
Iodide of Potassium	2 oz.

A tablespoonful three times a day.

292.

Wine of Ipecac	1 oz.
Spirits of Turpentine	1 "
Castor-oil	1 "
Molasses	½ pint.
Warm Water	½ "

Mix.

293.

Thoroughwort	1 oz.
Senna	1 "
Lobelia	½ drachm.
Cayenne	10 grains.
Epsom Salts	1 tablespoonful.
Molasses	½ pint.
Boiling Water	1 "

Make a strong decoction of the herbs, and then add the salt and the molasses.

294. **Pneumonia, Typhoid Fever.**

Bicarbonate of Soda	½ oz.
Compound Infusion of Gentian . .	4 "
Tincture of Colombo	1 "
Syrup of Orange Peel	½ "

A tablespoonful three times per day.

295.

Sulphate of Quinine	1 scruple.
Alcohol	4 oz.
Sulphuric Acid	5 drops.
Madeira Wine	1 quart.

Two wineglassfuls per day.

296.

Rose Water	6 oz.
Syrup of Orange Peel	1 "
Muriated Tincture of Iron . . .	1 "

For adults, one teaspoonful in a wineglass of water after each meal.

297. **Rheumatism.**

Diluted Acetic Acid	1 drachm.
Tinct. Jalap	15 minims.
Tinct. Orange	1 drachm.
Camphor Mixture	10 drachms.

A draught to be taken two or three times per day.

298. **Rheumatism and Cell-Dropsy.**

Tinct. Black Cohosh	1 oz.
Iodide of Potassium	2 drachms.
Syrup of Ipecac	1 oz.
Spring Water	2 "

A teaspoonful three or four times per day.

299. **Rheumatism.**

Peppermint Water	1½ oz.
Wine of Colchicum Root	½ "
Sulphate of Morphia	1 grain.
Magnesia	1 scruple.

One teaspoonful three or four times per day.

300. **Ringworm.**

Pulverized Sulphate of Copper . . .	10 grains.
Ext. Spanish Flies	5 "
Lard	1 oz.

Rub into the scalp or part affected.

301.

Castor-oil	2½ pounds.
Pure Alcohol	2½ pints.
Pulv. Spanish Flies	½ oz.
Oil of Bergamot	2½ "
Otto of Roses	20 drops.

A superior preparation for keeping the hair from falling, and to prevent dandruff. Let the mixture stand for a few days and then filter.

302. **Small-Pox, Angina Pectoris.**

Spirits of Mindererus	2 oz.
Sweet Spirits of Nitre	1 "

One teaspoonful every three hours.

303.

Comp. Tinct. of Cardamoms . . .	2 oz.
Comp. Tinct. of Lavender . . .	2 "
Comp. Tinct. of Gentian	2 "

One teaspoonful at a time, as occasion may require.

304.

Pulv. Gum Arabic	1 scruple.
Soft Water	2 oz.
Sweet Spirits of Nitre	½ "
Tinct. Veratrum Viride . .	20 drops.

Half a teaspoonful every hour.

305.

Aromatic Spirits of Ammonia　.　.　.	2 drachms.
Ether　..　.　.　.　.　.　.	1 drachm.
Laudanum　.　.　.　.　.　.	20 drops.
Spirits of Camphor .　.　.　.　.	1 drachm.

　　　Half a teaspoonful as often as required.

306.　　　　　　**Scrofula and Leucorrhœa.**

Sesquioxide of Iron, ⎫
Extract of Hemlock, ⎭ of each　.　.　1 drachm.

　　　Make into twenty-four pills.　Two twice per day.

307.　　　　　　**Scrofula.**

Oxysulphuret of Antimony, ⎫
Chloride of Mercury, ⎭ of each　.　½ drachm.

Ammonia　.　.　.　.　.　.	·1　"
Balsam of Peru　.　.　.　.　.	sufficient.

　　　Make thirty pills.　Take one or two every night.

308.　　　**Scrofulous and Cancerous Affections.**

Chloride of Baryta .　.　.　.　.	1 oz.
Muriatic Acid .　.　.　.　.　.	4 drops.

　　　Four drops, increasing to ten drops, twice per day.

309.　　　　　　**Sore Throat.— Malignant.**

Hydrochloric Acid .　.　.　.　.　1½ drachms.

Decoction of Cinchona, ⎫
Compound Infusion of Roses, ⎭ of each .　3½ oz.

Honey of Roses　.　.　.　.　.　1　"

　　　Make a gargle.

310.

Infusion of Sage	8 oz.
Hydrochloric Acid	1½ drachms.
Syrup of Mulberry	2 oz.

Make a gargle.

311. Putrid Sore Throat.

Hydrochloric Acid	15 minims.
Infusion of Cinchona	4 oz.
Honey of Roses	1 "

Make a gargle.

312. In Common Sore Throat.

Decoction of Quince Seed	7 oz.
Syrup of Sage	1 "
Hydrochloric Acid	20 to 30 minims.

Make a gargle.

313. Tonics.

Decoction of Cinchona	10 drachms.
Aromatic Confection	1 scruple.
Compound Tincture of Cinchona . .	1 drachm.

A draught to be taken every fourth hour.

314.

Infusion of Gentian (compound) . .	9 drachms.
Compound Tinct. Gentian . . .	2 "
Orange Syrup	1 drachm.

A draught to be taken twice per day.

315.

Infusion of Roses	10 drachms.
Aromatic Sulphuric Acid	15 minims.
Tincture of Rhatany	1 drachm.
Mix with Diluted Spirits	2 pints.
Syrup of Red Poppy	1 drachm.

A draught to be taken three times per day.

316.

Infusion of Orange	6 oz.
Aromatic Tincture of Rhatany, ⎫ of each .	1 "
Syrup of Ginger, ⎭	

Three tablespoonfuls three times a day.

317.

Hydrochloric Acid, diluted, ⎫ of each .	5 minims.
Nitric Acid, ⎭	
Compound Mixture of Gentian . .	10 drachms.

Draught three times per day. Also, laxative in its effects.

318.

Sulphuric Acid, diluted	12 minims.
Infusion Cascarilla	10 drachms.
Orange Syrup	1 drachm.

Draught to be taken three times per day.

319.

Sulphuric Acid, diluted	40 minims.
Compound Spirits Ether	2 drachms.
White Sugar	$\frac{1}{2}$ oz.
Green Peppermint Water	6 "

A fourth part four times per day.

320.   **Tonic in Dyspepsia.**

Powdered Rhubarb	1 drachm.
Magnesia	1½ drachms.
Powdered Ginger , ...	1 scruple.
Peppermint Water	1 pint.

Mix. Dose, one-half ounce.

321.   **Toothache.**

Tannic Acid	1 drachm.
Mastich	1 "
Sulphuric Ether	1½ oz.

To be introduced, on cotton, into a hollow tooth.

322.

Powdered Resin	½ drachm.

Chloroform sufficient to dissolve the Resin.

Steep a small piece of cotton in the preparation, and place it in the hollow of the tooth.

323.

Canada Balsam	1 drachm.
Slaked Lime	1 "

Pressed into the cavity.

324.

Oil of Cloves	1 drachm.
Cajeput Oil	1 "
Powdered Opium, } of each . . .	½ scruple.
Camphor,	

Rectified Spirit, sufficient to dissolve.

Apply as above, or to the swollen gums.

325.   **Typhus, and Other Fevers.**

Powdered Scammony, ⎫
Powdered Aloes,   ⎬ of each   .   .   2 grains.
Chloride of Mercury, ⎭
     Make two pills.   To be taken at once.

**326. Ulcerated Mucous Membranes, Mouths, etc.; and In-
          flammation of Mouth in Infants.**

Sugar of Lead .   .   .   .   .   .   3 grains.
Soft Water   .   .   .   .   .   1 oz.
     As a wash.

327.

Sulphate of Copper .   .   .   .   .   ½ drachm.
Soft Water   .   .   .   .   .   .   1 oz.
     To be applied twice a day to the ulcers in gangrene of
the mouth.

328.

Creosote .   .   .   .   1 drachm.
Alcohol   1   "
     To be applied with a camel's-hair pencil to the gangre-
nous ulcers of the mouth, after running a lancet through the
sloughs.

329.

Acid Nitrate of Mercury .   .   .   .   ½ drachm.
Soft Water   .   .   .   .   .   .   1 oz.
     To be injected into the throat with the shower-syringe, or
applied to the ulcers with a pencil.

330. For **Curdy Patches in Children's Mouths.**

Hydrochloric Acid	1 drachm.
Honey	1 oz.

331. **Vomiting.**

Ext. Belladonna	6 grains.
Pulv. Ipecac	10 grains.
Confection of Roses	2 grains.

Make thirty pills. One pill twice per day.

332.

Compound Infusion of Orange . . .	10 drachms.
Green Peppermint (Spirit) . . .	1 drachm.
Liquor of Potass	10 drops.
Carbonate of Magnesia	1 scruple.
Tinct. Hyoscyamus	$\frac{1}{2}$ drachm.
Extract Hops	8 grains.
Syrup of Ginger	1 drachm.

Mix. Make a draught.

333.

Sesquicarbonate of Ammonia . . .	2 drachms.
Gum Tragacanth	1 scruple.
Distilled Water	7 fluid oz.

Mix. A teaspoonful every hour.

334. **Vomiting.**

Creosote	6 minims.
Pulv. Tragacanth	$\frac{1}{2}$ drachm.
Camphor Mixture	6 oz.

A sixth part to be taken for a dose.

335. **For Wind and Belching.**

Powdered Chamomile	$\frac{1}{2}$ scruple.
Long Pepper Powder . . . :	3 grains.
Powdered Aloes	1 grain.

A powder, to be taken every night.

336. **Worms.**

Garlic Bulbs	4 drachms.
Milk	8 oz.

Boil gently and strain, to make an enema.

337.

Decoction Aloes	$1\frac{1}{2}$ oz.
Extract Liquorice	2 drachms.
Wine of Aloes	2 "

One or two teaspoonfuls twice a day.

338.

Oil of Almonds, } of each . . .	$\frac{1}{2}$ oz.
Distilled Water, }	
Sesquicarbonate of Ammonia (Liquor) .	20 minims.

A draught. To be taken every morning on an empty stomach.

339. **To prevent the Breeding of Worms.**

Hydrochloric Acid, diluted . . .	2 drachms.
Infusion of Quassia	$7\frac{1}{2}$ oz.
Syrup of Orange	$\frac{1}{2}$ oz.

Take one-fourth twice per day.

340. <center>**To Expel Tape-Worms.**</center>

Gamboge	$\frac{1}{2}$ scruple.
Sulphate of Iron	6 grains.
White Sugar	1 scruple.
Peppermint Water	3 drops.

Make a powder. Prepare six similar doses. One to be taken every four hours until expelled.

APPENDIX C.

DOMESTIC RECEIPTS.

Whooping-Cough Syrup.

Onions and garlics sliced, of each 1 gill; sweet oil, 1 gill; stew them in the oil in a covered dish, to retain the juices; then strain and add honey, 1 gill; paregoric and spirits of camphor, of each one-half oz., bottle and cork tight for use. Dose for a child of two or three years, 1 teaspoonful three or four times daily, increasing according to age.

Diarrhœa Tincture.

Compound tincture of myrrh, 6 oz.; tincture of rhubarb and spirits of lavender, of each 5 oz., tincture of opium, 3 oz.; oils of cinnamon and anise, with gum camphor and tartaric acid, each ½ oz. Dose, 1 teaspoonful in half a teacupful of warm water sweetened with loaf sugar. Repeat after each passage.

Cathartic.

Jalap and peppermint leaf, each 1 oz., senna, 2 oz., powder finely and sift through gauze. Bottle it, and keep it corked. Mix a good teaspoonful of the powder, and an equal quantity of sugar, into a wineglassful of boiling water; when cool, stir and drink. Repeat every three hours until operation.

Small-pox.

To prevent Pitting the Face. — A great discovery has recently been made by a surgeon of the English army in China, to prevent pitting or marking the face. The mode of treatment is as follows : —

When, in small-pox, the preceding fever is at its height, and just before the eruption appears, the chest is thoroughly rubbed with croton oil and tartar-emetic ointment. This causes the whole of the eruption to appear on that part of the body to the relief of the rest. It also secures a full and complete eruption, and thus prevents the disease from attacking the internal organs. This is said to be now the established mode of treatment in the English army in China, by general orders, and is regarded as perfectly effectual.

Female Laxative Pill.

Aloes, macrotin, and cream of tartar, of each 2 drs. ; podophyllin, 1 dr. ; make into common-sized pills by using oil of peppermint, 15 to 20 drops, and thick solution of gum mucilage. Dose, one pill at bed-time, and sufficiently often to keep the bowels just in a healthy condition.

If the aloes should not agree with the patient, the following may be used: —

Female Laxative and Anodyne Pill.

Macrotin and rhubarb, of each 10 grs. ; extract of hyoscyamus, 10 grs. ; Castile soap, 40 grs. ; scrape the soap and mix well together, forming into common-sized pills with gum solution. Dose, one pill, as the other, or sufficiently often to keep the bowels in proper condition, but not too free. The hyoscyamus tends to quiet the nerves without constipating the bowels.

To soothe and quiet the nervous system and pains, if very violent, when the courses commence, or during their progress, make the following : —

Pill for Painful Menstruation — Anodyne.

Extract of stramonium and sulphate of quinine, of each 16 grs. ; macrotin, 8 grs. ; morphine, 1 gr. ; make into 8 pills. Dose, one pill, repeating once or twice only, 40 to 50 minutes apart, if the pain does not subside before this time.

The advantage of this pill is that costiveness is not increased, and pain must subside under its use.

Tea — Injection for Leucorrhea.

When the glairy mucous discharge is present, prepare a tea of hemlock, inner bark, and witch hazel (often called spotted alder) leaves and bark ; have a female syringe sufficiently large to fill the vagina, and inject the tea, twice daily ; and occasionally, in bad cases, say twice a week, inject a syringe of the following : —

Injection for Chronic Female Complaints.

White vitriol and sugar of lead, of each $\frac{1}{4}$ oz. ; common salt, loaf sugar, and pulverized alum, of each $\frac{1}{2}$ dr. ; soft water 1 pt. Simmer all over a slow fire for 10 or 15 minutes ; when cool, strain and bottle for use, keeping well corked. Inject as mentioned in the paragraph above, holding the syringe in place for a minute or two at least. This injection is valuable in diseases of the generative organs for males and females.

STIMULANTS.

STIMULANTS are medicines which increase and sustain the action of the heart. They are often useful in diseases characterized by great prostration, and in convalescence from fevers, etc. Brandy, whiskey, wine, etc., are stimulants, and may be used in appropriate cases, in proper quantities, three to six times a day. The dose of brandy or whiskey should not, as a general rule, exceed one to two table-spoonfuls. If wine be used, about two ounces may be given.

ALCOHOLIC STIMULANTS should not be continued longer than they are required *as medicines;* for the daily use of spirituous liquors, by persons in health, is very injurious, and tends to undermine the constitution and bring on serious and even fatal diseases of the brain, liver, and kidneys.

Wine Whey.

Take of good milk, one-half pint; heat to the boiling point; then add of sherry wine, one gill; strain, and add of white sugar one ounce, and a little nutmeg.

There are few mild stimulants more employed or more useful, than wine whey. The dose must be regulated by the circumstances of the case. From one gill to a pint may be taken during the day.

Egg-Nog.

Take the White and Yolk of	4 eggs.
White Sugar	1 oz.
Beat well together, then add	
Sherry Wine	2 oz.
Water	4 oz.

Grated nutmeg to taste. The above is sufficient for about four doses. Stimulant and nutritious.

Milk Punch.

Take of good Brandy	.	.	.	.	1 gill.
Fresh Milk .	.	.	.	.	$\frac{1}{2}$ pint.
White Sugar	.	.	.	.	1 oz.
A little grated nutmeg.					

Mix. A table-spoonful or more may be taken every hour or two in low fevers, and other diseases characterized by great prostration.

Mixture of Carbonate of Ammonia.

Take of Carbonate of Ammonia	.	.	2 drachms.
White Sugar	.	.	.	3 "
Powdered Gum Arabic	.	.	3 "
Spearmint Water	.	.	.	$\frac{1}{2}$ pint.

Mix. Give a table-spoonful every two hours.

ANTHELMINTICS.

ANTHELMINTICS are medicines which are taken to destroy and bring away worms from the alimentary canal. They are frequently termed vermifuges.

Fluid Extract of Pinkroot and Senna with Santonin.

| Take of Fluid Extract of Pinkroot and Senna | 2 oz. |
| Santonin | . | . | . | . | . | 16 grains. |

Mix. Give a child, two years old, a teaspoonful night and morning, until purging takes place. This is very effectual for removing the common round-worm.

Mixture of the Extract of Male Fern.

Take of Ethereal Fluid Extract of Male

Fern	½ drachm.
Syrup of Gum Arabic . . .	1 oz.

Mix. Shake well before using. Take one-half at bed-time, and the remainder early in the morning, on an empty stomach. Used for tape-worm. If the worm does not come away in six or eight hours after the last dose, take some mild purgative, as castor-oil.

For Pin-Worms. — Injections are the most certain of all remedies for *ascarides*, or " pin-worms."

Injection of Salt Water.

Take of Common Fine Salt . . .	1 oz.
Warm Water	1 quart.

Mix. One-half or even the whole may be injected into the bowel of an adult, and retained fifteen minutes, if possible. It should be repeated once or twice a day for a week or two. When used for children, let the quantity be in proportion to age.

OINTMENTS.

Simple Ointment.

Take of Fresh Lard	3 oz.
White Wax	1 or 2 "
Oil of Sweet Almonds . .	1 "

Melt together, and stir until cool. If for winter use, one ounce of white wax will be sufficient, but if for warm weather, use two ounces. A few drops of oil of rose added when nearly cool, will give it a fine odor, but this is not essential. Useful in dressing wounds, burns, scalds, blisters, cuts, and sores of almost every description.

Resin Ointment.   (Basilicon Ointment.)

Take of Powdered Resin	1 oz.
Yellow Wax	½ "
Fresh Lard, or Simple Ointment .	2 "

Melt with a gentle heat, strain through flannel, and stir constantly until cool.   Useful in dressing indolent ulcers, sores, wounds, etc., where something more stimulating and adhesive than simple ointment is required.

Compound Sulphur Ointment.

This can be bought at the apothecaries.   Used for diseases of the skin ; itch, ringworm, etc.

Simple Sulphur Ointment.

Take of Lac Sulphur	½ pound.
Fresh Lard	1 "
Oil of Bergamot	2 drachms.

Mix well.   Used in itch.   In making simple sulphur ointment, the sublimed sulphur is generally used, but I prefer the precipitated or lac sulphur.

Astringents for External Use.

Tannin, powdered matico leaves, alum, etc., are often useful to check bleeding from superficial wounds.   In bleeding from the nose they may be used in solution, or the dry powder may be used as snuff.

BLISTERS

Are those articles which produce a serous or watery discharge beneath the cuticle, — the *blister* of common language.

Blistering Plaster.

This may be made by spreading blistering ointment on leather, cloth, or adhesive plaster. It may be of any size required, from one to six or eight inches square. It should remain on the part until a good blister has formed ; then remove it, and dress the blister with simple ointment, or sweet oil. This is sometimes termed a " Fly-blister," the active ingredient being Cantharides, — Spanish flies.

AQUA AMMONIA, MUSTARD, and some other articles, will " blister," but the ordinary *fly-blister* is preferable in most cases.

Mustard Plaster.

Take of Powdered Black Mustard . . 1 oz.

Water sufficient to make a paste or poultice.

The uses of the mustard plaster are too well known to require description. It is sometimes made by adding an equal quantity of flaxseed or Indian meal ; this should always be done for children, or persons of delicate skin. It should be kept on as long as it can be borne, which will not generally be longer than half an hour.

Ground Black Pepper.

If mustard is not at hand, this may be taken as a substitute.

Onion Poultice.

Onions are often used in cases of croup, and in diseases of the chest in children, as revulsives and antispasmodics.

Having been partially roasted, mashed, and spread between two folds of thin muslin, they may be applied over the chest, and permitted to remain as long as they retain their warmth and moisture.

EYE WASHES.

Sulphate of Zinc Eye Wash.

Take of Sulphate of Zinc .   .   .   .   2 grains.
      Rose Water .   .   .   .   .   1 oz.

  Mix.   Drop a little into the eye, two or three times a day, when sore or inflamed.

Acetate of Zinc Eye Wash.

Take of Acetate of Zinc .   .   .   .   2 grains.
      Rose Water .   .   .   .   .   1 oz.

  Mix.   Use three or four times a day, for sore eyes.

Sugar of Lead Eye Wash.

Take of Sugar of Lead .   .   .   .   2 grains.
      Rose Water .   .   .   .   .   1 oz.

  Mix.   Excellent for inflamed eyes.   It should be used two or three times a day.

WASH FOR THE EAR.

Goulard's Extract and Rose Water.

Take of Goulard's Extract (Liquid Subace-
        tate of Lead) .   .   .   .   ½ drachm.
      Glycerine   1   "
      Rose Water .   .   .   .   .   2 oz.

  Mix.   This may be dropped into the ear until the cavity is full, and be allowed to remain there for a few minutes. Use morning and night.   *For running from the ear.*

Solution of Nitrate of Silver.

Take of Nitrate of Silver .   .   .   .   10 grains.

       Distilled Water .   .   .   .   1 oz.

Mix. Used in ulceration, attended by discharges of matter (pus) from the ear. Dip a camel's-hair brush into the solution, and apply every second or third day. Between the applications use the preceding recipe.

FOMENTATIONS.

FOMENTATIONS, used warm, or hot, are good counter-irritants. They may be made by dipping a piece of flannel, folded several times, in warm or hot water, allowing it to remain three to five minutes, and then wringing it nearly dry. It should be of sufficient size to well cover the part to which it is applied. A piece of oiled silk or a dry cloth may be placed over it, to prevent evaporation. Fomentations should generally be about as hot as the patient can bear, where active counter-irritation is indicated, and must be renewed every ten or fifteen minutes. In pleurisy, congestion of the lungs, lung fever, and inflammation of the bowels, oil of turpentine may be gently applied to the parts, before applying the hot cloths; or it may be sprinkled upon them.

MISCELLANEOUS RECEIPTS.

Tooth-Ache Drops.

Take of Mastic   .   .   .   .   .   1 drachm.

       Chloroform .   .   .   .   .   1 oz.

Mix. Moisten with the solution cotton enough to fill the cavity of the tooth, and press it well in.

Another.

Take of Oil of Cloves   .   .   .   .   1 drachm.

Chloroform .   .   .   .   .   1 "

Mix. Use as directed above. Oil of cloves alone is excellent.

Another.

Take of Gum Camphor   .   .   .   .   ½ oz.

Strongest Alcohol   .   .   .   1 "

Mix in a mortar, and rub until the camphor is dissolved. Use as above.

Tooth Powder.

Take of Prepared Chalk   .   .   .   .   1 oz.

Powdered Orris Root   .   .   .   1 "

Powdered Castile Soap .   .   .   2 drachms.

Mix. Use after breakfast and supper.

Another.

Take of Powdered Orris Root .   .   .   1 oz.

Powdered Peruvian Bark .   .   1 drachm.

Powdered Gum Myrrh   .   .   ½ "

Powdered Cinnamon Bark .   .   1 "

Bicarbonate of Soda   .   .   .   1½ drachms.

Mix. Use as above.

PREPARATIONS FOR THE HAIR.

Hair Oil.

Take of Castor Oil   .   .   .   .   .   2 oz.

Glycerine   .   .   .   .   .   2 drachms.

Cologne Spirit   .   .   .   .   2 oz.

Oil of Bergamot .   .   .   .   1 drachm.

Oil of Rose .   .   .   .   .   4 drops.

Mix.

Another.

Take of Olive or Castor Oil . . . 3½ oz.

      Glycerine . . . . . 2 drachms.

      Oil of Jessamine . . . . 1 drachm.

      Oil of Orange . . . . 1 "

Mix.

APPENDIX D.

VARIOUS METHODS OF BATHING.

It would be difficult to exaggerate the importance of ablution as a means of preserving and maintaining health. In the earliest ages it was strictly enjoined on all persons, young and old ; and want of cleanliness was punished as a crime. In both the Jewish and Mohammedan religions constant ablutions (four or six times a day) were prescribed as an essential part of religious duty. The ancient philosophers, too, continually inculcated in their orations and writings the absolute necessity and beneficial influences of cleanliness. Aristotle describes it as one of the half virtues. Cicero taught his disciples to " preserve health by attention to the body and temperance in living," and the inhabitants of the East have for the last six thousand years looked upon bathing and washing, not only as an imperative duty but a positive necessity and indescribable luxury, far more important to life than even food itself. And, *theoretically*, it occupies an equally prominent position in the sanitary and hygienic laws of the most civilized communities of the present age, England and America ; but, unfortunately for the health, comfort, and physical and mental development of the masses of the people, there is, in this much-boasted nineteenth century, far too much theory, and comparatively little practical application of these first principles of cleanliness, especially among the more educated classes, to whom we would naturally look for a strict adherence to Nature's laws. Dr. Guy, an eminent member of the Medical Faculty in England, remarks that "a want of personal cleanliness is more frequently chargeable against persons of education than might at first sight be thought possible." The fact is, it is one thing to know what is right, and another to conscientiously act upon that knowledge. The practice of daily ablution of the whole body is observed by a very small number of the community, despite its acknowledged necessity.

In thickly populated towns and cities, especially, many hundreds of thousands of lives might be saved if soap and water were but freely used. Cholera, typhoid fever, small-pox, and skin diseases of all kinds, would be effectually and permanently driven from the land if daily ablution were systematically and compulsorily practised. Want of cleanliness has caused more misery, desolated more homes, originated more crime, and peopled many more church-yards, than all the wars that have occurred since the creation of the world. There are many

Absurd Prejudices against frequent Bathing

among all classes of the people; one would almost think that an hydrophobic mania had seized civilized communities, — for they seem to be frightened of the sight of or contact with water, so far as the ablution of the entire body is concerned (except in the warmest weather). A ludicrous illustration of the reluctance to perform this operation, occurs to us at this moment.

Dean Swift's Satire upon Uncleanliness.

While this eminent satirist was staying at his episcopal residence near Dublin, one of his servants (who had evidently not washed his hands and face for some days) approached him.

" James," said the dean, " when did you last wash yourself?" " Yesterday morning, your reverence," the servant
replied. The dean put on a very solemn countenance, and
said: " I should seriously advise you to discontinue such a
useless and unnecessary practice, for two or three reasons.
1st, on the score of economy: soap and water cost money,
and washing involves an expenditure of time, money, and
trouble. 2d, as a matter of personal comfort: you must
bear in mind that the dirt on your skin will keep you warm ;
and that if you wash yourself one morning you must repeat
the operation daily, else you will get dirty again ; so that I
would, if I were you, let it remain." It need scarcely be
said that the servant, who was constitutionally lazy, took
the advice literally, and refrained from washing himself
afterward.

Beneficial Effects of Daily Bathing.

In our chapter on Hygiene we have defined the various
methods of bathing, and the benefits directly derivable from
the constant adoption of the practice ; but we desire specially to call the attention of our readers to the fact that
warm and towel and sponge baths, or even wash-tub
baths, persistently adopted, not only keep the body in a
vigorous, healthy condition, but in our variable climate are
absolute preventives against catching cold, and ensure to
the skin the proper and unimpeded performance of its functions. Persons in ordinary health should use either of the
baths we have named, at least twice each week, if they wish
to keep perfectly clean.

The Functions of the Skin.

The pores of the skin are the great avenues of egress and
ingress for the human body, — the direct means by which we

exhale the noxious mephitic elements, and absorb or inhale the regenerating influences of pure water, pure air, etc. There are more than two thousand of these respiratory tubes to the square inch, and as a person of average size and height has about two thousand five hundred square inches of surface, the number of tubes in the skin of one person is *seven millions*, each about a quarter of an inch in length, making an aggregate length of tubing in the human skin of *twenty-eight miles*. Now, imagine the irreparable injury occasioned by the continued obstruction of those delicate ducts or tubes. If the organs of respiration and inspiration are prevented from performing their office, the whole of the complicated machinery of man is necessarily deranged and debilitated, and an interminable series of chronic diseases originated. But the evil does not rest here : the mental faculties are proportionately deteriorated, dwarfed, and demoralized as bodily cleanliness is neglected. The poet of Nature, Thomson, says : —

> Even from the body's purity the mind
> Receives a secret, sympathetic aid.

Directions for Bathing.

In all descriptions of bathing, great care should be taken to avoid imparting a " shock " to the system by sudden plunging, or other application of warm or cold water to the body. Water below 75° is considered cold, and, if applied to persons in ordinary health, is a powerful and efficient tonic, adding greatly to the strength, vigor, and compactness or firmness of the body. With the exceptions elsewhere named, it is always advisable to keep the temperature of the bath as near 75° as practicable or consistent with the comfort of the patient.

The Sponge Bath

is the simplest, most agreeable, and certainly the most efficacious method of bathing, if practised daily in the morning. Those who are feeble should expose only a part of the body at a time, quickly sponging and drying it, and proceeding in this way until the entire surface has been subjected to the bracing influence of water and friction. The only apparatus necessary for this is a good-sized sponge, a basin, and a towel.

The Shower Bath

requires great care and discrimination in its use. Either warm or cold water may be used, or both successively, but especial caution is necessary to let the *shower* be more after Nature's pattern — gentle, regular, equal in quantity, and not in sudden spasmodic spirts. It will then prove, what it was originally intended to be, a means of invigoration and stimulation, which the feeblest invalid might bear without injury.

The Cold Bath

and its application has been treated of in the earlier part of this work, and is only referred to here to intimate to our readers that it is the most easily arranged of any of the modes of bathing (the sponge-bath excepted.) An ordinary wash-tub, with a coarse towel and a flesh-brush or horse-hair glove, is all that is required to render it a pleasurable, healthful, and invigorating preparation for the labors and occupations of the day, and an efficient preservative against infection and the ordinary ailments of humanity.

The Sitz-Bath.

We have referred in various parts of this volume to the use of the Sitz-bath (or *Sitting Bath*) as a remedial agency, and as an invaluable tonic for the stomach, liver, bowels, spine, uterine organism, in eruptive fevers, etc. An ordinary wash-tub is all that is required, sufficiently large to permit of the immersion of the body just above the hips; soap may or may not be used at the discretion of the operator; but the bath should never be continued for less than ten or more than thirty minutes at a time.

The Leg-Bath,

especially designed for the cleaning of open wounds, ulcers, and persons afflicted with scrofulitic and other eruptions of the skin, should consist of an ordinary wooden tub, the water warmed to blood-heat, and, when necessary, medicated with the preparations indicated in the treatment for those diseases; but, in every case, pure, clean water is an essential and indispensable element in the cure of all cutaneous trouble.

Special Directions for Bathing.

In rubbing and drying the body, the muscles and nerves should be thoroughly manipulated, especially those of the arms, legs, and spinal column. The friction, either by hand or towel, should be upward, not *downward*, and the motion over the abdominal and thoracic viscera (the abdomen, chest, and trunk) *upward* and across — a kind of circular motion. By this means the ligaments through which the uterine organism is held in position would be materially strengthened, and *displacements* and *weaknesses*, in most instances, easily corrected or entirely averted.

Bathing in Advanced Age

should be confined to warm water, the temperature ranging between 95° and 105°. A warm bath of this kind, every other day, would, if the person be otherwise healthy, prolong her life from ten to twenty years.

Cold-Water Foot-Baths

(with or without salt) are invaluable as remedies in rheumatism, rush of blood to the head, inflammation of the eyes, hemorrhages, gout, and other inflammatory or poisoned conditions of the blood, and especially in fevers (whether remittent or intermittent, contagious or simple.)

Injections.

In the treatment of the various diseases, and also in the chapter on Hygiene, we have recommended the cleansing of the internal organism by injection, or the judicious use of *syringes*. Now, as the kind of syringe used is of the most vital importance, we deem it our duty to caution our readers to exercise the greatest possible care in their selection of the instrument. There are a great many in the market, each of which possesses some peculiar excellence, but at the same time there are others which are worthless and should never be used. We consider the use of the old style *glass syringes*, in uterine difficulties, to be attended with considerable danger, from their liability to break or chip, and cause laceration of the parts. Many metallic syringes are equally objectionable, from the oxidation and other chemical changes to which they are liable, and also from the fact that it is very difficult to regulate the force of the injection, so as to avoid stricture or other injury to the adjacent organs. Our own preference is for a syringe or injecting apparatus made of vulcanized rubber, which should be entirely under the control of the patient, easily cleansed, and simple in construction, such as " Fairbanks'," which are manufactured with glass, metal and hard rubber tubes ; but we should decidedly recommend the hard rubber as being the most cleanly, durable, and least likely to get out of order. If any of our readers should desire to purchase through us, we have made special arrangements with the manufacturers whereby we can supply them —

Small size, No. 1	$2.00
Medium size, No. 2	2.50
Large size, No. 3	3.00

A large and varied selection of the best and most approved instruments may always be obtained at Messrs. Codman & Shurtleff's, and other reputable and eminent surgical mechanicians, whose reputation would be a sufficient guarantee for the reliability of any instrument they might recommend.

GLOSSARY.

Abdomen. The lower belly, containing the stomach, intestines, liver, kidneys, etc.

Abnormal. Irregular; unhealthy.

Abortion. Miscarriage, or the expulsion of the foetus before the seventh month.

Abscess. A collection of pus.

Acute. Sharp; a disease of recent date, or which terminates in a short time.

Adipose. Fatty.

Albumen, albuminous. A substance resembling the white of an egg.

Aliment. Nourishment; food.

Alimentary canal. The passages from the mouth to the anus, through which the food passes.

Alvine. Relating to the intestines.

Amenorrhœa. Suppression of the menses.

Anatomy. The knowledge of the structure of the human body.

Anæmia. Debility.

Anthelmintic. Whatever procures the evacuation of worms.

Aorta. The large artery from the left ventricle of the heart.

Aperient. Whatever gently opens the bowels.

Anus. Termination of the rectum externally.

Atrophy. A wasting away.

Auricle. A cavity of the heart; an ear.

Belladonna. The deadly night-shade; a narcotic.

Bile. The gall; a bitter fluid secreted by the liver.

Bronchia. The smaller branches of the windpipe.

Bronchial. Belonging to the windpipe.

Bronchitis. Inflammation of the bronchii.

Cæsarean operation. Cutting the child out of the womb.

Cardiac. The heart.

Caries. Mortification of the bones.

Carotid. An artery of the neck.

Cartilage. The *gristle* attached to bones.

Cartilaginous. Having the appearance of cartilage.

Catamenia. The menses.

Cataplasm. A poultice.

Catarrh, Catarrhal. An increased discharge from the nose.

Cathartic. Whatever produces an evacuation from the bowels.

Catheter. A hollow instrument to introduce into the bladder for the purpose of drawing off the urine.

Caustic. Whatever *burns* or destroys the texture of a part.

Cellular. Having little cells.

Cellular membrane. The *fatty* membrane immediately beneath the skin.

Cerebellum. The lower part of the brain.

Cerebral, cerebrum. The brain.

Cervical. Belonging to the neck.

Chest. The cavity of the body which contains the lungs.

Chlorosis. The green-sickness.

Chlorotic. Affected by chlorosis.

Chronic. Applied to diseases of long standing.

Chyle. The milk-like fluid resulting from the digestion of food, and from which the blood is formed.

Clavicle. The collar-bone.

Clyster. An enema or injection.

Coagula. The clot of the blood.

Coccyx. A small bone belonging to the pelvis.

Colon. The largest intestine.

Coma. A propensity to continued sleep.

Conception. Impregnation of the ovum of the female.

Congenital. Existing at birth.

Congestion. Distended with blood.

Conjugal. Relating to marriage.

Conjunctiva. The membrane that covers the anterior part of the eyeball and the inner surface of the eyelids.

Constipation. Costiveness.

Counter-irritation. Artificial irritation in a part distant from that diseased.

Cranium. The skull or upper part of the head.

Craniotomy. The operation of perforating the cranium.

Cutaneous. Relating to the skin.

Defecate. To purge; to evacuate the bowels.

Demulcent. A medicine of a soft, mild and viscid character.

Derivatives. Counter-irritants.

Diagnosis. The art of distinguishing diseases.

Diaphragm. The midriff, or muscle that divides the chest from the abdomen.

Diarrhœa. An increased discharge from the bowels.

Diathesis. Peculiarity of constitution.

Dietetic. Relating to the food.

Digestion. The process by which the food is converted into chyle.

Diluents. Those substances which increase the fluid portion of the blood.

Diuretic. That which increases the flow of urine.

Dorsal. Relating to the back.

Douche. Water poured from a height upon any part of the body.

Duodenum. The intestine next to the stomach.

Dysentery. A discharge of bloody mucus from the bowels,

Dysmenorrhœa. Painful menstruation.

Dyspepsia. Indigestion.

Dysuria. Suppression of the urine.

Edematous. Dropsical swelling.

Embryo. A germ; the fœtus before the seventh month.

Emetic. That which produces vomiting.

Emmenagogue. That which promotes the discharge of the menses.

Emollient. Softening; relaxing.

Emulsion. A soft or oily substance.

Enema, pl. *enemata*. A clyster or injection into the rectum.

Engorgement. Distention; fulness.

Epidemic. A generally prevalent disease.

Epigastric. Over the stomach.

Epilepsy. The falling-sickness.

Ergot. The *spur* of rye.

Erotic. Passionate; lustful.

Erysipelas. St. Anthony's fire.

Esophagus. The gullet or passage for food.

Exacerbation. An increase in the violence of a disease.

Excrement. The fœces.

Exhale. To emit.

Expectorant. That which increases the discharge of mucus from the lungs.

Fœces. The excrementicial portion of the food.

Farinaceous. Of flour or meal.

Fauces. The cavity behind the tongue.

Febrile. Feverish.

Femoral. Belonging to the thigh.

Femur. The thigh.

Fetid. Offensive to the smell.

Fibre. A filament or thread.

Fibrine. A fibrous substance obtained from blood and animal matter.

Fistula. An ulcerous cavity with a narrow opening.

Flex. To bend.

Fœtus. The child in the womb.

Follicle. A small gland.

Function. The natural action of any organ.

Fungus. Proud flesh,

Gangrene. Mortification.

Ganglion. A knot; applied to the enlargement of nerves.

Gastric. Appertaining to the stomach.

Gastritis, gastralgia. Inflammation and pain of the stomach.

Gelatin. Known as jelly.

Gestation. Pregnancy.

Glottis. The opening of the larynx.

Hæmatemesis. Vomiting of blood.

Hæmoptysis. Spitting blood.

Hectic. A fever arising from internal suppuration.

Hemiplegia. Palsy of one half of the body.

Hemorrhage. A bleeding.

Hemorrhoids. Bleeding piles.

Hepatic. Belonging to the liver.

Hernia. A rupture.

Humor. A fluid of the body.

Hygiene. Relating to the regimen, diet, etc., of the sick.

Ilium, Iliac. Relating to the haunchbone or region of the body near it.

Induration. A hardening.

Inferior The lower part.

Insalivation. Mixing the food with saliva.

Integument. An external covering.

Intercostal. Between the ribs.

Issue. An artificial ulcer, kept open for the purpose of producing a discharge.

Jugular. Belonging to the throat.

Labia. The lips; applied to the external female genitals.

Lactation. Nursing.

Larynx. The superior opening of the windpipe.

Laxative. A gentle purge.

Leucorrhœa. The whites.

Ligament. A strong membrane connecting the bones.

Ligature. A thread.

Liniment. An oily fluid.

Liquor amnii. The fluid in which the fœtus floats.

Lochia. The cleanings, or *show*, after labor.

Lumbar. Relating to the loins.

Lymph. A thin white, or rose-colored fluid found in the lymphatic vessels.

Lymphatic. Relating to lymph; a cold, relaxed habit of body.

Marasmus. Wasting away.

Marital. Relating to a husband.

Masticate. To chew.

Meatus urinarius. Opening of the urethra.

Meconium. The excrement found in the intestines of the child at birth.

Membrane. A thin expanding and elastic substance.

Menorrhagia. Immoderate flow of the menses; flooding.

Menses, menstruation. The monthly discharge from the vagina of females.

Micturition. Voiding of urine.

Morbific. Diseased.

Mucilage, mucilaginous. Of a gummy or slimy character.

Mucus. A viscid fluid secreted by mucous membranes.

Narcotic. Anodyne; a medicine that produces sleep.

Nasal. Relating to the nose.

Nates. The fleshy parts on which we sit.

Nerve. A long white cord transmitting sensation.

Neuralgia. Pain in a nerve.

Nitrogen. A component of the atmosphere.

Normal. Regular; healthy.

Nucleus. That about which something is formed.

Nutrition. Nourishing.

Nymphomania. Uncontrollable desire in women for coition.

Œdema. A swelling; dropsical.

Ophthalmia. Inflammation of the eye.

Opiate. That which procures sleep.

Organic. Belonging to a part of the body; a structural disease.

Organism. Organized structure.

Os. A bone; the mouth.

Osseous. Bony.

Ovaria, Ovarium, Ovary. The organs that contain the female ova.

Oxygen. A gas which forms the vital part of the air.

Paralysis. The palsy.

Parenchymatous. Relating to the cellular substance.

Parturient, parturition. Childbirth.

Pathology, pathological. Diseased.

Pectoral. Belonging to the chest.

Pelvis. The bony cavity below the abdomen.

Penis. The male organ.

Percussion. The act of striking against the abdomen.

Peritoneum. The membrane that surrounds the whole contents of the abdomen and pelvis.

Peritonitis. Inflammation of the peritoneum.

Perspiration. The vapor that passes off from the skin.

Pessary. An instrument to support the uterus by being introduced into the vagina.

Phthisis. Pulmonary consumption.

Physiology. The science of life.

Placenta. The afterbirth.

Plethora. An excess of blood.

Pleurisy. Inflammation of the membrane that surrounds the lungs.

Post mortem. After death.

Pregnancy. Being with child.

Prolapsus. A slipping-down.

Puberty. Ripe age.

Pubes. The part covered with hair above the genitals.

Pulmonary. Belonging to the lungs.

Purgative. Whatever produces increased discharges from the bowels.

Pus. A matter found in abscesses and sores.

Rachitic. Rickety.

Rectum. The last portion of the large intestines, terminating in the anus.

Regimen. The diet, etc., of the sick.

Reproduction. Producing anew.

Resolution. Termination of inflammation without disorganization.

Sacrum. A bone belonging to the pelvis.

Saliva. The fluid that is secreted in the mouth.

Salivation. An increased flow of saliva.

Sanguine. Bloody.

Scirrhus. Cancer.

Scrotum. The skin that covers the male testicles.

Sedative. Whatever diminishes animal energy.

Serum, serous. The watery part of the blood.

Simulate. To appear like; counterfeit.

Spine, spinal. The backbone.

Stethoscope. An instrument for making out internal diseases by the abnormal sounds.

Strumous. Scrofulous.

Strychnine. A stimulant; poisonous substance.

Suppurate. The formation of pus.

Sutures. The union of the bones of the cranium.

Tendon. A white, hard cord by which a muscle is attached to a bone.

Tenesmus. A continual inclination to go to stool without a discharge.

Therapeutical. Means employed to cure disease.

Thorax, thoracic. The chest; relating to the chest.

Tissue. The texture of an organ.

Trachea. The windpipe.

Trunk. The body apart from the limbs.

Tubercle, tuberculous. Scrofula; a hard, indolent tumor in the lungs and other glandular parts.

Ulceration. A purulent solution of any part of the body.

Umbilical cord. The navel string or cord which connects the child with the placenta.

Umbilicus. The navel.

Uterus. The womb.

Vascular. Relating to the veins.

Veins. The vessels which return the blood to the heart.

Venery, venereal. Intercourse of the sexes.

Venous. Relating to the blood of the veins.

Vertebræ, vertebral. The spinal column or backbone.

Vertex. The crown of the head.

Vertigo. Dizziness.

Vesicle. An elevation of the scarf-skin, containing a watery fluid.

Vicarious. Instead of.

Virus. The *poison* of contagion.

Viscera. Applied to the internal organs of the body.

Viscid. Sticky; ropy.

Vulva. The passage from the external generative organs to the womb.

ALPHABETICAL INDEX.

ALPHABETICAL INDEX.